ENERGY AND ENVIRONMENTAL ENGINEERING

Communications in Environmental Science

eISSN 2333-8032

Volume 1

PROCEEDINGS OF THE INTERNATIONAL CONFERENCE ON ENERGY AND ENVIRONMENTAL ENGINEERING (ICEEE 2014), SEPTEMBER 21–22, 2014, HONG KONG

Energy and Environmental Engineering

Editor

Yijin Wu
*Asia Pacific Environmental Science Research Centre, Hong Kong,
People's Republic of China*

CRC Press is an imprint of the
Taylor & Francis Group, an **informa** business

A BALKEMA BOOK

CRC Press/Balkema is an imprint of the Taylor & Francis Group, an informa business

© 2015 Taylor & Francis Group, London, UK

Typeset by MPS Limited, Chennai, India

All rights reserved. No part of this publication or the information contained herein may be reproduced, stored in a retrieval system, or transmitted in any form or by any means, electronic, mechanical, by photocopying, recording or otherwise, without written prior permission from the publishers.

Although all care is taken to ensure integrity and the quality of this publication and the information herein, no responsibility is assumed by the publishers nor the author for any damage to the property or persons as a result of operation or use of this publication and/or the information contained herein.

Published by: CRC Press/Balkema
 P.O. Box 11320, 2301 EH Leiden, The Netherlands
 e-mail: Pub.NL@taylorandfrancis.com
 www.crcpress.com – www.taylorandfrancis.com

ISBN: 978-1-138-02665-0 (Hardback)
ISBN: 978-1-315-75211-2 (eBook PDF)

Table of contents

Preface   IX
Organizing Committee   XI

Energy science and technology

Study on industrial structure and energy efficiency: An empirical analysis of China   3
X.L. Liao

Research into different pretreatment methods to enhance the enzymatic saccharification of Green Bamboo   7
J.J. Liu

Research in the specialized construction of new energy science and engineering   13
L. Yang, W.Y. Zhang, D.C. Jin, L. Wang & Z.B. Bao

Catalytic hydrogenation liquefaction of lignite in supercritical water   17
M.L. Wang, Q.Q. Guan, Q.L. Zhang, P. Ning, J.J. Gu, S.L. Tian, Q.L. Chen & R.R. Miao

Study on the power supply reliability of a nuclear power plant with GO-FLOW methodology, based on Simulink   21
J. Wang, J. Jia, J. Zhao, D.C. Liu, L. Wang & C. Cheng

Power supply safety assessment of a nuclear power plant, based on fuzzy comprehensive evaluation   27
C. Cheng, J. Zhao, D.C. Liu, J. Jia, S.S. Liang & X. Lin

The electric curtain method of dust removal and measurement of particle size and density on solar cells   33
C.D. Zhou, H.B. Xiao, J. Zhang & S.H. Sun

Passive control of floating wind turbine for vibration and load reduction based on TMD   39
M.W. Ge, S. Wang & H.W. Xiao

Environmental science and engineering

Controlled synthesis of manganese oxides with different morphologies and their performance for catalytic removal of gaseous benzene   47
W.X. Tang, S.D. Li & Y.F. Chen

The isolation, identification and biochemical reducing pathway of Cr (VI)-removal bacterium *Brevibacillus Parabrevis* from sludge biosystems   53
Y. Zhou, Y.B. Xu, S.H. Xu, X.H. Zhang, J.X. Xu, J.S. Luo, W.L. Gao & Y.J. Deng

Effect of operating parameters on the photocatalytic oxidation disinfection of swimming pool water   59
C.W. Kan, Y.L. Pan & H. Chua

Study on photochemical degradation of sulfamethazine in an aqueous solution   63
H.H. Xiao, G.G. Liu, Z.M. Chen & R.H. Wu

Study on landscape diversity characteristic of Xilin Gol grassland, based on remote sensing technology   69
M.X. Huang, G. Bao, Y.H. Bao & F.H. Zhang

Plant diversity and community stability in two subtropical karst forests in Southwest China   75
Z.H. Zhang, G. Hu & B.Q. Hu

Evaluating the smart land use of small towns in the context of Chinese new-type urbanization: A case study of Shanglin County in South China 79
B.Q. Hu, J.M. Wei, G. Hu & Z.Q. Zhang

Assessing how rainy or snowy weather conditions change residents' home based and return-related travel behaviour 85
J.L. Li, X.H. Li & D.W. Chen

Establish evaluation system of haze pollution 93
Y. Shen, L.Q. Li, H.D. Ma & Z.Y. Wang

Performance and real-time control of a novel intermittent SBR based on simulating photovoltaic aeration for organics removal 101
F.S. Ma, B.H. Zhou, Y. Qin, X. Du & Y.J. Yuan

Study on a multi-objective optimisation model for multi-reservoir ecological operation 109
L.N. Liu & W.L. Liu

Legal thinking and countermeasures of environmental pollution in the course of rural urbanisation 113
G.L. Zhang

Motivation, electrical engineering and automation

The modelling and simulation of a doubly-fed induction motor 119
Z.Y. Pei, Z.J. Wang & H. Xiao

Parameter inversion of horizontal multilayer 123
J.T. Quan, L. Ruan, X. Tong, X.S. Wen, Z.H. Pan & Q. Yang

Prediction of the performance and exhaust emission of a diesel engine based on Boost 129
X.G. Yang, G.Y. Han, H. Wang, L. Zhang & L.Y. Huang

The study on the simulation model for power system 135
J.G. Zhu & C.L. Wang

Design of a self-adaptive fuzzy PID control system for a supercritical once-through boiler 139
D. Zhang & D.M. Xi

The development and utilization of resources

A study on the comparison of the several typical processes for dealing with vanadium titanium magnetite resources 145
J. Qin, G.G. Liu, Z.J. Li & J.L. Qi

Internal architecture analysis and modeling of point-bar—A case study of PI2 formation, X area 149
J.Y. Wu, S.M. Shi, X. Feng, X.X. Wu, X.L. Du & Y. Bao

Influence of mined-out area on blast vibration effect in mine's underground stope 155
G.S. Zhong & L.P. Ao

Overburden movement in a shallow coal seam with great panel width 159
Y. Li, S.H. Hao, Y.J. Guo, W.J. Hu, B. Jiang & X.Y. E

Optimisation of flocculation conditions for aqueous extracts of Apocynum Venetum leaves using chitosan by response surface methodology 163
H.S. Yang, W.J. Gao, Y.M. Luo & L. Wu

Mechanisms of multi-thermal fluid huff and puff recovery for offshore heavy oil reservoirs and its influential law 169
Q.J. Du, J. Hou, J.C. Wang, L.N. Shi, G.F. Zhang, X.N. Li & S.P. Li

Discuss on application and development of bamboo and rattan materials in low carbon furniture 177
J. Zhang, W.D. Bai, J. Xu & W.Q. Jiang

Analysis of geological conditions of coal-bed methane and tight gas, and favourable areas selection in the Upper Paleozoic of the Ordos Basin 181
P. Liu, W.F. Wang, L. Meng & S. Jiang

Theory and practice of sustainable development

The current situation and outlook on environmental protection and sustainable development in China L.S. Zhang, Y. Feng, G.H. Li & X.G. Liu	189
Research on experimental methods of structural progressive collapse resistance Y.H. Huo & S.R. Ding	195
Contrastive analysis on anchorage length of pre-embedded and post-embedded bars Y.H. Huo & S.R. Ding	201
Empirical analysis of an urban development model in the western minority areas, taking Linxia City as an example of the coordination degree method J. Xu	207
Empirical analysis of main industrial capacity utilization in Gansu Province, focusing on an overcapacity situation J. Xu	211
Research on business model innovation for China's environmental industry Y.F. Li & D. Li	217
The establishment of a public participation system in regulatory detailed planning—a study based on the interactive model of government and NGOs D. Li, J.F. Chen & Q. Chen	223

Other related topics

Input-output analysis of CO_2 emissions embodied in international trade and the analysis of geopolinomic structure implications Q.N. Tang, Z. Wang, J.B. Xue & X.N. Cong	231
Dynamic simulation and experimental investigation on the performance of a refrigeration system under frosting conditions at low temperatures H.J. Qin, W.Z. Li, B. Dong & W.Y. Zhu	239
A study of the physical properties of soil under different ecological restoration measures in degraded red soil Y.Y. Li	247
Modelling of electric vehicle penetration in China based on innovation diffusion theory X.M. Ou, Q. Zhang, X. Zhang & X.L. Zhang	251
A study on the readjustment evaluation of a marine nature reserve based on the analytic hierarchy process W.H. Lu, Y. Wu & X.Q. Xiang	255
A study on regional comparison of farmland resources' value and innovative strategies for farmland protection compensation in China X.P. Zhou, Q. Wang & D. Chai	263
The application of the ETKF data assimilation method in the Lorenz-96 system J.C. Wang, J. Hou, Q.J. Du & R.X. Xu	271
Robust suboptimal control of a nonlinear enzyme-catalytic dynamical system in microbial continuous culture of glycerol to 1,3-propanediol J. Wang & Z.L. Xiu	277
Development and application of deep foundation pit internal bracing support technology in Nanchang Y.H. Li, Z.Y. He & X.P. Wang	281
Author index	287

Preface

It is our pleasure to welcome you to the 2014 International Conference on Energy and Environmental Engineering (ICEEE 2014), held September 21–22, 2014, in Hong Kong. A major goal and feature of ICEEE 2014 is to bring academic scientists, engineers and industrial researchers together to exchange and share their experiences and research results about most aspects of Energy and Environmental Engineering, and discuss the practical challenges encountered and the solutions adopted.

Human beings have only lived on this planet for several million years, but no species has ever changed the planet like we have. In our process of making the world a better place, we have kept advancing our understanding of natural laws and to our joy, we have succeeded a lot with our understanding of these laws. One of our greatest successes in understanding and applying the natural law is our knowledge of Energy and Environmental Engineering.

In the past twenty years, Energy and Environmental Engineering has become involved in many varied applications throughout the world, with multiple products and rapid market services. It has not only provided industries with new methods, new tools and new products from Energy and Environmental Engineering to operation and management process, but also is changing the manners, thinking styles and working environments of people in the manufacturing field. Today, scientists are still having disagreements over certain issues, and the final settling of these issues will surely advance our technology another big step. For this reason, our gathering and sharing of our ideas at ICEEE2014 will definitely not be in vain.

Serving as an important and influential platform for authors to publish manuscripts in excellent international proceedings and exchange new ideas face to face, ICEEE 2014 promises to be both stimulating and informative with a wonderful array of keynote and invited speakers from all over the world. Delegates will have a wide range of sessions to choose from and will face a difficult decision in deciding which sessions to attend.

The subjects covered several areas and were gathered into six themes for publication:

(I) Energy science and technology
(II) Environmental science and engineering
(III) Motivation, electrical engineering and automation
(IV) The development and utilization of resources
(V) Theory and practice of sustainable development
(VI) Other related topics

This conference can only succeed as a team effort, so the editors want to thank the international scientific committee and the reviewers for their excellent work in reviewing the papers as well as their invaluable input and advice.

In the end, we would like to thank the organization staff, the members of the program committees and reviewers. They have worked very hard in reviewing papers and making valuable suggestions for the authors to improve their work. We also would like to express our gratitude to the external reviewers, for providing extra help in the review process, and the authors for contributing their research results to the conference. Special thanks go to the publisher CRC Press/Balkema (Taylor & Francis Group).

We wish all attendees of ICEEE 2014 an enjoyable scientific gathering in Hong Kong.

The Editor

Organizing Committee

KEYNOTE SPEAKER AND HONORARY CHAIR

Gerald Schaefer, *Loughborough University, UK*

GENERAL CHAIRS

Wei Deng, *American Applied Sciences Research Institute, USA*
Ming Ma, *Singapore NUS ACM Chapter, Singapore*

PROGRAM CHAIRS

Prawal Sinha, *Department of Mathematics & Statistics, Indian Institute of Technology Kanpur, India*
Honghua Tan, *Wuhan University, China*

PUBLICATION CHAIR

Yijin Wu, *Huazhong Normal University, China*

INTERNATIONAL COMMITTEE

C.W. Kan, *Hong Kong Polytechnic University, Hong Kong*
Jiayi Wu, *Northeast Petroleum University, China*
Haijie Qin, *Dalian University of Technology, China*
Xinlin Liao, *Wuhan University, China*
GuoSheng Zhong, *Huizhou University, China*
Zeyang Pei, *Shanghai DianJi University, China*
Xunmin Ou, *Tsinghua University, China*
Huahua Xiao, *Guangdong University of Technology, China*
Y.L. Pan, *Hong Kong Polytechnic University, Hong Kong*
Yang Li, *China University of Mining and Technology, Beijing, China*
M.X. Huang, *Inner Mongolia Normal University, China*
Wenhai Lu, *Ocean University of China, China*
Xiaoping Zhou, *Beijing Normal University, China*
Jianguo Zhu, *Three Gorges University, China*
Lei Yang, *Tianjin Agricultural University, China*
Junlong Li, *Southeast University, China*
Chuande Zhou, *Chongqing University of Science and Technology, China*

Energy science and technology

Study on industrial structure and energy efficiency: An empirical analysis of China

XinLin Liao
School of Economics and Management, Wuhan University, Wuhan, China

ABSTRACT: This paper studies the relationship between industrial added value and energy efficiency in China over the period 2004–2012 by means of gray relational analysis. The empirical results show that some industry sub-sectors are high in industrial value but low in energy efficiency. China should pay more attention to the development of 'win-win' industrial sub-sectors which are able to both enhance industrial value and energy efficiency.

1 INTRODUCTION

As the main body of energy consumption in industry is the key to the whole energy efficiency in China, over the years China's industrial growth has relied heavily on high input and high energy consumption. The industrial sectors play a dominant role in total energy consumption, and industrial energy consumption accounts for about 70% of the total consumption. The rapid increase of imports and exports, especially in the processing trade, promoting the development of steel, nonferrous metal, building materials, chemical and other high energy consumption industrial sectors, is also an important reason for the rapid rise in the total energy consumption in China recently.

China is coming to the environmental constraint boundary of resources which is hindering sustained economic growth and the smooth progress of industrialization. China tries to pursue the win-win strategy of industrial structure and energy efficiency, so as to realize the long-term, steady and rapid growth of the national economy, and to make greater contribution to the world economy. During the past ten years, China has succeeded in reducing her overall industrial energy consumption, and has also made significant efforts in energy efficiency improvement in order to achieve the greatest progress in production capacity. Energy is the basis of a modern industrial economy. China is facing severe challenges from energy supply and environment protection due to heavy dependence on energy.

Many foreign and domestic scholars have given sustained attention to China's industrial structure and energy efficiency, a brief literature review as follows. Sinton and Levine (1994) studied China's industrial sectors data over the period 1980–1990 and believed that the improvement of energy efficiency was the main reason for the decline of energy intensity. Garbaccio and Jorgenson (1999) decomposed the reduction of energy use in China into structural change and technical change. Their main conclusion was that the structural change actually increased the use of energy while the technical change within industry sectors accounted for most of the fall in the energy output ratio between 1987 and 1992. Han and Wei (2004) divided the change of energy intensity into a structure share and an efficiency share and quantitatively analyses the structure share and efficiency share of the change in Chinese energy intensity. They thought that the decrease of Chinese energy intensity was mainly due to the increase in energy utilizing efficiency in the secondary industrial sectors.

Karen, Gary, Liu et al. (2004) analysed the panel data of large and medium-sized industrial enterprises in China, and the result was that the main reasons for the decline of energy efficiency were the reduction in coal consumption in the industrial sector, technological development, investment, and industrial restructuring. He and Zhang (2005) thought that the increasing of China GDP energy efficiency in recent years was mainly due to industry, particularly the heavy and chemical industries. Fisher-Vanden (2006) found that production structural changes were one of the major factors for the decline of energy efficiency in China. Hu and Wang (2006) insisted that the main influencing factors of China's regional energy efficiency include energy structure, industrial structure, technology, and so on. Guo, Chai and Xi (2008) analyzed the relationship between the primary energy structures, technique, management, and the energy use per unit of GDP, based on the data of the energy statistics from 1980 to 2004, and found that the major factor for the decline of energy efficiency was included in production structure.

Zhang, Liu and Cheng (2009) discussed the energy efficiency and its fluctuations in industrialized countries, and analysed the reasons for China's energy efficiency change and its development trends by using measurement analysis. By using the error correction model, they analysed the changing trend in energy efficiency under the conditions of the increasingly heavy industrial structure in China. They thought that industrial structure is the main reasons why China's energy efficiency had fallen after 2000, and insisted that with the formation and development of heavy industries, China's energy efficiency will continue to fall. Zhang (2010) studied the regional energy efficiency differences in China, and insisted that the energy efficiency in the eastern region is the highest, the middle region second, and the western region is third. He suggested that there was much more development space for energy efficiency and its optimization in the industrial structure.

Wang, Zheng and Wu (2011) thought that China's industrial restructuring had inhibited energy efficiency improvement, and the influence of industrial structural change was greater than the evolution of energy consumption structure, so the impact of industrial restructuring and upgrading was the key to improve energy efficiency and carry out energy saving policies. Zhu, Cao and Luo (2013) studied the characteristics of energy efficiency, industrial structure, and the energy consumption structure of China, and suggested that in the last forty years China's tertiary industrial sector had played a greater role than its second in improving energy efficiency. He insisted that the energy problem had already become an important restricting factor in China, and that the evolution of the industrial structure is one of the important factors to affect energy efficiency, and that the inhibition of secondary industry on energy efficiency improvement had not yet appeared in China, which indicated that China's economy will still rely on extensive and high energy consumption for a long time.

In this paper, I go through the main studies of previous literature and explore the relationship between industrial structure and energy efficiency in China, so as to determine which industrial sectors should be the priority development ones from the perspective of energy efficiency. My goal is to identify the key industrial sectors of 'win-win' which both increasing industrial output and energy efficiency.

2 MODEL AND ANALYSIS

The gray system theory deals with a system containing insufficient information, and it can effectively avoid subjective bias. In this paper, the gray system theory is adopted to evaluate the relationship between energy efficiency of China's industrial sectors and their industrial added value by means of establishing a gray associated sequence and calculating a gray correlation coefficient. A Gray Relational Analysis (GRA) model is a kind of impact measurement model of two series with named reference series and comparable series.

In this paper, take energy efficiency as the reference series, and industrial added value as the one

Table 1. The Energy Efficiency and the Gray Correlation Degree.

Sectors	Energy efficiency	Sort	Correlation degree	Sort
S_1	0.074	1	0.877	6
S_2	0.157	2	0.921	2
S_3	0.204	3	0.834	24
S_4	0.359	4	0.724	35
S_5	0.411	5	0.855	14
S_6	0.434	6	0.842	23
S_7	0.611	7	0.895	3
S_8	0.652	8	0.845	21
S_9	0.667	9	0.843	22
S_{10}	0.628	10	0.821	30
S_{11}	0.681	11	0.941	1
S_{12}	0.687	12	0.856	13
S_{13}	0.714	13	0.876	7
S_{14}	0.765	14	0.855	15
S_{15}	0.771	15	0.741	34
S_{16}	0.786	16	0.833	25
S_{17}	0.791	17	0.890	4
S_{18}	0.827	18	0.817	31
S_{19}	0.918	19	0.875	8
S_{20}	0.983	20	0.847	20
S_{21}	0.999	21	0.852	17
S_{22}	1.123	22	0.808	33
S_{23}	1.222	23	0.814	32
S_{24}	1.374	24	0.832	27
S_{25}	1.386	25	0.833	26
S_{26}	1.403	26	0.827	28
S_{27}	1.529	27	0.823	29
S_{28}	1.666	28	0.864	11
S_{29}	1.780	29	0.851	18
S_{30}	1.906	30	0.864	12
S_{31}	2.226	31	0.851	19
S_{32}	2.408	32	0.855	16
S_{33}	3.428	33	0.672	36
S_{34}	4.386	34	0.866	10
S_{35}	5.755	35	0.880	5
S_{36}	6.303	36	0.874	9

to compare. The energy efficiency data come from the average energy consumption (Tons of standard coal/yuan) of industrial added value of China's 36 industrial sectors from 2004 to 2012, the compare series are the industrial added value of the above-mentioned thirty-six sub-sectors. The data comes from China Statistical Yearbooks (CSY, 2005–2013). By the methodology of GRA, I summarize the gray relational degrees of China's industry energy efficiency and industry value added in Table 1, and I take also China energy efficiency data (in ascending order) in Table 1, so as to find out the "win-win" sub-sectors.

China's industrial sector has traditionally been disaggregated into 39 sub-sectors, however, I drop three sub-sectors of 'other mining industry', 'arms and ammunition manufacturing' and 'waste of resources and waste materials recycling industry' because gross output data for these sub-sectors will not be officially released for some years. Such practice should not have any significant impact on the final results due to the minimal gross output share of these sub-sectors in the whole industry.

The remaining thirty six industrial sub-sectors are by name as follows: Tobacco manufacturing (S_1), Oil & gas manufacturing (S_2), Transportation equipment manufacturing (S_3), Coal mining & extraction (S_4), Instrumentation & culture equipment (S_5), Electrical machinery & equipment (S_6), Pharmaceuticals manufacturing (S_7), Universal machinery & equipment manufacturing (S_8), Special equipment manufacturing (S_9), Textiles & garments shoes, caps manufacturing (S_{10}), Beverage manufacturing (S_{11}), Leather, furs, feathers manufacturing (S_{12}), Food manufacturing (S_{13}), Communication & electronic equipment (S_{14}), Nonferrous metals mining & extraction (S_{15}), Furniture manufacturing (S_{16}), Printing and copying (S_{17}), Metal products (S_{18}), Ferrous metals mining & extraction (S_{19}), Cultural, educational, sporting articles (S_{20}), Chemical raw materials & products (S_{21}), Agricultural by-product processing (S_{22}), Nonmetal mining & extraction (S_{23}), Plastic products (S_{24}), Nonferrous metals processing (S_{25}), Wood, bamboo processing (S_{26}), Water processing & supplying (S_{27}), Rubber products (S_{28}), Chemical fibre manufacturing (S_{29}), Textiles (S_{30}), Electric power & hot power (S_{31}), Papermaking & paper products (S_{32}), Gas producing & supplying (S_{33}), Oil processing, coking (S_{34}), Nonmetal mineral products (S_{35}), Ferrous metals smelting & rolling (S_{36}).

3 CONCLUSION

Table 1 indicates that the five sub-sectors, i.e. Tobacco manufacturing (S_1), Oil & gas mining (S_2), Pharmaceuticals manufacturing (S_7), Beverage manufacturing (S_{11}), Food manufacturing (S_{13}), are the 'win-win' industry sub-sectors which are both high in energy efficiency and in industrial added value. These kind of industrial sub-sectors should be given priority in development.

Some sub-sectors rank bottom in gray correlation degree but they rank top in energy efficiency, such as Transportation equipment manufacturing (S_3), Coal mining & extraction (S_4), Instrumentation & culture equipment (S_5), Electrical equipment (S_6), Universal equipment manufacturing (S_8), Special equipment manufacturing (S_9), Textiles & garments shoes, caps manufacturing (S_{10}). Although they cannot play an important role in creating value, their energy efficiency is high and attention should also be paid to them.

REFERENCES

Fisher-Vanden, K., Jefferson, G. H. & Jingkui, M. 2006. Technology development and energy productivity in China, *Energy Economics* 28(5): 690–705.

Garbaccio, R. F., Ho M. S. & Jorgenson D. W. 1999. Why has the Energy Output Ratio Fallen in China? [J]. *Energy Journal* (3): 63–91.

Guo, J., Chai, J. & Xi Y. M. 2008. Analysis of Influences between the Energy Structure Change and Energy Efficiency in China, *China Population, Resources and Environment* 18(4): 41–46.

Han, Y. Y. & Wei, M. 2004. Research on Change Features of Chinese Energy Intensity and Economic Structure, *Statistics and Management* (1): 1–6.

He, J. K. & Zhang, X. L. 2005. Analysis on the impact of the structural changes in the manufacturing industry on the rising of efficiency of GDP resources and its trend, *Environmental Protection* (12): 37–41.

Hu, J. L. & Wang S. C. 2006. Total-factor energy efficiency of regions in China, *Energy Policy* 17(5): 3206–3217.

Karen, F. V., Gary, H. J., Liu, H. M. & Tao, Q. 2004. What is driving China's decline in energy efficiency? *Resource and Energy Economics* 26(1): 77–97.

Sinton, J. E. & Levine. M. D. 1994. Changing Energy Intensity in Chinese Industry; the Relative Importance of Structural Shift and Intensity Change. *Energy Policy* (22): 239–255.

Wang, Q., Zheng, Y. & Wu, S. D. 2011. Mechanism of Energy Efficiency Response to Industrial Restructuring and Energy Consumption Structure Change, *Acta Geographica Sinica* 66(6): 741–749.

Zhang, W. S. 2010. Energy Efficiency, Industrial Structure and Regional Economic Development in China, *Journal of Shanxi Finance and Economics University* (7): 63–69.

Zhang, Y. X., Liu, J. & Cheng, J. H. 2009. Changing Trends and Adjusting Policies of China's Energy Efficiency: An Empirical Analysis from the Perspective of Heavy Industrial Structure, *Chinese Journal of Management* 6(6): 818–822.

Zhu, S. Q., Cao, W. D. & Luo, J. 2013. Research on Regional Differences of Mechanism of Energy Efficiency Response to Industrial Restructuring in China, *Human Geography* 28(6): 118–124.

Research into different pretreatment methods to enhance the enzymatic saccharification of Green Bamboo

JianJun Liu
School of Mechanical & Electronic Engineering, Sanming University;
Sanming Engineering Research Center of Mechanical CAD;
Post-doctoral Research Station of Fujian Sanming High-tech Industrial Development Zone

ABSTRACT: Bamboo is a kind of annual output of high biomass material. Green Bamboo has a 12,000 kg yield per hectare, and is very suitable for the production of cellulosic ethanol. We use Green Bamboo as a kind of biomass; it will simplify the processes and reduce production costs. It is pretreated with hot water, liquid ammonia, and steam explosion. Then cellulase and amylase for hydrolysis are added. Analysis is made from the samples of hydrolysis solution and filtrate by HPLC (high performance liquid chromatography), using the Biorad Aminex HPX-87H column for glucose, xylose, and arabinose. The results showed that the glucan/xylan conversion was greatest under hot water conditions for ten minutes. They were 28.45% and 45.55% respectively, compared with untreated biomass where they increased by 111.07% and 224.23%. The research conclusion has important significance for biomass energy conversion and utilization.

1 INTRODUCTION

Biomass energy is the human use of fire, the direct application of energy. With the progress of human civilization, the application of biomass energy research and development, after several twists and turns, before and after the second world war, European wood energy reached research peak, and then with the development of the petroleum chemical industry and coal chemical industry, the application of biomass energy gradually tended to be low[1]. By the 1970s, triggered by the global energy crisis because of the war in the Middle East, renewable energy, including wood energy development and the utilization of research, caused people to look at it again[2][3]. People deeply recognized that oil, coal, natural gas, and other fossil energy resources had limitations and environmental pollution problems. Relevant data is introduced, according to the already proven reserves and demand, to the middle of the 21st century, and with the world's oil and natural gas resources drying up, and use of coal, not only its own capacity being limited, as a result of combustion produces a large amount of SO_2, CO_2, causing serious environmental pollution[4]. Increasingly serious environmental problems, which have aroused the international community's common concerns, are closely related to the energy problem, which is one of common focus of the world today. There are data showing that the use of fossil fuels is the main cause of air pollution. 'Acid rain, 'greenhouse effect', and so on have brought catastrophic consequences to people on earth[5]. However, the use of nature's gift of biomass energy almost does not produce pollution; almost no SO_2 is produced, in use process of CO_2 and need to absorb a large number of CO_2 in the process of plant growth in quantity balance, known as CO_2 neutral fuel. Biomass energy is renewable and will not dry up, at the same time; it plays an important role in the protection and improvement of the ecological environment and is one of the ideal renewable energy sources.[6]–[10].

By the end of 2009, China's remaining economic recoverable reserves of oil were 1.478 billion t, and China's crude oil output was 189.4 million t, and with mining at the current rate, the existing reserves can only last eight years. On the other hand, consumption of fossil fuels generates hundreds of millions of t SO_2, NO_x and billions of t of CO_2 emissions, thus, no matter from the point of view of energy security or environmental protection, the search for clean alternatives to fossil fuel and renewable energy is imperative. Cellulosic ethanol fuel, compared with gasoline, can reduce greenhouse gas emissions by 86%[10]. The U.S. government in 2007 proposed a 20% fuel alternative plan[11], funded by the U.S. department of energy in 2009~2009, investment construction was made in six pilot studies of plant cellulose ethanol[12], and it can be seen that the use of lignocellulose as a raw material and the use of enzymatic hydrolysis fermentation technology for ethanol are priority strategies in the development of bioenergy in developed countries.

Green Bamboo[13] belongs to the *gramineae bambusoideae* bamboo species, widely distributed in southern ZheJiang, FuJian, Taiwan[14], GuangXi and GuangDong and other places, which is one of the key national forest regions of FuJian province, a provincial forest area of 100 million mu, covering 57.3% of the province and being the largest in China, in terms of the development of energy agriculture, energy, forestry resource conditions. About 13,300 hm² flicker existing bamboo resources in FuJian province, are mainly distributed in the east of NingDe city, XiaPu county (city), NanPing, SanMing county, north central FuZhou region, QuanZhou, ZhangZhou and other regions in the south.

It is mainly composed of plant cell walls, basic components, such as cellulose, hemicellulose and lignin, cellulose and has a kind of linear polysaccharide, with multiple molecules parallel arranged in filamentous insoluble tiny fibres, hemicellulose is mainly composed of xylose, a small amount of arabinose, galactose, mannose, lignin is benzene propane, as the basic unit of polymer aromatic compounds and derivatives. These complex spatial structure allows plants to avoid attack by microbes and various physical and chemical factors. When using cellulose enzyme hydrolysis biological resources to make ethanol, wood cellulose enzyme must contact adsorption onto cellulose substrates to make a reaction. The presence of lignin hinders the accessibility of cellulose of enzyme, and cellulose crystal structure and surface state, multi-component structure, the protective effect of cellulose and lignin cellulose, covered by hemicellulose, flickers the bamboo structure and chemical composition of the factors such as difficult to hydrolysis[15]. Therefore it must undergo pretreatment to separate the cellulose, hemicellulose, and lignin to open and cut off their crystalline structure of the hydrogen bond damage, reduce the degree of polymerization[16], according to statistics, pretreatment of the production cost as much as thirty cents a gallon of ethanol[17], is 1/4~1/3 of the cellulose ethanol production cost.

Flicker is bamboo research and is mostly focused on its economic aspects such as the green bamboo shoots grow flicker processing, bamboo fibre paper, making raw materials, bamboo processing Chinese medicinal materials, flicker and bamboo revetment dike to beautify the environment, etc. Flicker and the use of bamboo as raw material to produce cellulosic ethanol in the world is still in its infancy, but its development prospect is considerable[18]. Flicker in bamboo as raw material, the methods of hot water, liquid ammonia, steam explosion for digestion experiment was carried out after pretreatment of material, using high performance liquid chromatography (HPLC) analysis of sugar content, the comparison of different pretreatment methods on the influence of various enzyme solutions of sugar conversion rate, for the further utilization of hetian, and flickering the bamboo as a raw materials to produce fuel ethanol.

2 MATERIALS AND METHODS

2.1 Experimental materials

Biomass has been adopted in the experiment for three years: flicker of raw bamboo, collected mostly from the county of FuJian Province. An entire bamboo tree (the part on the ground, including the bamboo itself and the bamboo leaves) is cut into less than 1 cm long flakes, dried at 40°C to a moisture content lower than 15%, with the tiny plants pulverized and shattered into 50~100 mesh powder. After mixing and measuring the moisture content it is set in a in the rear plastic seal at −20°C in the refrigerator.

For an HPLC analysis of a reference substance, such as glucose or xylose, it is sourced from Sigma Company, with a purity of more than 99%.

2.2 Methods

The experimental analysis method, with reference to the National Renewable Energy experiment regulations, formulated a Laboratory Analytical Procedure, with some appropriate improvement in experiments within the relevant regulations.

2.2.1 Acid hydrolysis

The acid hydrolysis method to determine the biomass of carbohydrates and insoluble lignin components and carbohydrate composition is determined by LAP002, an insoluble lignin, determined with LAP003.

With the weight known, the first steps in the acid hydrolysis of the moisture content of biomass at about 0.3 g, is added 72% sulfuric acid 3 ml, keep 2 h in 30 DHS C water bath, adding demonized water 84 ml (diluted to 4% sulfuric acid), in the sterilization pot 121 DHS C keep 1 h. After vacuum filter to filter on the solids drying to constant weight, using HPLC to determine the sugar content, and calculate the dextran (cellulose, xylan and araban and lignin content.

2.2.2 Hot water pretreatment

Hot water treatment technology refers to the material added to the hot water under high pressure (by pressing to prevent water evaporation at more than 100 degrees Celsius), heating to a predetermined temperature T (depending on the power of the heater, for

Figure 1. Reaction kettle of Hot water.

Figure 2. Equipment of AFEX Pretreatment.

Figure 3. Equipment of Steam explosion Pretreatment.

about 40 mins), recording the preheating time at maximum pressure, and keeping the time T (timer). Then the reactor flux of cooling water for cooling below 60°C. Samples after take out the sample, vacuum filter, filter into the evaporating dish (plate), the sample in 45°C oven to keep 12 h, or take out after weighing, and the determination of moisture content, determine the rate of processing, sample – 20 DHS C refrigerator to be processed.

2.2.3 AFEX pretreatment

Ammonia Fibre Expansion (AFEX) pretreatment process is: firstly to process the raw materials into the reactor, then placed into a suitable amount of liquid ammonia, heated to a predetermined temperature (about 20~30 mins, heating depending on the power of heater), then kept at a certain amount of time (control) at a predetermined temperature, and then ammonia was quickly released, taking out the processed raw material in a dry ventilation cabinet at room temperature for 12 h. Weighed and measured, the moisture content, determined, the raw material yields, after pretreatment, into the cold standby – 20 DHS C.

2.2.4 Steam explosion pretreatment

Steam method refers to the pressure of 0.69 MPa to 4.83 MPa, under the condition of water will be lignocellulose material heated to 200°C~240°C, lasts 30 s~20 min, make the space inside the high pressure water vapour permeability to the raw material in the hydrolysis of hemicellulose, and lignin can be softened and part of the degradation. Then all of a sudden after depressurization, lignocellulose material surged instantly out of the gas phased medium instantly surged, raw material in high temperature liquid bumping formation flash quickly, have a huge energy cause blowout of cellulose crystal and fibre bundles, the raw material from the intercellular layer dissociated cells into single fibre, increase the susceptibility.

2.2.5 Enzyme hydrolysis

Enzymatic hydrolysis by reference to the relevant procedures, LAP009 reaction amount from the original 10 ml to 15 ml. Said take 0.15 g glucan equivalent of biomass in the enzyme solution bottle (20 ml sample bottles), in turn, add distilled water, sodium citrate buffer solution (after balance pH value of 4.8), antibiotics (tetracycline and cyclohexanol imide), by premixed 1 hour after adding cellulose enzyme, beta glycosidase enzymes, amylase and xylosidase. Related enzyme added amount of microcrystalline cellulose and starch mixed optimization simulation; see the instructions in the result analysis. Enzymatic hydrolysis conditions are a temperature setting of DHS 50C, shaking table speed is 150r. min^{-1}. Enzyme solution sampling time of 24 h, 72 h.

Quantitative analysis by HPLC method, chromatograph for Agilent Technologies, 1200 Series, using the differential detector. Chromatographic Column for the HPX – 87H Ion Exclusion Column 300 mm × 7.8 mm, mobile phase 0.05 mol·L^{-1} dilute sulphuric acid, flow rate of 0.5 mL·min L^{-1}, Column temperature of 45°C.

Table 1. The analysis of the components of green bamboo (dry basis).

Components	Green bamboo (%)
Glucan	40.87 ± 0.37
Xylan	14.68 ± 0.14
Lignin	34.15 ± 0.03
Ash	3.71 ± 0.10
Extract*	6.48

*Using 95% ethanol soxhlet extraction.

3 RESULTS AND DISCUSSION

3.1 Raw material composition analysis

The composition of raw materials such as shown in table 1, carbohydrates and lignin content, through the analysis of the acid hydrolysis ash obtained from muffle furnace heating analysis. Glucan containing cellulose and starch, cellulose content is 40.87%, hemicellulose content is 14.68%, total 55.55%, the analysis of the components of the bamboo both at home and abroad are basically identical.

3.2 Hot water pretreatment effects on digestion simple sugars yield

According to the basic pretreatment conditions of bamboo, we will hot water temperature control at 190°C, holding time, respectively for 30 min, 20 min, 10 min, and 5 min. Enzyme solution results as shown in figure 4.

Place the cursor on the T of Title at the top of your newly named file and type the title of the paper in lower

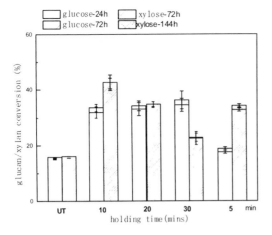

Figure 4. Different retention time in hot water pretreatment glucan/xylan enzymatic hydrolysis conversion rate.

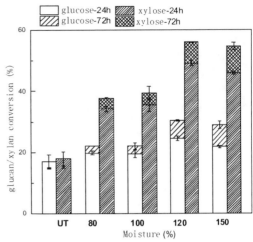

Figure 5. Different moisture content in ammonia explosive pretreatment glucan/xylan enzymatic hydrolysis conversion rate.

case (no caps except for proper names). The title should not be longer than 75 characters). Delete the word Title (do not delete the paragraph end).

Without hot water treatment of biomass, glucose conversion rate is 13.48%, xylose conversion rate is 14.05%; the transformation effect is the worst. When the heat preservation time of 30 min, glucan conversion rate is highest, 34.56%. But the amount of total sugar conversion is highest at the time of heat preservation for 10 min, biomass glucose conversion rate is 28.45%, xylose reached the highest conversion rate is 45.55%, less preprocessing enzymolysis effect of glucose and xylose conversion rate increased by 111.07% and 224.23% respectively. 30 min time of heat preservation materials but its high conversion rate of glucose and xylose conversion rate than other processing materials decreased obviously, the reason may be that under high temperature and high pressure retention time is too long, resulting in a large number of hemicellulose degradation, which is similar to steam explosion pretreatment. We are to deal with after the filtrate of composition analysis, almost no found simple sugars, can decompose the furfural and other substances.

3.3 Ammonia explosive pretreatment effects on digestion simple sugars yield

Because we select flicker of bamboo raw plant for 3 years, the crystal structure of density is higher, the ammonia pretreatment at low temperature the effect is not obvious, so we choose the temperature of 190°C, the ammonia with dry biomass loading ratio of 2:1, the processing time for 10 min, relative water were 80%, 100%, 120% and 150%. Enzyme solution results as shown in figure 5.

Untreated biomass enzymatic hydrolysis of glucose conversion rate is 17.05%, xylose conversion rate of 17.88%. The moisture content of 120% ammonia explosive biomass conversion rate as high as 24.499% glucose, xylose conversion rate is 49.13%, less preprocessing effect of enzymatic hydrolysis of glucose and xylose conversion rate increased by 43.64% and 174.48% respectively.

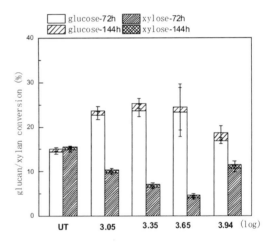

Figure 6. Steam explosion pretreatment of different steam explosion intensity glucan/xylan enzymatic hydrolysis conversion rate.

3.4 Steam explosion pretreatment effects on digestion simple sugars yield

According to the research of the Overend and others, the introduction of the concept of steam explosion intensity R_0: $R_0 = t \times \exp(T - 100)/14.75$, where t = pretreatment time; T = steam temperature. In this experiment, by steam explosion intensity index log (R_0) steam explosion on the strength of the material processing. We investigated materials for 3.05, 3.05, 3.35, 3.94 four intensity of steam explosion pretreatment. Enzymatic hydrolysis results are shown in figure 6.

10

Untreated biomass glucose conversion rate was 14.4%, the conversion rate of xylose is 15.03%. Relatively the best effect is 3.65 steam explosion intensity is 190°C temperature to maintain 10 min, the digestion of glucose conversion rate is 23.71%, relatively unprocessed materials increased by 64.5%. In addition, the xylose conversion rate is 6.5%; it decreased. The reason caused a large number of hemicellulose degradation, resulting in a decline in the xylan content in the raw material of the after pretreatment, because under the effect of high temperature, the water had the effect of acid, generated by the acid as a catalyst and accelerated the decomposition of hemicellulose.

4 CONCLUSIONS

4.1 Text and indenting

Experiments on different pretreatment process of the flicker through the bamboo, like a conclusion:

(1) hot water treatment pretreatment conditions for 10 min, the heat preservation time biomass conversion rate of 28.45% glucose, xylose reached the highest conversion rate is 45.55%, less preprocessing effect of enzymatic hydrolysis of glucose and xylose conversion rate increased by 111.07% and 224.23% respectively. Hot water treatment effect is relatively good, glucose and xylose conversion rates are relatively high. The accurate control of the temperature in the process of the experiment is the main problem, if the temperature is too high and produce aldehyde substances may inhibit fermentation, reduce the production of ethanol.

(2) the ammonia explosive pretreatment processing in condition of 110°C, the ammonia and dry biomass base load ratio of 2:1, the processing time is 10 min, relative water cut is 120%, the enzymatic hydrolysis of glucose conversion rate as high as 24.49%, xylose conversion rate is 49.13%, less preprocessing effect of enzymatic hydrolysis of glucose and xylose conversion rate increased by 43.64% and 174.78% respectively. Ammonia pretreatment can effectively remove lignin mainly, but compared with the method of steam, halfcellulose is dissolved. It can effectively destroy the crystal structure of cellulose, improve digestion rate. Small hydrolyzes inhibition of fermentation, ammonia can be recycled, pollution is small. Equipment investment cost is low, the operation is simple, but the energy consumption. Generally speaking, this kind of technology is still the most promising pretreatment. At present, the use of this technology has not been reported in China.

(3) ammonia pretreatment in steam explosion intensity of 3.65, digestion of glucose conversion rate is 23.71%, relatively unprocessed materials increased by 64.5%. Steam explosion method has the advantage of low energy consumption, can be batch also can continuous operation, the main raw materials and crops straw, suitable for hardwood faults is the number of xylose loss, effect of cork is poorer, and produce the material that is harmful to the fermentation pretreatment intensity, the greater the cellulose enzyme hydrolysis, the easier, but also by hemicellulose get sugar is less, and the fermentation the more harmful substances. However, the operation involves high pressure steam blasting equipment, the investment cost is higher, and produce fermentation inhibitor.

(4) pretreatment method for biomass species selective, structure and composition of green bamboo factors limit the lignocellulose cellulase for plants in cellulose enzyme solution, it is a kind of biomass pretreatment is difficult, to optimize the pretreatment method or screening is very important.

ACKNOWLEDGEMENT

The paper is supported by the Key Program of Agriculture Research of Science and the Technology Department of Fujian (NO.2012N0023), the Foundation of Educational Research of Young Teachers Fu'jian Educational Committee (JA. 13293), the Foundation of Educational Research of Young Teachers Fu'jian Educational Committee. (JB. 13183), and the Foundation of Teaching Reform of Higher Education of SanMing University (J1309/Q).

REFERENCES AND NOTES

A. P. C. Faaij, Energy Policy 34, 322 (2006).
U.S. Department of Energy, Office of the Biomass Program, 30×30 Workshop, Washington, DC, 1 to 2 August, 2006 (www.30x30workshop.biomass.govtools.us/).
Office of the Biomass Program (OBP), "Multi-Year Program Plan, 2007–2012" (OBP, U.S. Department of Energy, Washington, DC, 2005) (http://www1.eere.energy.gov/biomass/pdfs/mypp.pdf).
Biofuels Research Advisory Council, "Biofuels in the European Union: A Vision for 2030 and Beyond" (2006); available online (www.biomatnet.org/publications/1919rep.pdf).
J. Houghton, S. Weatherwax, J. Ferrell, "Breaking the Biological Barriers to Cellulosic Ethanol: A Joint Research Agenda," Biomass to Biofuels, Rockville, MD, 9 December 2005 (http://genomicsgtl.energy.gov/biofuels/b2bworkshop)
U.S. Department of Agriculture–National Resource Conservation Service, National Resources Inventory: 2001 Annual NRI (NRCS, USDA, Washington, DC, 2003) (www.nrcs.usda.gov/technical/land/nri01/nri01lu.html).
R. D. Perlack et al., "Biomass as Feedstock for a Bioenergy and Bioproducts Industry: The Technical Feasibility of a Billion-Ton Annual Supply"(Oak Ridge National Laboratory Report TM-2005, under contract DOE/GO-102005-2135, Oak Ridge, TN, (2005).
D. Tilman, J. Hill, C. Lehman, Science 314, 1598 (2006).
S. Nonhebel, Renew. Sustain. Energy Rev. 9, 191 (2005).
Wang M., Hong MW, Hou H. Life-cycle energy and greenhouse gas emission impacts of different corn ethanol

plant types [J]. Environmental Research Letters, 2007, 2(2):9–22.

Office of the Biomass Program (U.S. Department of Energy, Washington, DC.). Biomass Multi-Year Program Plan[R]. 2007.

Service R.F. Cellulosic ethanol: biofuel researchers prepare to reap a new harvest [J]. Science, 2007, 315 (5818): 1488–1491.

QingFang Zheng, YiMing Lin. Flora of FuJian [M]. Fuzhou, FuJian Science and Technology Press, 1995: 29–96.

Himmel M E, Ding SY, Johnson D K, et al. Biomass recalcitrance: engineering plants and enzymes for biofuels production [J]. Science, 2007, 315(5818): 804–807.

Mosier N, Wyman C, Dale B, et al. Features of promising technologies for pretreatment of lignocellulosic biomass [J]. Bioresource Technology, 2005, 96: 673–686.

Teymouri F., Perez L.L., Alizadeh H., Dale B. E. Optimization of the ammonia fiber explosion (AFEX) treatment parameters for enzymatic hydrolysis of corn stover [J]. Biorsource Technology, 2005, 96: 2014–2018.

Mosier N, Wyman C, Dale B, et al. Features of Promising Technologies for Pretreatment of Lignocellulosic Biomass [J]. Bioresource Technol, 2005, 673–686.

Yu G, Yano S, Inoue H, Inoue S, e t al .Pretreatment of Rice Straw by a Hot-compressed Water Process for Enzymatic Hydrolysis[J] . Appl Biochem Biotech, 2010, 539–551.

Research in the specialized construction of new energy science and engineering

Lei Yang, WeiYu Zhang, DengChao Jin, Li Wang & ZhenBo Bao
College of Engineering and Technology, Tianjin Agricultural University, Tianjin, China

ABSTRACT: New energy science and engineering, as major bachelor degrees, related to the national strategies of new industries, was first started in 2010. Thus, it is still in an exploratory stage of a specialty construction plan and a talents-training program. Explaining the background of the development of new energy science and engineering, based on the social development of Tianjin's regional economies, this research puts forward the plan for the construction of new energy science and engineering training skills.

1 INTRODUCTION

New energy, also known as unconventional energy, as various forms of energy except conventional energies, includes the energy just beginning to develop or being studied positively and evangelized such as solar power, geothermal energy, wind energy, ocean energy, biomass energy, fusion energy, etc. With common characteristics of less pollution and large reserve, there is great significance for solving the problems of environmental pollution and resource exhaustion (especially fossil fuel). Moreover, it is highly important for national energy security that new energies have the features of renewability and uniform distribution (uneven distribution of conventional energy like oil is the root cause of war).

2 NECESSITY OF ESTABLISHING NEW ENERGY PROFESSIONS IN COLLEGES IN CHINA

Although China is a big energy producer and consumer, energy resources are extremely limited and per capita consumption of energy is far less than the global average. According to the relevant data, per capita possession of economically recoverable reserves (101t) of coal in our country is no more than 46% of the average amount of world consumption (222t). The per capita occupation rate of recoverable hydroelectric potentiality (1955 KWh) is 81% of the world average (2423 KWh). The remaining recoverable reserves of oil account for 6.8% of the world average. The per capita recoverable reserves of gas is merely 1.5% of the world average. To meet the energy needs for economic development, and our country is obliged to export the external oil, participating in the competition for energy markets all over the world. Due to fast economic development, the increasing needs for energy year by year, the decrease of energy reserves day by day and the constant exploitation, thus, all lead to the obvious imbalance between supply and demand of energy.[1] In addition, it has been the primary choice that the program of new energy substitution is carried out in our country to get rid of the scarcity of conventional energies and the stress generated by more serious environmental problems caused by energy development.

On the basis of the anticipation by United Nations Organization and the International Energy Agency, human sustainable development can't do without new energy exploitation, the same as the high-speed sustainable development of our country' economy. Currently our country is the second energy producer and consumer around the world. The increasing supplies of energy support mainly the development of the economy and society. To fulfil economic sustainable development, the Twelfth Five-Year Plan gives priority to the energy development of conventional energies, new energies, energy conservation and emission reduction, etc. Energy has been the important basis of the national economy developing rapidly in our country. According to the national medium and long-term program, this is the critical period to develop new energy sources and renewable energy from 2000 to 2020. Before 2020, there has been a general objective for discovering renewable energy: Increasing the proportion of the consumption of renewable energy as a proportion of the total amount of energy consumed, solving the problems of giving access to electricity for people who are without electric power in remote districts and who experience a shortage of vital rural fuels, transforming organic waste into energy, and pushing ahead with industrial development of renewable energy techniques. In accordance with *medium and long-term plans for renewable energies*, it is focusing on water power, biomass energy, wind energy and solar energy in future. By 2020, there are several goals to achieve: 300 million kilowatts from wind, 30 million kilowatts from biomass, 1.8 million kilowatts from solar energy, 300 million square metres of solar water heater, 4.4 billion cubic metres of biogas every year, 50 million tons of fuels generated by biomass, and 10 million tons oil fuels, replaced by non-biomass liquid fuels. For those, approximately two billion yuan will be invested in the development of renewable energy, and about 1.5 billion yuan in the period between now and 2020. According to investment stimulating employment formulas used in relevant departments, the amount of per million fixed capital investment to stimulate employment remains between 297–706. 1.5 billion yuan provides 7110 thousand jobs at the average of 474 people/million. Therefore, during the 12th Five-Year Plan, the needs of talents in the field will rise sharply. With a strategic emerging industry profession in 2011, new energy science and engineering is aimed at training professionals in the proper exploitation of energy resources and a highly efficient and clean utilization of new energies, taking up the heavy responsibility of relieving the tension of needs of new energy talents.[2],[3]

The establishment and education of new energy courses in our country's colleges lags behind those of developed countries. As the times require, the Department of Education examined and approved eleven colleges such as Zhejiang University, Central South University, Jiangsu University, to set up new energy science and engineering courses in July 2010 in order to train professionals for the emerging priorities.

3 OVERALL GOAL AND MEASURE OF SPECIALTY CONSTRUCTION

On the basis of the full investigation of professionals related to new energy at home and abroad, the new energy industry characteristics and the demands of enterprises and society to the knowledge structure and ability structure of professionals, analyzing the requirements of social and economic development of our country, at the same time, concentrating on the demand for energy professionals in the social development of Tianjin's regional economy, combined with the college's own subject characteristics, the plan of new energy professional talent training is established, which mainly includes construction goals, scientific and reasonable curriculum systems and feasible teaching plans, etc.[4]

3.1 Construction goals

The advanced education thinking (e.g. Cognitive Flexible Theory) has been introduced with the consideration of the need for energy professionals for the development of Tianjin's regional economy. Firstly, centring round the 'General simulation work training system', the humanistic education is strengthened by hierarchically training, enabling students to have a strong sense of dedication and responsibility, a good ethics, work ethics and legal consciousness. Meanwhile, optimizing the specialty's structure to improve its whole leverage, practical talents training base is erected by the further cooperation between the universities and enterprises to strengthen effective links between the specialty chain and industrial chain. The quality of skills is enhanced all-around by a trinity of talents training alliances and collaborations with enterprises, colleges and research institutes with more competitiveness in teaching conditions. Teachers, training modes, training programs, course systems and teaching contents, teaching methods to achieve the goal of advanced education theory, tangible training target, leading teaching reform, teachers optimized and excellent teaching achievements.

3.2 Construction thoughts and implementing scheme

3.2.1 Orientation serving social development in Tianjin's regional economies

Centring the Agricultural Technology Innovation Project and Agriculture Facilities Upgrade Project put forwarded by Tianjin, an engineering technology platform of the waste recycling and green power efficiently used in the productive process of "large-scale agriculture (suburb farming and urban greening)" with metropolitan agriculture characteristics. In addition, on this basis, the good situation of the win-win and interactive developing are achieved by combining personnel training and local needs to cultivate innovation and compound talents qualified for Tianjin's agriculture and industry circles.

3.2.2 Goal: Cultivating innovation and compound talents

Innovation and compound talents are urgently demanded in recent times, as well as the premise of training excellent energy engineers. To this end, a new education system with art education blending and profession crossing is erected for the education reform, essentially primary education, primary profession education and professional education, with general simulate post training throughout it, to explore a hierarchical personnel training mode of wide specification, good foundation, extraordinary ability, high quality and preeminent individuality with the ability training as the master line.

3.2.3 Method paying equal to theory and practical teaching

The flexible teaching system and the teacher communication platform are established, adhering to the reform ideas of human-rooted, student-centric, and the goal of innovation and practice, breaking bottlenecks of primary education, primary profession education and professional education. First, qualities are paid more attention including thoughts, orientation, methodological basis and the comprehensive basis of overall control while emphasizing knowledge accumulation of basic skill, professional foundation and primary education, interests and curiosity in their specialties are enhanced under the stimulation of primary education. Second, it is a main approach, vigorously implementing the practical program of simulative posts, students' scientific and technological innovative activities, seamless cooperation project between school, enterprises and research institutions, to enable students to solve practical problems by the theory, personality development by training hierarchically as the primary measures to accelerate adaptation of the society and career.

3.3 *Measures of curriculum system reform*

There is a proper proportion between all kinds of courses to optimize curricular design and content, balancing spirit cultivation of humanity and spirit cultivation of science, public primary course, professional basis and professional course, curriculum system and the training target for the integrative optimization of curriculum system. Main course systems are initially formed which consist of municipal selected courses, school-level selected courses and high quality courses to perfect curriculum system, reflecting the features of subject development, the professional characteristics, the harmonization between course curriculum design and social needs, and the great connection among the courses.

New Energy Science and Engineering belongs to interdisciplinary field, coming down to extensive subjects such as Physical Engineering, Energy Source Engineering, Power Engineering, Electronic Engineering, Automatic Control, Material Science, Mechanical Engineering and Civil Construction. The setting of curriculum not only enriches students' basic theories and specialized knowledge, but also experiment skills and the capabilities of experiment design. It is appropriate to enhance practical training and experimental curriculum to not only cultivate manipulative ability and innovation ability, but also have a better knowledge of the subject's forefront information and development tendency, forming a relatively systematic curriculum system. The system is made up of six parts: Liberal education, public foundation, professional basic course, compulsory course, elective course and practice work.

4 CONCLUSION

New Energy Science and Engineering is emerging as a strategically important component of specialized undergraduate courses in universities, of which training program's design and content are keeping pace with the development of new energy science. A scientific, advanced and developmental curriculum system has been established, based on the professional training aims of tracking dynamically to create training patterns. Specialty construction cultivates positively high-grade specialized talents for the development of new energy industries, based on the needs of businesses and society, remaining relationship with industry, subject docking to industry and major docking to professions.

ACKNOWLEDGMENTS

This work was financially supported by the Tianjin Normal undergraduate teaching quality and teaching reform research project (C03-0805) and Tianjin Agricultural University Research and Education Reform Project (12-A5-12).

REFERENCES

Qin Yuanping. Thinking of Energy Sustainable Development in Our Country's New Situation, *Energy of China*, Vol. 28 No. 6, pp. 45–46, 2006

Ren Dongming. New Energy Development and System Innovation in China, *Sino-Global Energy*, vol. 16, pp. 31–36, 2011

Chen Xuejun. Suggestions on New Energy Science and Engineering, *Bulletin of Chinese Academy of Sciences*, vol. 20, pp. 451–455, 2005

Han Xinyue, He Zhixia, Wang Qian, Ji Hengxong. Research on Cultivating Professional Talents of New Energy Science and Engineering, *China Electric Power Education*, No. 5, pp. 9–11, 2013

Catalytic hydrogenation liquefaction of lignite in supercritical water

MinLi Wang, QingQing Guan, QiuLin Zhang, Ping Ning, JunJie Gu, SenLin Tian, QiuLing Chen & RongRong Miao
Faculty of Environmental Science and Engineering, Kunming University of Science and Technology, Kunming, China

ABSTRACT: Catalytic hydrogenation liquefaction of lignite in supercritical water with catalyst Pt/C, Pd/C, Ru/C and Ni/SiO$_2$/Al$_2$O$_3$ was investigated. The results indicated that the oil yields were dramatically improved by four heterogeneous catalysts, especially with Ru/C and Ni/SiO$_2$/Al$_2$O$_3$, It s also found that four catalysts have some different selectivity to the liquefied products, especially the oil generated by Pd/C with large amounts of long-chain alkane, which was considered to improve the energy density of oil. These results indicate that a one-step liquefaction treatment strategy at a low temperature 380° represents an efficient option for the hydrothermal catalytic upgrading of lignite.

Keywords: catalytic hydrogenation liquefaction; lignite; Supercritical water; catalysis

1 INTRODUCTION

Lignite is considered as low rank coal (high ratio of C/H and large aromatic content) with high moisture content (10–40%), sometimes as high as 66% and a low heating value only 10–20 MJ/kg. It is mined worldwide, China having an abundant amount of lignite reserves of about 128 billion tons, mainly located in Inner Mongolia, the north-eastern provinces, and Yunnan province. Although production is large, it is unsuitable for long-term storage, uneconomic to be transported, and widely used due to its unique natural properties.

Water at temperature and pressure around its thermodynamic critical point (T$_c$ = 647 K, P$_c$ = 22.1 MPa) is miscible with organic compounds and gases, so that a single homogeneous fluid phase can be achieved. Since Supercritical water liquefaction takes place in water around its critical point, liquefaction reactions are rapid as the mass transfer rates are dramatically increased. Supercritical water liquefaction (abbreviated as SCWL) is a promising method for converting wet substances like algal biomass and lignite et al. into liquid fuels, because it obviates the need (capital, energy and time) for feedstock dewatering and drying as is needed for other conversion methods such as gasification and pyrolysis. Furthermore, the oil from the SCWL has lower oxygen content and a higher energy density than the oils produced from fast pyrolysis [1].

There are a lot of works about SCWL [2, 3] and few reviews are available. However, a systematic study on the liquefaction of lignite with catalytic hydrotreatment was not found in the literature. Moreover, the liquefied products of crude oil are common in high heteroatoms (O, N, S) content and low heating values. Our previous work [4] also examined a quantity of acid in the crude oil of lignite. Thereof further chemical or physical upgrading is needed. Up to now, numerous upgrading strategies that improve crude oil quality and reduce O, N and S content were reported to produce a fungible transportation fuel, including hydrotreatment (e.g., hydrodeoxygenation (HNO), hydrodenitrogenation (HDN) and hydrodesulfuration (HDS), and hydrocracking [5, 6]. Of these proposed crude oil upgrading strategies, catalytic hydrotreatment is a promising one, which involves the treatment of crude oils, heavy oils, and coal tar et al. with hydrogen in the presence of the catalyst [7–10]. However, those studies were a two-step process of the raw material rather than a one-step process and were also time-consuming.

In this work, lignite liquefactions of four heterogeneous catalysts (Pd/C, Pt/C, Ni/SiO$_2$/Al$_2$O$_3$ and Ru/C) were preformed firstly, and we considered a one-step processing approach, wherein the catalysis, hydrogenation and liquefaction occurred simultaneously in the reactor. The effect of different catalysts on gaseous and coke yields, and products composition was compared firstly. All upgrading experiments were conducted at low temperature 380°C, 30 min, 2 Mpa H$_2$ and 0.3 g/cm^3 water density, and 50 wt% catalyst loading.

2 MATERIALS AND METHODS

The lignite was got from Zhaotong (Yunnan, China), Table 1 lists its elemental and chemical composition along with other properties. The four heterogeneous

Table 1. Characteristics and compositions of Zhaotong lignite (%).

Proximate analysis w_{ad}/%				Ultimate analysis w_{ad}/%				
M	A	V	FC	C	H	N	S	O
27.8	14.33	34.3	23.57	55.13	4.91	1.24	0.85	29.06

catalysts (Pd/C, Pt/C, Ni/SiO$_2$/Al$_2$O$_3$ and Ru/C) were purchased from Sigma-Aldrich. Freshly deionized water was prepared in the lab.

316-stainless steel mini-batch (10 mL) reactors were used for liquefaction experiments, which were depicted in detail in our previous works [11, 12]. Reactions were carried out by placing the reactors vertically in a Techne Fluidized Sand Bath (model SBL-2). In a typical process, lignite and the catalyst were loaded into a reactor and then deionized water was added. After the desired reaction time, the reactors were removed from the sand bath and cooled. The gaseous products were analysed by an Agilent Technologies model 6890N gas chromatograph (GC) with a thermal conductivity detector (TCD). The liquefied products were analysed by GC-MS (PE SQ 8T-680), equipped with an Elite-5MS capillary column.

The yield of each product was calculated by using the following equations:

$$\text{Yields of oil (wt \%)} = \frac{\text{weight of oil}}{\text{weight of lignite loaded}} \times 100\%$$

$$\text{Yields of gas (wt \%)} = \frac{\text{weight of gas - initial weight of H}_2}{\text{weight of lignite loaded}} \times 100\%$$

$$\text{Yields of coke (wt \%)} = \frac{\text{weight of coke}}{\text{weight of lignite loaded}} \times 100\%$$

At least duplicate independent runs were conducted under identical conditions to determine the uncertainties in the experimental results. Results reported herein represent the mean values for the two independent trials.

3 RESULTS AND DISCUSSION

Fig. 1 shows the experimental results. The gas yields were mainly unreacted H$_2$ together with lower yields of CO$_2$ and CH$_4$ hydrocarbon. The gas yield of Ni/SiO$_2$/Al$_2$O$_3$ (17.6 wt%) and Ru/C (16.9 wt%) was almost double the yield of Pd/C, Pt/C catalyst and three times the condition of no catalyst, which indicated Ni/SiO$_2$/Al$_2$O$_3$ and Ru/C can enhance the gasification of lignite in supercritical water at lower temperature. Guan et al. [13] also reported that Ru/C was an excellent catalyst to convert wet microalgae into a fuel-rich gas, containing H$_2$ and/or CH$_4$.

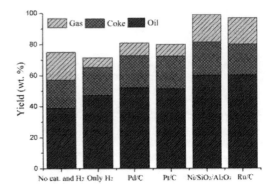

Figure 1. Effect of different catalysts on product yields (380°C, 50 wt% catalyst loading, 30 min, 2 MPa H$_2$, 0.3 g/cm^3).

The coke yields for those four catalysts are close, which is slightly higher than that of no catalyst. However, it obvious that a better mass balance (80–97 wt%) is achieved with the loading of a catalyst rather than an absence of catalyst (70–75 wt%). Duan [14] suggested that increased water density can realize a very well control to the coke formation, which is also consistent with the results observed in the partial oxidation of bitumen in SCW [15]. The most likely reason was the loss of lignite due to its tendency to stick to the reactor wall and tubing. Besides, high water density may lead to the decrease of the mass balance, and an average 80% of the mass balance was obtained in our experiment, which suggested that some water soluble, formed in the liquefaction, remained in the aqueous phase.

Fig. 1 also obviously showed that those four heterogeneous catalysts can be improved by the efficient use of lignite catalytic liquefaction. Moreover, the activity of those catalysts in liquefaction is the following order: Ru/C > Ni/Al$_2$O$_3$/SiO$_2$ > Pd/C > Pt/C. Ru/C and Ni/SiO$_2$/Al$_2$O$_3$ is the most efficient one of catalysts tested with oil yields of 60.23 wt% and 60.13 wt%. They also proved to be effective on the hydrothermal liquefaction of a microalga [16]. High catalytic hydrogenation activity promoted hydrogen incorporation into the lignite with H wt% in the liquefaction, as Bai et al. [17] confirmed in the study of algal oil upgrading. Water density also is a key factor affecting the yields of oil, since our experiment conducted at high water density of 0.3 g/cm^3, water is more favourable to participate in the liquefaction reaction, besides water with a good solubility of potential coke precursors so that to extract and transport them from the catalyst pores, thus retarded the coke formation and proceeded the catalytic reaction.

Although the oil yields obtained from several of the catalytic hydrogenation reactions are very similar, the colour and apparent viscosity of the upgraded oils did vary depending on the catalyst type (Fig. 2.). The oil obtained from treatment with the noble metals Pd/C, Pt/C, Ru/C flowed easily, and was much less viscous than the oils from the Ni/SiO$_2$/Al$_2$O$_3$ and the other two un-catalysed treatments. Oil produced with Pd/C, Pt/C and Ru/C shows a pale yellow, while a red-brown colour appears with the Ni/SiO$_2$/Al$_2$O$_3$ catalyst, and yellowish-brown with no catalyst.

GC-MS was applied to gain insight into the molecular compositions of the treated oil under different conditions. Fig. 3 compares the total ion chromatograms (TIC) of oils produced from the hydrothermal processing of the lignite with no catalyst and H$_2$, no catalyst and with added Pd/C, Pt/C, Ni/SiO$_2$/Al$_2$O$_3$ and Ru/C. The intensity scale for the chromatograms differs for the different oils. The general appearance of the TIC under the two liquefaction conditions without catalyst (a and b) altered little. However, the number and size of peaks of the TIC were significantly changed under four conditions of catalytic hydrogenation. This outcome suggests that catalysts play a vital role in the upgrading of pretreated oil.

The identified compounds of the liquefied oil components are categorized into the following groups: phenols, long-chain alkanes, ketones, fatty acid, and a small amount of olefins, aromatics, alcohols, lipids and other substances. Moreover, their relative amounts are given in Table 2. We use the sum of the areas of each individual peak in each category to represent their total relative amounts in different conditions.

Table 2 shows that long-chain alkanes and phenols are predominant in liquefied oils. Catalytic hydrogenation of lignite liquefaction significantly decreases the phenols, ketones and increases the long-chain alkanes in oils, while showing little effect on the fatty acid amides contents. N-containing and S-containing compounds were substantially not detected and this suggests that denitrogenation and desulfurization occurred during liquefaction. Of the four upgraded oils, the one produced from the Pd/C and Ru/C catalysed reaction contained the highest proportion of long-chain alkanes and lowest proportion of fatty acid. Those long-chain alkanes, mostly

Figure 2. Samples of liquefied oil treated under different conditions.

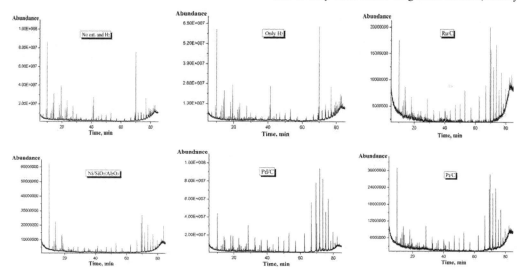

Figure 3. Total ion chromatograms for (a) oil processed with no catalyst and H$_2$ (b) oil processed with H$_2$ only; (c) oil processed with Ru/C; (d) oil processed with Ni/Si$_2$O$_3$/Al$_2$O$_3$; (e) oil processed with Pd/C; and (f) oil processed with Pt/C.

Table 2. Main components of oils under different conditions (% of total peak area by GC-MS).

Oils	Phenols	Long-chain alkanes	Fatty acid	Ketones	Alcohols	Olefins	Aromatics	Lipids
The peak time/min	8–30	40–75	70.02	8–30	30–75	84	19	70.02
No cat. and H$_2$	53.79	15.73	15.7	7.7	7.09	–	–	–
Only H$_2$	51.37	13.83	14.53	7.08	11.2	–	–	–
Ru/C	23.9	67.1	9	–	–	–	–	–
Ni/SiO$_2$/Al$_2$O$_3$	42.2	38.32	9.8	–	1.86	5.85	1.97	–
Pd/C	19.24	73.25	–	–	2.74	–	–	4.76
Pt/C	23.83	65.12	11.05	–	–	–	–	–

straight chain alkanes, which ranged from C14-C44, with C14-C20 as the most abundant, dominated the chromatogram for all the upgraded oils, which were considered the main contributor to the high energy density values.

4 CONCLUSIONS

A one-step process of catalytic hydrogenation liquefaction of lignite was dramatically improved by four catalysts. The results indicated Ru/C, Ni/Al$_2$O$_3$, Pd/C and Pt/C can serve as efficient catalysts for lignite liquefaction. The main gas products are mainly CO$_2$, CH$_4$ and H$_2$. Coke formation can be well controlled at higher water density, but lower mass balance will be obtained. As to the yields of oil, the Ru/C catalyst is the most efficient one of the catalysts tested. A 60.23 wt% oil yield was gained in SCWL. However, for oil quality, the Pd/C catalyst may be better, since producing an upgraded oil, with low viscosity and a high energy density, makes it suitable for the production of transportable fuels.

Improving the yield and quality of oil simultaneously is our main goal in a one-step process of catalytic hydrogenation liquefaction and thus using alloy catalyst in SCWL will be our next step.

ACKNOWLEDGEMENTS

This work is supported by the High Technology Talent Introduction Project of Yunnan in China (2010CI110), the Science and Technology Major Project of Yunnan Province (2012ZB002) and the National Natural Science Foundation of China (21307049). Foundation from Yunnan Natural Science Foundation (2013FZ032) and Kunming University (14118583) are also appreciated.

REFERENCES

Brown M.T., Duan P.G., Savage P.E. Hydrothermal liquefaction and gasification of Nannochloropsis sp. Energy Fuels 2010; 24: 3639–3646.

Duan P.G., Savage P.E. Hydrothermal liquefaction of a microalga with heterogeneous catalysts. Ind Eng Chem Res 2011; 50: 52–61.

Miao X., Wu Q., Yang C. Fast pyrolysis of microalgae to produce renewable fuels. J Anal Appl Pyrol 2004; 71: 855–63.

Guan Q.Q., Chen H.H., Ning P., Tian S.L., Gu J.J., Hu H.C., and Chai X.H. Rapid Determination of Total Acid Content of Oils Resulting from Sub/Supercritical Water Liquefaction of Lignite by Headspace Gas Chromatography. Energy & Fuels 2013; 8.

Wildschut J., Melián-Cabrera I., Heeres, H.J. Appl. Catal. B: Environ. 2010; 99: 298–306.

Tran N.H., Bartlett J.R., Kannangara G.S.K., Milev A.S., Volk H., Wilson M.A., Fuel 2010; 89: 265–274.

Li J., Yang J.L., Liu Z.Y. Hydrogenation of heavy liquids from a direct coal liquefaction residue for improved oil yield. Fuel Processing Technology 2009; 90: 490–495.

Nikulshin P.A., Tomina N.N., Pimerzin A.A., Kucherov A.V., Kogan V.M. Investigation into the effect of the intermediate carbon carrier on the catalytic activity of the HDS catalysts prepared using heteropolycompounds. Catalysis Today 2010; 149: 82–90.

Duan P.G., Savage P.E. Catalytic hydrotreatment of crude algal bio-oil in supercritical water. Applied Catalysis B: Environmental 2011; 104: 136–143.

Yeh T.M., Dickinson J.G., Franck A., Linic S., Thompson Jr. L.T., Savage P.E. Hydrothermal catalytic production of fuels and chemicals from aquatic biomass. J. Chem. Technol. Biotechnol 2013; 88: 13–24.

Guan Q.Q., Savage P.E., Wei C.H. Gasification of alga Nannochloropsis sp. in supercritical water. J. of Supercritical Fluids 2011; 61: 139–145.

Guan Q.Q., Wei C.H., Savage P.E. Kinetic model for supercritical water gasification of algae. Phys. Chem. Chem. Phys 2012; 14: 3140–3147.

Guan Q.Q., Wei C.H., Savage P.E. Hydrothermal gasification of Nannochloropsis sp. with Ru/C. Energy Fuels 2012; 26: 4575–4582.

Duan P.G., Bai X.J., Xu Y.P., Zhang A.Y., Wang F., Zhang L., Miao J. Catalytic upgrading of crude algal oil using platinum/gamma alumina in supercritical water. Fuel 2013; 109: 225–233.

Sato T., Adschiri T., Arai K., Rempel G.L., Ng F.T.T. Upgrading of asphalt with and without partial oxidation in supercritical water. Fuel 2003; 82: 1231–1239.

Duan P.G., Savage P.E. Hydrothermal liquefaction of a microalga with heterogeneous catalysts. Ind Eng Chem Res 2011; 50: 52–61.

Bai X.J., Duan P.G., Xu Y.P., Zhang A.Y., Savage P.E. Hydrothermal catalytic processing of pretreated algal oil: A catalyst screening study. Fuel 2014; 120: 141–14.

Study on the power supply reliability of a nuclear power plant with GO-FLOW methodology, based on Simulink

Jun Wang, Jun Jia, Jie Zhao, DiChen Liu, Li Wang & Chen Cheng
School of Electrical Engineering, Wuhan University, Wuhan, China

ABSTRACT: This paper builds a Simulink simulation interface of power supply reliability calculation of the nuclear power plant with GO-FLOW methodology, based on Simulink technology. Considering the impact of the shared signals on the system, a simulation model is established, and the reliability of the power supply is analysed with an example of a power supply system from the nuclear power plant. Simulation results show that the simulation system can be efficiently and accurately applied to the reliability calculation of a complex power supply system of the nuclear power plant.

Keywords: GO-FLOW Methodology; Reliability; nuclear power

1 INTRODUCTION

The GO-FLOW methodology is a new method of analysis and processing of system reliability, which is derived from the GO methodology. The GO-FLOW methodology was first proposed in 1997 by Takeshi Matsuoka and earlier applied to the reliability design of ships. After more than ten years of research and development by domestic and foreign scholars, this methodology has been successfully applied to the cooling system of boiling water reactor core, the water supply system of a pressurized water reactor [1] and the reliability analysis of a power supply of nuclear reactor plant [2]. Currently, on the basis of GO-FLOW, domestic and foreign scholars add 'periodic task', 'k-out-of-n' and other extended operators with operation rules to this methodology. The GO-FLOW algorithm with its processing framework [3], including common cause failure and the corresponding computer-aided calculation system [3,4], is also proposed. Compared to the GO methodology, GO-FLOW can describe operating characteristics of the dynamic system better and calculate the reliability of the system more accurately.

Simulink is a graphical tool of programming and modelling, which is based on signal flow. It is mainly used for dynamic simulation of actual system analysis and has a strong interaction with other parts of Matlab. Simulink is suitable for the joint simulation of multiple modules and has been widely used to control modelling and signal processing.

In this paper, taking the power supply system of nuclear power plants as an example, a full operational subsystem is established and packaged as Simulink functions according to modelling rules of the GO-FLOW methodology, which includes processing modules of shared signals. According to the electrical simplified design, a corresponding GO-FLOW model is also established to calculate its reliability.

2 ESTABLISHMENT OF GO-FLOW MODULES UNDER SIMULINK

2.1 *Calculation models*

2.1.1 *Operator22: OR gate*

OR gate has a number of inputs and outputs. The output S_i exists when there is at least one input signal. The model is given as follows:

$$R(t) = 1 - \prod_{i=1}^{n}\left(1 - S_i(t)\right) \quad (1)$$

With two inputs, as an example, the Simulink model is shown in Figure 1. Add, Product and Constant are used to achieve the transfer function expression in the model. The multiple-input model can be equivalent to a number of two-input models in series.

Shared signals often exist in the GO-FLOW chart of engineering systems, so the calculation problem of logic gates, including shared signals, cannot be ignored. Shared signals are mainly generated in the system of backup equipment. The interaction of each backup component ensures that the calculation has to consider correction factors and cannot be implemented by independent probability. Therefore, the calculation model of OR gate with consideration of shared signals is established and shown in Figure 2 and Figure 3, respectively.

Funding: The National Natural Science Funds Fund (51347006, 51307123)

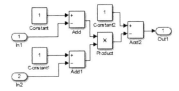

Figure 1. Model of OR gate.

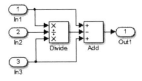

Figure 2. Model of two-input OR gate with shared signals.

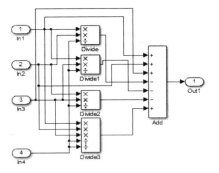

Figure 3. Model of three-input OR gate with shared signals.

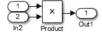

Figure 4. Model of AND gate.

Figure 5. Model of two-input AND gate with shared signals.

2.1.2 Operator30: AND gate

The operator of AND gate represents the logic and relationship of a plurality of signals and its output value is the probability that all inputs are present. The formula is as follows:

$$R(t) = \prod_{i=1}^{n} S_i(t) \tag{2}$$

The Simulink model is shown in Figure 4.

The calculation model of AND gate considering shared signals is similar to the OR gate. A model, with correction of shared signals, should be built when shared signals exists. The two-input AND gate and three-input AND gate, with shared signals, are shown in Figure 5 and Figure 6 respectively.

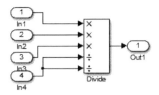

Figure 6. Model of three-input AND gate with shared signals.

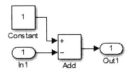

Figure 7. Model of NOT gate.

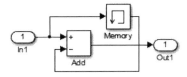

Figure 8. Model of differentiator

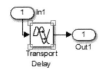

Figure 9. Model of delay

2.1.3 Operator23: NOT gate

The NOT gate operator is mainly used to calculate the probability of logic not. The component is shown in Figure 7.

2.1.4 Operator24: Differentiator

The differentiator is used to simulate the calculation of the differential relationship of inputs. Its Simulink model is shown in Figure 8.

2.1.5 Operator28: Delay

The Delay operator is mainly applied to the construction of the dynamic system where delay exists. It is used in the calculation of signal delay. The system structure is shown in Figure 9.

2.1.6 Operator40: Periodic task

The relationship of inputs and outputs of a periodic task is shown as follows:

$$R(t) = \begin{cases} 1, t < t_i \\ S(t), t_i \leq t \leq t_j \\ S(t_j), t > t_j \end{cases} \tag{3}$$

The arithmetic logic is shown in Figure 10, where the function of a Simulink custom module is achieved

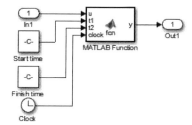

Figure 10. Model of periodic task.

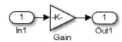

Figure 11. Model of two-state element.

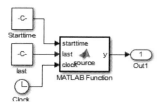

Figure 12. Model of signal generator.

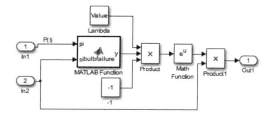

Figure 13. Model of component fault.

by judging the current task with the system timed to determine the output value.

2.2 Component model

2.2.1 Operator21: Two-state element
The operator is used to represent components with only two states of normal operation and fault, such as resistors, lines and diodes. Assuming the fault rate is P_g, the relationship between inputs and outputs is as follows:

$$R(t) = P_g \cdot S(t) \qquad (4)$$

The mathematical expression shows that the probability of an output equals to the probability of an input multiplied by the probability of fault rate, which can be constructed with the gain module in Simulink. The corresponding model is shown in Figure 11.

2.2.2 Operator25: Signal generator
The signal generator is mainly used to generate control signals of each module, which can be either discrete or continuous. In practical application of GO-FLOW, the signal generator is often used to generate step signals. This paper sets up two parameters for the operator, which is the start time and the duration time of the step signal. The Simulink model is shown in Figure 12. The custom module of Matlab Function is used to judge whether the system time is within the period set by the user.

2.2.3 Operator35: Component fault
Component fault is the most common operators among the basic operators of GO-FLOW. It is mainly used to describe the components which fail in operation. The model of component fault has one primary input signal with a number of branch input signals and one output signal. Assuming the fault rate of component is constant, the output value is as follows:

$$R(t) = S(t) \cdot exp\left\{-\lambda \cdot \sum_i \sum_{t_k \le t} P_i(t_k) \times min[1, S(t_k)/S(t)]\right\} \qquad (5)$$

As seen from the equation above, the calculation of $R(t)$ in every moment needs to consider the function value of $P(t_k)$ and $S(t_k)$ in every moment since t_1, which means there is an array store all values of $P(t_k)$ and $S(t_k)$. In a common Simulink module it is difficult to achieve what is required, so the custom function of Simulink by M-file is applied to the construction of the model. Coupled with the constant, product, MathFunction modules, the Simulink subsystem is shown in Figure 13. The MathFunction module calculates the next output of $S(t)$ and $P(t)$ by recording the current values.

2.3 Switch model

2.3.1 Operator26: Normally closed valve
Normally a closed valve is turned off in the normal state and turns on when the input signal exists. The operator consists of a main input, a control input, and an output. When the on signal comes, the probability of the on state is P_g. If the on signal doesn't come, the probability of the on state is P_p. The output value is as follows:

$$\begin{cases} R(t) = S(t) \cdot O(t) \\ O(t_1) = P_p \\ O(t) = O(t') + [1 - O(t')] \cdot P(t) \cdot P_g \end{cases} \qquad (6)$$

where $O(t')$ = probability of the on state before t_1. t_1 = start time.

When establishing the Simulink operator, the memory component is applied to store the value of the previous time and other common mathematical calculation modules are used to build transfer function expressions. The Simulink model is shown in Figure 14. In addition, the can also be expressed as a normally open valve.

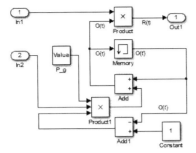

Figure 14. Model of normally closed valve.

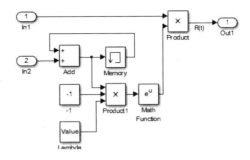

Figure 16. Fault model of normally open valve.

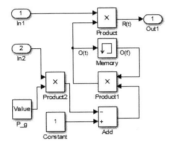

Figure 15. Model of normally open valve.

2.3.2 *Operator 27: Normally open valve*

Similar to the normally closed valve, the normally open valve is on in normal operation and turns off when there is a control signal. The operator also consists of a main input, a control input and an output. When the on signal comes, the probability of the off state is P_g. If the on signal does not come, the probability of the off state is P_p. The output value is as follows.

$$\begin{cases} R(t) = S(t) \cdot O(t) \\ O(t_1) = 1 - P_p \\ O(t) = O(t')[1 - P(t) \cdot P_g] \end{cases} \quad (7)$$

The Simulink model is shown in Figure 15. The operator can also be expressed as normally open valve.

2.3.3 *Operator37: Fault of normally open valve*

The operator model mainly represents the valve when it fails in the open state. The operator consists of a main input, a number of branch input and an output. Assuming the fault probability of this component is constant λ, the mathematical expression is as follows.

$$R(t) = S(t) \cdot exp\left\{-\lambda \cdot \sum_i \sum_{t_k \le t} P_i(t_k)\right\} \quad (8)$$

The component of Memory and Add is used to achieve the function of cumulative summation in the process of the construction of Simulink models. The model is shown in Figure 16.

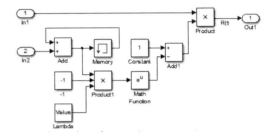

Figure 17. Fault model of normally open valve.

2.3.4 *Operator38: Fault of normally closed valve*

Similar to the fault of normally open valve, this operator, whose fault probability is constant λ, consists of a main input, a number of branch inputs and an output. The mathematical expression is as follows:

$$R(t) = S(t) \cdot \left\{1-exp\left[-\lambda \cdot \sum_i \sum_{t_k \le t} P_i(t_k)\right]\right\} \quad (9)$$

The model is shown in Figure 17.

2.3.5 *Operator39: valves which can open and close*

This operator, with an integrated operator 26 and operator 27 is a model of valves that can open and close frequently. The operator consists of a main input, two branch inputs, and an output. The probability of the successfully open state is Po and the probability of the successfully close state is Pc.

When the open signal P_1 comes, the output value is as follows:

$$R(t) = S(t) \cdot \left\{O(t') + [1-O(t')] \cdot P_1(t) \cdot P_g\right\} \quad (10)$$

When the close signal P_2 comes, the output value is as follows:

$$R(t) = S(t) \cdot O(t')[1-P_2(t) \cdot P_g] \quad (11)$$

where $O(t')$ = probability of the open state before the time t.

The constructed Simulink subsystem is shown in Figure 18. P_1 and P_2 is processed by Constant, Add and Product, to prevent the logic that two signals fail

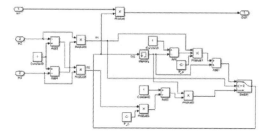

Figure 18. Model of valve which can open and close.

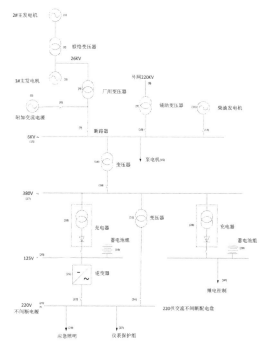

Figure 19. Power supply system of critical load in nuclear power plant.

simultaneously. The Memory component is used to store $O(t)$ and generate the signal $O(t')$. The Switch component, in the logic diagram, is applied to achieve the optional output of the two expressions of P_1 and P_2. When $P_1 = P_2 = 0$, the two expressions are equivalent and expressed as $R(t) = S(t)*O(t')$, so each output can be selected.

The GO-FLOW methodology is mainly applied to reliability analysis in the industry field. The following demonstrates the application in the reliability analysis of a practical system with two examples.

3 EXAMPLES

This paper takes the power supply system of critical load in a nuclear power plant [5] as an example to analyse its reliability by the GO-FLOW methodology.

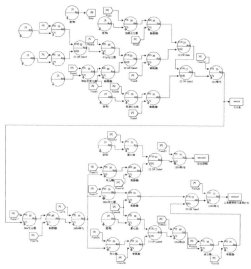

Figure 20. GO-FLOW chart of Power supply system of critical load in nuclear power plant.

Table 1. Reliability of each component of the electrical system.

Unit name	Fault frequency	MTTR	Reliability
Generator	2.059e−02	1.060e+02	0.999751
220 kV	2.300e−03	1.000e+02	0.999973
Contact transformer	5.694e−03	1.000e+01	0.999993
Additional AC Power	8.468e+00	3.351e+00	0.996771
Auxiliary transformer	5.694e−03	1.000e+01	0.999994
Breaker	4.643e−04	7.200e+01	0.999996
6 kV AC bus	3.679e−03	5.000e+00	0.999998
Auxiliary transformer	1.139e−02	1.000e+01	0.999987
Diesel generators	1.743e+02	5.000e+00	0.909516
6 kV transformer	5.694e−03	1.000e+01	0.999994
380 V AC bus	3.679e−03	5.000e+00	0.999998
380 V transformer	1.139e−02	1.000e+01	0.999987
Charger	8.760e−02	3.351e+00	0.999967
Battery	6.549e−02	2.000e+00	0.999985
125 V DC bus	4.643e−4	7.200e+01	0.999996
Inverter	3.914e−02	3.000e+00	0.999987
220 V AC bus	3.679e−03	5.000e+00	0.999998

where MTTR = mean time to repair.

The corresponding GO-FLOW chart is as follows. The operator 35 is used to describe electrical equipment of the distribution network in nuclear power network and the operator 26 is used to describe the control signal which are necessary in a simulation of the circuit breaker in the distribution network.

According to fault records of the nuclear power plant, the reliability of each component of the electrical system is shown in Table 1.

The calculation method of component reliability is as follows:

$$A = \frac{MTBF}{MTBF + MTTR} \quad (12)$$

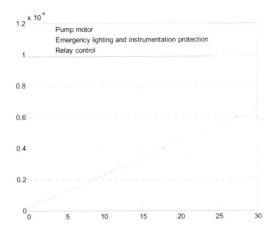

Figure 21. Reliability of critical load in nuclear power supply which varies with time.

where MTBF = mean time between failures of components. The calculation method of MTBF is as follows.

$$\text{MTBF} = \frac{365 \times 24}{\text{failure frequency}} \quad (13)$$

MTTR can be checked in the table.

4 SIMULATION RESULTS

The reliability data is imported into the corresponding Simulink model to make fixed-step simulation in the step of one year with the algorithm of ode3. The simulation time is 30 years. The reliability of the critical load in a nuclear power supply varies over time, as is shown in Figure 21. In order to improve the effective bits of data, the reliability of simulation results is represented by the fault rate of equipment, where

$$\text{Fault rate} = 1 - \text{reliability} \quad (14)$$

5 CONCLUSION

(1) With the GO-FLOW methodology, the reliability calculation model of a nuclear power plant is established on the basis of Simulink. Examples demonstrate that the model can precisely calculate power supply reliability of a nuclear power supply and improve the efficiency of analysis.
(2) The GO-FLOW calculation component, considering shared signals, is established. In comparison with the traditional GO-FLOW methodology, the model with shared signals, is closer to engineering practice, thereby increasing the accuracy of calculation results.
(3) GO-FLOW components with the time-varying characteristics are used to establish the dynamic reliability model of a nuclear power plant to reflect the characteristic that system reliability changes with time.

REFERENCES

G.J Wu, Y. Wang, Y.L. Shang. Reliability analysis of repairable system based on GO-FLOW methodology. Nuclear Power Engineering. Vol. 33 (2012), p. 25-29.

Q.M. Zhen. A study of reliability monitoring system based on GO-FLOW for nuclear power. 2009.

J. Lin. Research on the principles and the computer aided technology of the GO-FLOW. 2003.

Takeshi Matsuoka, Michiyuki Kobayashi. The GO-FLOW reliability analysis methodology-analysis of common cause failures with uncertainty. Nuclear Engineering and Design, Vol. 175 (1997), p. 205–214.

J.Q. Zhang, G.S. Wang. Research on practicability of safety design criteria of power system in nuclear power plant. Nuclear standard measurement and quality, Vol. 2. (2010), p. 17–21.

… Energy and Environmental Engineering – Wu (Ed.)

Power supply safety assessment of a nuclear power plant, based on fuzzy comprehensive evaluation

Chen Cheng, Jie Zhao, DiChen Liu, Jun Jia, ShanShan Liang & Xue Lin
College of Electrical Engineering, Wuhan University, Wuhan, Hubei, China

ABSTRACT: A Power Supply Safety Assessment (PSSA) model of nuclear power plant based on fuzzy comprehensive evaluation combined with the auxiliary electrical system structure of a nuclear power plant is proposed to implement the goal of quantizing power supply safety risk in a nuclear power plant. The auxiliary electrical system of a nuclear power plant and the fuzzy comprehensive evaluation were introduced in this paper firstly, then assessing matrix is established based on fuzzy comprehensive evaluation building assemble of auxiliary electrical system safety element and its weight coefficient and index assemble. Moreover, the matrix is applied to the safety assessment of nuclear power plant power supply. Lastly, the example of analytical results demonstrates the feasibility and effectiveness of the method in the article and it provides a theoretical basis to improve the supply safety of nuclear power plants.

Keywords: Nuclear Power Plant; Auxiliary Electrical System; Fuzzy Comprehensive Evaluation; Safety Assessment

1 INTRODUCTION

With the large-scale development of China's nuclear power, nuclear safety has drawn increasing attention. Among the many issues affecting the safety of nuclear power, reliable power supply for protecting the safety of nuclear power plants plays a crucial role. The vast majority of nuclear power plants are required to work in a reliable electricity supply environment to achieve the plant's normal operation and safe shutdown in the event of an accident and to ensure sufficient depth shutdown [1]. Therefore, the risk analysis for the powered security of electric power systems in nuclear power plants plays a great significance in the safe operation of nuclear power. Therefore, academic and engineering research has been active with reliability studies of the power supply system of nuclear power plants.

The literature [2] discusses whole plant power-off of nuclear power plants and puts forward some optimization methods to heighten the ability to prevent accidents. Literature [3] analyzes a station blackout accident in a nuclear power plant, using fault tree approach, and illustrates the main failure mode and probabilities and the importance of occurred events for the station blackout accident. GO methodology is used in a reliability analysis of the main electrical connections in a nuclear power plant and the result of quantitative calculation is compared with the result of fault tree approach in Literature [4]. The correctness and advantages of GO methodology is verified in a reliability analysis of the main electrical connections.

The PSSA model of a nuclear power plant, based on fuzzy comprehensive evaluation, is proposed in this passage to implement the goal of risk assessment of a station blackout accident in a nuclear power plant and a quantization of the safety risk. Fuzzy evaluation is a kind of method that can establish the model of a multi factor system and comprehensive evaluation is widely used in fuzzy neighbourhood. However, this application in power systems is less, and only present in the boiler operation economic evaluation, while scheme selection of flue gas desulfurization and denitrification and other fields have had some preliminary study [5–8]. Because the PSSA is complicated and unique for nuclear power plants compared to others, the PSSA mode established by this method is able to conduct multi-factor comprehensive investigation effectively. In addition, it provides a new evaluation standard for the power supply risk evaluation of nuclear power plants.

2 INTRODUCTION OF POWER SUPPLY SYSTEM OF NUCLEAR POWER PLANT

An auxiliary electrical system of a nuclear power plant is composed of a power supply and a distribution system, as shown in Figure 1. The power supply includes: main generator, main external power grid, aided external power grid and emergency diesel generator set. A station blackout mainly refers to the total loss of 6.6 kV. The so-called loss of 6.6 kV distribution system refers to emergency AC disc LHA and LHB cannot supply

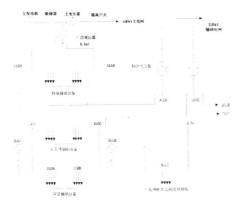

Figure 1. Schematic of Power System in Nuclear Power Plant.

power (?), in addition to the failure of the AC disc itself, while the loss of both external power supplies and loss of two emergency diesel generators, will result in a total loss of AC 6.6 kV. Power supply is provided, under normal circumstances, by the 400 kV main power supply. The auxiliary power load will be borne by a power plant islanding if the main power is lost. The auxiliary power supply will be borne by an auxiliary power supply of 220 kV if the islanding operation fails. When all external power is lost and the islanding operation fails, the diesel generator is started to take the important load of LHA and LHB of electricity.

The combination of various power supply failure modes, common cause failure of bus LHA, as well as bus LHB, and the failure of auxiliary transformer almost constitute all failure modes of the station blackout. Currently, the event importance and event reliability, which cause station blackout accidents, have been discussed in much of the literature, which has laid the foundation for the research of this article.

3 METHOD OF FUZZY COMPREHENSIVE EVALUATION

Comprehensive evaluation is a reasonable evaluation process when evaluating things affected by many factors. It uses the principle of fuzzy linear transformation and considers each factor of the evaluated things [9–11]. Fuzzy comprehensive evaluation is divided into single-stage and multi-stage, this paper uses the method of two-stage fuzzy comprehensive evaluation for this complex system of power supply in nuclear power plants.

The main steps of the application of multi-level fuzzy comprehensive evaluation model for comprehensive evaluation to multi factors are as follows:

a) To set up the hierarchical level of factor sets of evaluation object

The factor set U contains all the evaluation factors. The factors of U are divided into several categories according to a criterion. Factors of similar nature are generally divided firstly into classes and a class of factors is comprehensive judged and then an evaluation result is comprehensive judged on the high level of 'class' elements.

The factors of U are divided into one group, that is:

$$U = \{U_1, U_2, \cdots, U_l\} \qquad (1)$$

The above formula is satisfied:
$U = \cup_{i=1}^{l} U_i$; when $i \neq j$, $U_i \cap U_j = \varnothing$; for $\forall i \in \{1, 2, \ldots, l\}$, there is $U_i \neq \varnothing$.

For each U_i, there is $U_i = \{u_{i1}, u_{i2}, \ldots, u_{in}\}$, among them n represents the number of factors included in the factors set of I.

This shows that factor sets are divided into different levels. U is the factors set on the top level and $U_i (i = 1, 2, \ldots, l)$ are factor sets on low level.

b) To establish evaluation set and weighting coefficients

The comments to all the appraisal objects that may appear were included in an evaluation set. Risk reviews are given one by one by the experts to each risk factor. The risk reviews are divided into m levels and that evaluation set is $V = \{v_1, v_2, \ldots, v_m\}$ applied to evaluation of the factors of any layer and any kind. The definition of $A_i = (a_{i1}, a_{i2}, \ldots, a_{in})$ is the weight coefficient set that each factor is relative to V of U_i and $a_{i1} + a_{i2} + \cdots + a_{in} = 1$ is satisfied. a_{in} are assigned according to the importance of each factor of $U_i \cdot A = (a_1, a_2, \ldots, a_l)$ is the weight coefficient set that each sub factor set U_i is relative to V and $a_1 + a_2 + \cdots + a_l = 1$ is satisfied. a_l are assigned according to the importance of each factor of U.

c) To calculate the fuzzy evaluation matrix on each level

The fuzzy evaluation matrix R_i of single factor and that single factor U_{in} is relative to the safety risk assessment set V and is determined first. That is:

$$R_i = [r_{k1}, r_{k2}, \cdots r_{km}] \quad (k = 1, 2, \cdots, n) \qquad (2)$$

In the formula: $r_{k1} + r_{k2} + \cdots + r_{km} = 1$; among them r_{kj} represents the percentage of safety risk indicator v_j in the factor U_{in} of safety risk factor subset U_i.

Then the comprehensive evaluation result to factor subset U_i is determined by using compound operation according to a univariate fuzzy evaluation matrix R_i.

$$B_i = A_i \cdot R_i = [b_{i1}, b_{i2}, \cdots, b_{im}] \qquad (3)$$

Then the B of safety risk fuzzy comprehensive assessment matrix of senior factors U can be determined.

$$R = \begin{bmatrix} B_1 \\ B_2 \\ \vdots \\ B_l \end{bmatrix} = \begin{bmatrix} A_1 \cdot R_1 \\ A_2 \cdot R_2 \\ \vdots \\ A_3 \cdot R_3 \end{bmatrix} \qquad (4)$$

28

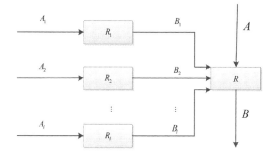

Figure 2. Model of two stage fuzzy comprehensive evaluation.

At last, safety risk U considering variety of security risk factors is computed.

$$U = A \cdot R \quad (5)$$

Figure 2 is a model of two stage fuzzy comprehensive evaluation.

4 POWER SUPPLY SAFETY ASSESSMENT OF LARGE-SCALE NUCLEAR POWER PLANT BASED ON FUZZY COMPREHENSIVE EVALUATION

4.1 The evaluation process

The model of comprehensive evaluation is established taking the power supply safety of large-scale nuclear power plant as an example. Moreover, the safety risk of this nuclear power plant is evaluated by comprehensive calculation.

Main factors set affecting the power supply security of nuclear power plant is $U = \{U_1, U_2, U_3, U_4, U_5\}$. Among them: U_1 represents failure of the main generator; U_2 represents failure of the external power grid; U_3 represents failure of the emergency diesel generators; U_4 represents failure of the buses; U_5 represents failure of the auxiliary transformer.

The sub-factors set of U_1 is $U_1 = \{u_{11}, u_{12}, u_{13}\} = $ {island operation failure, unit shutdown, blackout of external power grid, caused by unit shutdown}. The sub-factors set of U_2 is $U_2 = \{u_{21}, u_{22}, u_{23}\} = $ {common cause loss of external power grid, loss of aided external power grid, loss of main external power grid}. The sub-factors set of U_3 is $U_3 = \{u_{31}, u_{32}, u_{33}, u_{34}, u_{35}, u_{36}, u_{37}, u_{38}, \} = $ {running failure of LHP, running failure of LHQ, running common cause failure of LHP and LHQ, maintenance of LHP, maintenance of LHQ, starting failure of LHP, starting failure of LHQ, starting common cause failure of LHP and LHQ}. The sub-factors set of U_4 is $U_4 = \{u_{41}\} = $ {common cause failure of bus LHA and LHB}. The sub-factors set of U_5 is $U_5 = \{u_{51}\} = $ {failure of auxiliary transformer}.

There are ten experts, for comments set of all levels of different factors by five level mechanism $V = \{v_1, v_2, v_3, v_4, v_5\} = $ {very good, better, good, general, poor}. The main factors set and secondary weight coefficient are determined by events of a critical degree leading to a station blackout of the nuclear power plant. (The data of events critical degree is taken from 200 reactor-years of the operating experience feedback data of the French standard 900,000 kilowatts PWR power plant.) The factor sets the weight coefficient at all levels and a secondary evaluation matrix are shown in Table 1.

4.2 Calculation of comprehensive evaluation

Do the one-level comprehensive assessment to the factor subset using data in Table 1, obtain:

$$B_1 = A_1 R_1 = [0.86\ 0.07\ 0.07] \cdot \begin{bmatrix} 0 & 0 & 0 & 0.3 & 0.7 \\ 0 & 0 & 0.7 & 0.3 & 0 \\ 0 & 0 & 0.7 & 0.3 & 0 \end{bmatrix} = \quad (6)$$

$$[0\ 0\ 0.098\ 0.3\ 0.602]$$

Similarly, doing the one-level comprehensive assessment to other factor subsets can obtain:

$$B_2 = [0.152\ 0.632\ 0.2\ 0.01546\ 0] \quad (7)$$

$$B_3 = [0.00054\ 0.00728\ 0.27752\ 0.63776\ 0.0769] \quad (8)$$

$$B_4 = [0.7\ 0.3\ 0\ 0\ 0] \quad (9)$$

$$B_5 = [0.1\ 0.9\ 0\ 0\ 0] \quad (10)$$

Do the second-level comprehensive assessment to $U = \{U_1, U_2, U_3, U_4, U_5\}$ using B_1, B_2, B_3, B_4, B_5, obtaining the evaluation result of the factor set U is:

$$U = A \cdot R = [0.2566\ 0.2607\ 0.4112\ 0.0688\ 2.7 \times 10^{-3}] \cdot \begin{bmatrix} B_1 \\ B_2 \\ B_3 \\ B_4 \\ B_5 \end{bmatrix}$$

$$= [0.0883\ 0.1908\ 0.1914\ 0.3433\ 0.1861] \quad (11)$$

4.3 Analysis of the result

The matrix B is the comprehensive evaluation result of the security risk of the station blackout accident in a large nuclear power plant. Using three kinds of method analysis of the evaluation result as follows:

(1) Maximum membership degree law. Select the maximum value of the membership function of fuzzy appraisal set as the evaluation grade. Because $B(v_4) = 0.3433 = \max\{0.0883, 0.1908, 0.1914, 0.3433, 0.1861\}$, think the reliability level of station blackout accident be general by the maximum membership principle.

Table 1. Factors setting weight coefficients at all levels and a secondary assessing matrix.

Factors set at first level	Weight coefficient at first level	Factors set at second level	Weight coefficient at second level first level	v_1	v_2	v_3	v_4	v_5
Main generator fault	0.2566	Island operation failure	0.86	0	0	0	0.3	0.7
		Unit shutdown	0.07	0	0	0.7	0.3	0
		Blackout of external Power Grid caused by unit shutdown	0.07	0	0	0.7	0.3	0
External Power Grid fault	0.2607	Common cause loss of external power grid	0.76	0.2	0.8	0	0	0
		Loss of Aided external power grid	0.1546	0	0.1	0.8	0.1	0
		Loss of main external power grid	0.0854	0	0.1	0.9	0	0
Emergency diesel generator fault	0.4112	Running failure of LHP	0.3845	0	0	0.1	0.8	0.1
		Running failure of LHP	0.3845	0	0	0.1	0.8	0.1
		Running Common cause failure of LHP and LHQ	0.1644	0	0	0.9	0.1	0
		Maintenance of LHP	0.0185	0	0	0.9	0.1	0
		Maintenance of LHQ	0.0185	0	0	0.9	0.1	0
		Starting failure of LHP	0.0121	0	0.1	0.8	0.1	0
		Starting failure of LHQ	0.0121	0	0.1	0.8	0.1	0
		Starting Common cause failure of LHP and LHQ	0.0054	0.1	0.9	0	0	0
Bus fault	0.0688	Common cause failure of bus LHA and LHB	1	0.7	0.3	0	0	0
Fault of Auxiliary transformer	2.7×10^{-3}	Failure of Auxiliary transformer	1	0.1	0.9	0	0	0

(2) Fuzzy fractional step method. The result of comprehensive evaluation B is normalized. Make

$$b_k^{'} = b_k \bigg/ \sum_{j=1}^{n} b_j, k = 1, 2, \cdots, n, \text{ then } B^{'} = \begin{bmatrix} b_1^{'} & b_2^{'} & \cdots & b_n^{'} \end{bmatrix}$$

as the equivalent of the evaluation result. The intuitive explanation is: $b_k^{'}$ indicates rating scale v_k being the percentage of comprehensive evaluation result. $B^{'}$ shows the distribution of the percentage of the n-th evaluation grade.

The above normalized evaluation result of evaluation set is $B = [0.0883 \quad 0.1908 \quad 0.1914 \quad 0.3433 \quad 0.1861]$. From fuzzy fractional step method, it shows the reliability rating of station blackout accident in this nuclear power plant. The proportion of very good, better, good, general and poor is 8.83%, 19.08%, 19.14%, 34.33%, 18.61%.

(3) Weighted average method. The weighted average method uses membership degree b_j of each evaluation grade v_j as a proxy weight coefficient.

Take weighted average v' of each v_j as evaluation result $v' = \left(\sum_{j=1}^{n} b_j v_j \right) \bigg/ \sum_{j=1}^{n} b_j$, and compare v' with the elements in the comment V. The safety level of the system can be judged by the comparison of v' and the elements in the comment V. When the weighted average method is used, v_j should be quantized first to use the above formula. If the quantization of $V = \{v_1, v_2, v_3, v_4, v_5\}$ is $\{10, 8, 6, 4, 2\}$, the result of the weighted average is $v' = \left(\sum_{j=1}^{5} b_j v_j \right) \bigg/ \sum_{j=1}^{5} b_j = 5.3037$.

The result shows: the reliability level of station blackout accident of this nuclear power plant is between good and general.

5 CONCLUSION

Based on the safety-influencing factor of the auxiliary electrical system in a nuclear power plant, this passage proposed a way to evaluate it by using fuzzy comprehensive evaluation. It includes the introduction of the method and procedures. Through a quantitative calculation by a ten-people expert team, it got a brief and accurate result. The establishment of PSSA mode of a nuclear power plant and mathematical evaluation method provides a new way of thinking on the study of reliability of the auxiliary power system in nuclear power plants. It also has significance for the maintenance of the safety system in a nuclear power plant.

ACKNOWLEDGEMENT

Funding: The National Natural Science Funds Fund (51347006, 51307123).

REFERENCES

Li Zhe. 2011. Reliability Analysis of LOOP and SBO Events in Nuclear Power Plant Based on the GO Methodology. Beijing. Qinghua University.

Mei Qizhi. 1994. Reliability Analysis of Electric Power System for Large-scale NPP. Nuclear Power Engineering 15(6): 486–492.

Chen Haiyan, Li Xiaohua & Ke Guotu. 1998. Failure mode analysis for the station blackout of large-scale nuclear power plant. Chinese Journal of Nuclear Science and Engineering 18(4): 304–310.

Li Zhe, Lu Zongxiang & Liu Jingquan. 2010. Reliability Analysis of Nuclear Power Plant Bus Systems Arrangement Based on GO Methodology. Nuclear Power Engineering 31(3): 69–77.

Gu Yujiong, Dong Yuliang & Yang Kun. 2004. Synthetic Evaluation on Conditions of Equipment in Power Plant Based on Fuzzy Judgment and RCM Analysis. Proceedings of the CSEE 24(6): 189–194.

Chen Yafei, Gao Xiang, Luo Zhongyang, Ni Mingjiang & Cen Kefa. 2004. A Fuzzy Evaluation on Flue Gas Desulphurization Technologies. Proceedings of the CSEE 24(2): 215–220.

Man Ruoyan & Fu Zhong. 2000. Condition Assessment of Fossil-fired Power Plant Based on Fuzzy Comprehensive Evaluation. Proceedings of the CSEE 29(5): 5–10.

Jin Tao, Fu Zhongguang, Man Ruoyan & Yang Yongping. 2010. A Fuzzy Evaluation Method for Comprehensively Evaluating a Thermal Power Plant. Journal of Engineering for Thermal Energy and Power 25(5): 474–477.

Feng Xiaoan, Xie Hongbin & Liu Yanping. 2008. Security Evaluation of Power Network Information Systems Based on Fuzzy Comprehensive Judgement. Power System Technology 32(23): 40–43.

Cong Lin, Li Zhimin, Pan Minghui, Gao Kunlun & Pian Ruiqi. 2004. Information Security Evaluation in Power Systems Based on Fuzzy Comprehensive Judgment. Automation of Electric Power Systems 28(32): 65–69.

Liang Dingxiang & Chen Xi. 2009. Safety assessment mode of electric power information system based on fuzzing synthetical theory and its application. Power System Protection and Control 37(5): 61–64.

The electric curtain method of dust removal and measurement of particle size and density on solar cells

ChuanDe Zhou & HongBo Xiao
College of Mechanical and Power Engineering, Chongqing University of Science and Technology, Chongqing, China

Jie Zhang
College of Mechanical Engineering, Chongqing University, Chongqing, China

ShaoHua Sun
Changqing Oilfield Company, Oil Production Plant No. 7, Qingyang, China

ABSTRACT: As Solar Power Generation Systems are usually located in wild environments, dust particles are susceptible to accumulate on solar cell arrays. Electric curtains, embedded in the substrate on the surface of solar cells, are an effective way to remove dust, as they are able to produce a traveling wave field between two electrode grids, thus they can be used to manipulate charged particles to move in a direction vertical to the grid axis, so that the particles can be removed. The potential distribution and electric field intensity distribution was simulated by MATLAB. We found that the larger the diameter and density the particles have, the lower the acceleration the particles would have, which means that particles with different diameters and densities would have different motions when moving within the waves of the electric field, and this indicates that measuring the diameters and density of particles, which have accumulated on the solar cell, was essential for the design of the electric curtain within appropriate parameters. Therefore, we carried out the measurements of particle diameters and densities and the results showed that the average density of particles was 0.4526 kg/m^3 and most of the particles' size ranged between five and twenty μm.

1 INTRODUCTION

Solar Power Generation is a new electricity generation technique, composed of solar cell arrays, containing photovoltaic material, which is silent, without pollution or exhaust, and having no geographic restrictions or safety implications. Thus, it has become the main trend in energy generation technology (Gaofa et al. 2011, Guanjun et al. 2013).

Dust particles were susceptible to accumulate on solar cells, as Solar Power Generation Systems were usually located in wild environments. In addition, the power generation efficiency would be decreased by 40%, if the dust accumulation increased into 4 g/m^2 (Mazumdera et al. 2007, Chuande et al. 2013). Therefore, cleaning up dust particles effectively was quite essential.

The commonly used methods of dust removal were natural cleaning methods: mechanical cleaning, self-cleaning nano-film method, and the electric curtain method (Gaofa et al. 2011, Guanjun et al. 2013, Chuande et al. 2013). Natural cleaning method moved particles by using the wind, rain and gravitational force etc. to wash out particles, and it was simple, but lower in efficiency. Mechanical cleaning methods clean up dusts by using sweeping, blowing with high pressure airflow, vibrating or by an ultrasound method. The efficiency of these was higher than natural cleaning methods, but would cause some damage to the solar cells in the operation, and the tiny particles were unable to be removed as small particles adhered close to cells. Self-cleaning nano-film method means using self-cleaning nano-film, such as superhydrophilicity film to prevent dust accumulating on solar cells. Although it was effective, the high cost and immature preparation process made it not suitable for application in solar power generation industry.

The electric curtain method, chosen in the following study, means using an electric curtain, consisting of a series of parallel electrode grids, connected to a three phase AC source to produce a traveling wave electric field in order to levitate and repulse charged particles on the surface of solar cells (Atten et al. 2009, Calle et al. 2011, Calle et al. 2009, Chuande et al. 2013). In the electric field, the charged particles could be moved in a direction vertical to the grid axis under the Coulomb force, dielectrophoretic force, adhesion force, viscous force and gravitational force, when the appropriate curtain parameters such as voltage, frequency, grid diameter etc. were applied (Chuande et al. 2013).

The principle of electric curtain method was introduced and the electric potential distribution and electric field intensity distribution were simulated. Through the simulation of particles' acceleration, we found that the larger the diameter and density that particles have, the lower the acceleration particles would have, which means particles with a different diameter and density would have different motions in the traveling wave electric field, and indicate that measuring the diameters and density of particles, accumulated on the solar cell, was essential for the design of the electric curtain with appropriate parameters. Therefore, we measured the particles' diameter and density.

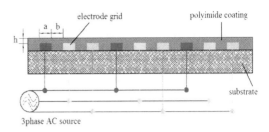

Figure 1. Schematic of 3-phase traveling wave electric curtain.

2 PRINCIPLE

The schematic design of a 3-phase traveling wave electric curtain is shown in Figure 1. The electric curtain, composed of a series of parallel electrode grids with grid width a and spacing distance b, was embedded in the substrate and the insulation material polyimide with thickness h was coated on the surface for break-down (Atten et al. 2009, Calle et al. 2011, Calle et al. 2009, Chuande et al. 2013). The motion of charged particles under the Coulomb force, dielectrophoretic force, adhesion force, viscous force and gravitational force in the traveling wave electric field can be expressed as (Calle et al. 2009, Chuande et al. 2013):

$$m\frac{d^2r}{dt^2} = qE\cos\omega t - 6\pi r\eta\frac{dr}{dt} - mg \quad (1)$$

where m, r, q denotes particle mass, radius and charge respectively; E was electric field intensity; η was viscosity in air.

In the electric field, the charged particles can be moved along the direction vertical to the grid axis up to the end of the curtain edge, if appropriate curtain parameters, such as voltage, frequency, grid width and spacing distance, were applied.

2.1 Potential-field distribution

The potential field distribution in the traveling wave field under applied voltage $U = u_0\cos(\omega t)$ can be expressed as (Atten et al. 2009, Qixia, et al. 2012 a,b):

$$\phi(x,y,t) = u_0\left[\varphi(x,y)\cos(\omega t) + \varphi\left(x - \frac{\lambda}{3}, y\right)\cos\left(\omega t - \frac{2\pi}{3}\right) + \varphi\left(x - \frac{2\lambda}{3}, y\right)\cos\left(\omega t - \frac{4\pi}{3}\right)\right] \quad (2)$$

where $\omega = 2\pi f$ is frequency of applied 3-phase AC source; λ is the period of electrode grid.

The potential field distribution, simulated by MATLAB, is shown in Figure 2, and we can find that the potential field is distributed periodically in x direction, whereas it decayed exponentially in y direction.

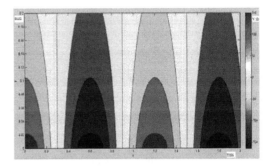

Figure 2. Distribution of potential-field.

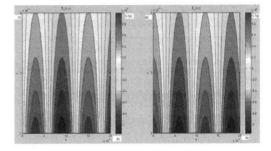

Figure 3. Distribution of electric-field intensity of the component Ex, Ey.

2.2 Electric-field distribution

According to the physical relationship between potential difference and electric intensity, electric distribution can be expressed as (Qixia, 2012 a,b,c):

$$E = -\nabla\phi(x,y,t) \quad (3)$$

The electric field intensity distribution simulated by MATLAB was shown in Figure 3, picture on the left, denoted Ex, the component of electric field intensity in x direction; and the right one denotes Ey, the component of electric field intensity in y direction. In addition, the figure showed that the electric field distributed periodically in x direction, whereas decayed exponentially in y direction.

As shown in Figures 4–5, traveling wave was moving forward in x direction with velocity $V = f\lambda$.

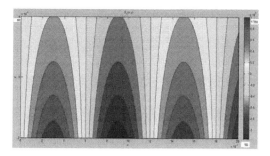

Figure 4. Electric field intensity distribution of Ex at t = 0.

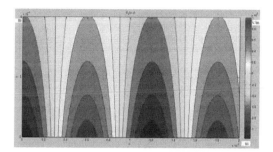

Figure 5. Electric field intensity distribution of Ex at t = T/4.

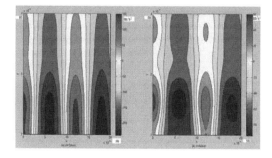

Figure 6. Comparison of dust acceleration with different particle diameters.

Figures 4–5 showed the electric-field distribution of Ex at time t = 0, and t = T/4 respectively, and we can find that the node in the electric field moved λ/4 forward in x direction (λ = 0.6 mm, T = 0.02 s in the simulation).

2.3 Influence of dust removal by particle size and density

Figures 6a, b showed the acceleration distribution of particles in diameter 35 μm and 60 μm respectively, which indicated that the larger the diameter, the lower the acceleration.

Figures 7a, b illustrates the acceleration distribution of particles, in density $\rho = 1.0 \times 10^3$ kg/m^3 and $\rho = 5.0 \times 10^3$ kg/m^3 respectively, which indicates that the larger the density, the lower the acceleration.

By the comparison of different diameters and densities in Figures 6 and 7, we found that particles with different diameters and different densities would have different motions in the traveling wave electric field. Therefore, measuring the diameter and density of particles, which had accumulated on the solar cell, was beneficial for the design of the electric curtain within appropriate parameters in order to remove dust from solar cell surface effectively.

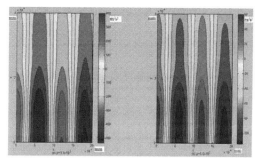

Figure 7. Comparison of dust acceleration with different particle density.

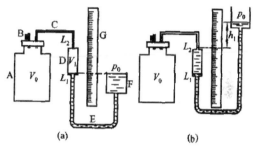

Figure 8. Schematic diagram of particle volume measurement.

3 MEASUREMENTS OF DUST DENSITY

According to the definition of density $\rho = m/V$, particle density can be measured by measuring the mass and volume of particle. Particle mass can be weighed precisely by high precision scales, and volume can be measured by the method introduced by the reference (Huixian 2004). As shown in Figure 8, jar A, volume V_0, was connected to wide pipe D with a thin pipe C; L_1 and L_2 were labels on pipe D to indicate the height of liquid level, and B was a sealing plug on jar A. The height of the liquid level in pipe D can be measured by ruler G, when water was put into funnel F.

Firstly, we measure the original height of liquid level h_1 in pipe D, and then measured the height h_2, after dust particles with weight m_0 were put into the jar A, according to reference (Huixian 2004), particle density can be calculated by formula: $\rho = m/[(1 - h_1/h_2)V_0]$.

The results of ten measurements of dust particle mass varied from two g to twenty g with mass interval of twenty g, as is shown in Figure 9. The ordinate was

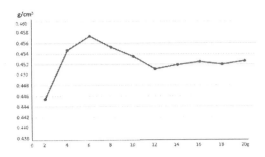

Figure 9. Curve of particle density results in measurements.

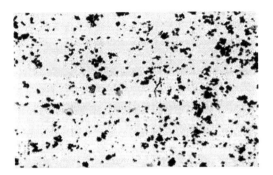

Figure 10. Photograph of particles under microscope.

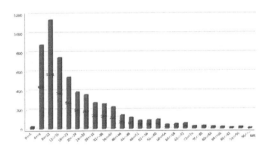

Figure 11. Distribution of particle size.

the measured density; abscissa was particle mass, and the average density was 0.4526×10^3 kg/m^3.

4 MEASUREMENTS OF PARTICLE SIZE DISTRIBUTION

As the particle size was small, a microscope was used to take a photograph after the particles had been scattered uniformly on the section, as shown in Figure 10. Then, RGB to grey conversion, binaryzation processing, edge detection, particle number statics and particle equivalent area calculation were worked out by image processing software *Open Source Computer Vision Library (OpenCV)*.

The result was shown in Figure 11, the ordinate denotes the number of particles; abscissa denotes particle size range. We can then find that about 58.3%, 26.5%, 9.4%, 5.3% and 0.5% of the particles' diameters were in the range of 5 to 20 μm, 20 to 40 μm, 40 to 60 μm, 60 to 100 μm and below 5 μm.

5 CONCLUSIONS

As the solar cell arrays were susceptible to accumulate particles, an electric curtain embedded in the substrate is a better choice for removing dust on solar cell surfaces, as it has various advantages, such as high efficiency, low cost, and does not damage the solar cell etc. The electric curtain connected to 3-phase AC source was able to produce a wave field between two electrode grids, thus can be used to manipulate charged particles on the solar cell surface in order to move in a vertical direction to the grid axis so that the particles can be cleaned up. The principle of electric curtain method of dust removal was introduced, and the potential distribution and electric field intensity distribution were simulated by MATLAB. We also found that the larger the diameter and density that particles have, the lower the acceleration the particles would have, which means particles with different diameters and different densities would have different motions when traveling in the waves of an electric field. This indicates that measuring the diameter and density of particles, which have accumulated on the solar cell, was essential for the design of an electric curtain within appropriate parameters. Therefore, we carried out the experiments with particle diameter and density measurements, and the results showed that the average density of particles was 0.4526×10^3 kg/m^3 and most of the particle size were in the range of five to twenty μm.

ACKNOWLEDGEMENT

This work has been supported by the National Natural Science Foundation of China (NSFC), Grant Number: 51205431. The authors wish to thank the National Natural Science Foundation of China.

REFERENCES

Atten, P. Hailong, P. & Reboud, J. L. 2009. Study of dust removal by standing-wave electric curtain for application to solar cells on mars. *Ieee transactions on industry applications* 45(1): 75–86.

Calle, C. I, Buhler. C. R. & McFall, L. et al. 2009. Particle removal by electrostatic and dielectrophoretic forces for dust control during lunar exploration missions. *Journal of Electrostatics* 67: 89–92.

Calle, C. I. Buhler. C. R. & Johansen, M. R. et al. 2011. Active dust control and mitigation technology for lunar and Martian exploration. *Acta Astronautica* 69: 1082–1088.

Chuande, Z. Gaofa, H. & Jie, Z. et al. 2013. Research on the motion pattern of dust particles and self-cleaning mechanism of solar cells under traveling wave electric curtain. *Journal of Hebei University of Science and Technology* 34 (20): 97–101. Chinese.

Gaofa, H. Chuande, Z. & Zelun, L. 2011. A review of self-cleaning method for solar cell array. *Procedia Engineering* 16: 640–645.

Guanjun, B. Linwei, Z. & Shibo, C. et al. 2013. Review on dust depositing on PV module and cleaning techniques. *Journal of Mechanical & Electrical Engineering* 30(8): 909–913. Chinese.

Huixian, C. 2004. Measurement of density of dissolvable material. *College physic* 23(1): 37–38.

Mazumdera, M. K. Sharmaa, R. & Birisa, A. S. et al. 2007. Self-cleaning transparent dust shield for protecting solar panels and other devices. *Particulate science and technology* 25: 5–20.

Qixia, S. Ningning, Y. & Xiaobing, C. et al. 2012a. Advance in lunar surface dust removal method by electrodynamic field. *Advance in mechanics* 42(25): 785–803. Chinese.

Qixia, S. Ningning, Y. & Xiaobing, C. et al. 2012b. Mechanism of dust removal by a standing wave electric curtain. *Science China* 55(6): 1018–1025.

Qixia, S. Ningning, Y. & Zhikun, X. et al. 2012c. Experimental study on efficiency of dust removal by standing wave electric curtain. Spacecraft engineering 21(3): 72–79. Chinese.

Passive control of floating wind turbine for vibration and load reduction based on TMD

M.W. Ge, S. Wang & H.W. Xiao
North China Electric Power University, Beijing, P.R. China

ABSTRACT: In order to achieve vibration and load reduction of floating wind turbines, a Tuned Mass Damper (TMD) is introduced into the nacelle. Taking the 5MW offshore floating wind turbine as a research object, the coupled simulation model, including aerodynamics, structure, hydrodynamics and control is based on a FAST program. Under the condition of Normal Wind Profile (NWP) and Extreme Operating Gusts (EOG), the effect of TMD on the structural control is studied. The results show that under EOG with high wind speed, a reduction of 15% on vibration and of 30% on load can be achieved using TMD, which is very significant for the structural control of floating wind turbines.

1 INTRODUCTION AND MOTIVATION

Offshore wind turbines have become a trend of wind power. So far, the offshore platform of a wind turbine is mainly confined to the bottom of mixed structures, such as gravity foundations, fixed pile foundations, multi-pile foundations, jacket foundations, and so on. However, the expense of these support structures increases sharply with the depth of the sea water. When the water depth exceeds 50 meters, the support structure will be very expensive. In deep water, floating support structures have become a hot topic concerning offshore wind power (Musial et al. 2004; Hendersen et al. 2004; Wayman et al. 2006). Vibration is one of the difficulties of floating-base technologies. Under the combined effects of the wind and waves, excessive vibration will be produced on the floating wind turbine, which will significantly increase the fatigue and ultimate load. Hence, the control of vibration is very important for the development of floating offshore platform technology.

In recent years, some scholars have tried to improve suspended load on the support structure through the pitch control of blades and the adjustment of motor torques. Although these methods can achieve some success, they need to increase the workload of the pitch institutions or the control system. It needs to use a variety of control methods to suppress the vibration of large-scale floating platforms under the combined action of water and wind. Passive control methods based on TMD is an effective way to control the structure and is widely used in the field of bridges, buildings, etc., which provide a good idea for the vibration reduction in floating wind turbines (Murtagh et al. 2007; Wilmink & Hengeveld 2006; Enevoldsen 1996; Calderon 2009).

In the present paper, the 5MW offshore floating wind turbine is taken as a research object. The coupled simulation model, including aerodynamics, structure, hydrodynamics and control is based on a FAST program. Under the condition of EOG, the effect of TMD on the structural control is studied.

2 INTRODUCTION OF FAST PROGRAM

FAST (Fatigue, Aerodynamics, Structures, and Turbulence) program is developed and currently maintained by the US Department of Energy's National Renewable Energy Laboratory (NREL). It is a complex simulator which can evaluate the extreme fatigue loads for the two-bladed horizontal wind turbines as well as the three-bladed wind turbines. The FAST program consists of a horizontal axis wind turbine and pneumatic subroutine of AeroDyn components. In 2005, Fast and AeroDyn passed the 'Germanischer Lloyd Wind Energies' assessment and was determined to be appropriate software of 'computing load for the design and certification of onshore wind turbines'.

The FAST program is used to create the whole model of a wind turbine and Simulink is used to design a controller. The wind is generated using the subroutine TurbSim. The aerodynamic calculation subroutine AeroDyn is used to compute the aerodynamic loads. However, the traditional FAST program lacks the freedom of the TMD system. Wind turbines, equipped with a TMD system, cannot be calculated through FAST without extension. Researchers in the University of Massachusetts have presented an improved version of the software "FAST-SC" (SC for structural control). In the modified version, they introduced the TMD model in the original FAST (Lackner & Rotea 2011(a); Lackner & Rotea 2011(b); Rotea et al. 2010). Hence, in this study, the effect of TMD on the vibration reduction of floating wind turbines is investigated using the program of "FAST-SC".

Table 1. The main parameters of the 5MW wind turbine.

Rated power (MW)	5
Wind rotor	Upwind, Clockwise
Number of blades	3
Control system	Pitch variable speed Yaw
Drive train system	Gear box
Diameter of wind rotor and Hub	126, 3
Height of Hub (m)	90
Cut-in, rated, Cut-out wind speed (m/s)	3, 11.4, 25
Tip Speed	80
Cone angle, tilt angle (degree)	2.5, 5
Mass of the wind rotor (T)	110
Mass of the nacelle (T)	240
Mass of the tower (T)	347.5
Size of the nacelle (m)	18*6*6
Platform	Floating platform
Location of the TMD	Nacell

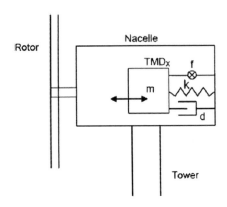

Figure 1. Configuration of TMD (Stewart & Lackner 2011).

3 INTRODUCTION OF FAST PROGRAM

3.1 The main parameters of the 5MW wind turbine

The simulation model used in this paper is the 5MW Barge offshore wind turbine prototype, designed by the US Department of Energy's National Renewable Energy Laboratory. The main parameters are shown in Table 1.

3.2 The model of aerodynamics

The aerodynamic load of the wind turbine is calculated by AeroDyn program, in which the parameters of the blade, and the wind model should be defined. As a widely used method, the Blade Element Momentum Theory (BEMT) is used in the computational model. In the theory, the induction factors are the main focus:

$$a = \left(1 + \frac{4\sin^2\varphi}{\sigma C_n}\right)^{-1} \quad (1)$$

$$b = \left(1 + \frac{2\sin^2\varphi}{\sigma C_t}\right)^{-1} \quad (2)$$

where ϕ = the inflow angle
σ = the solidity of the blade
C_n = axis force coefficient
C_t = Tangential force coefficient

3.3 The configuration of TMD

In the FAST-SC program two new freedoms for TMD system are added to the coupled dynamic model of the original structure. Figure 1 shows the configuration of the TMD system on the wind turbine structure.

3.4 The model of control system

The control system includes wind turbine yaw control, generator torque control, and pitch control. Yaw is controlled automatically to track the changes of wind direction, which ensures that the wind turbines can capture the maximum wind power.

Below rated wind speed, a variable speed turbine may try to stay at its optimum tip speed ratio wherever possible, by changing the rotor speed in proportion to the wind speed. This maximizes the power coefficient and hence the aerodynamic power available. This can be achieved in a steady state by setting the generator torque to be proportional to the square of the rotor or generator speed. The Optimal mode gain multiplies the square of the generator speed to give the required generator torque demand. It can be calculated as:

$$k_{opt} = \frac{1}{2}\rho\pi R^5 C_p \frac{1}{\lambda^3 N^3} \quad (3)$$

where k_{opt} = Optimal mode gain
ρ = air density
R = rotor radius
C_P = power coefficient at
λ = desired tip speed ratio
G = gearbox ratio

The torque demand is then given by:

$$Q = k_{opt}\omega^2 \quad (4)$$

where Q = generator torque demand
ω = generator speed demand

When the wind speed is greater than the rated wind speed, wind turbines require pitch control to maintain the operation at rated power. To ensure the stable operation of the generator, the generator torque should be adjusted timely:

$$Q = \frac{P}{\omega\eta} \quad (5)$$

where η = the efficiency of the generator
P = the power of the generator

3.5 *The model of wind condition*

In the present study, the extreme operating gust (EOG) will be taken as the wind condition to evaluate the effect of TMD for vibration reduction. To increase the versatility of this study, GL standard model is introduced here as a continuous stream condition. The model is as following

$$V(z,t) = \begin{cases} V(z) - 0.37 V_{gustN} \sin(3\pi t/T) \\ (1 - \cos(2\pi t/T)) & \text{for } 0 \le t \le T \\ V(z) & \text{for } t < 0 \text{ and } t > T \end{cases} \quad (6)$$

where D is the diameter of the wind rotor, $\beta = 6.4$, $T = 14$ s. V_{gustN} is calculated as the following:

$$V_{gustN} = \beta \left(\frac{\sigma_1}{1 + 0.1 \left(\frac{D}{\Lambda 1} \right)} \right) \quad (7)$$

$$\Lambda_1 = \begin{cases} 0.7 z_{hub} & \text{for } z_{hub} < 60m \\ 42m & \text{for } z_{hub} \ge 60m \end{cases} \quad (8)$$

The wind profile $V(z)$ represents the average wind speed over the ground height z as a function. In the case of standard grade wind turbines, wind profilers' normal law exponent is given by:

$$V(z) = V_{hub}(z/z_{hub})^\alpha \quad (9)$$

where $V(z)$ is the wind speed at the height z [m/s], z is the height [m] over the ground, z_{hub} is the height of hub [m], α is the exponent of wind profile. In this paper, the exponent is selected to be 0.2. The power spectral density shall meet the following requirements:

$$\sigma_1 = I_{15}(15m/s + aV_{hub})/(a+1) \quad (10)$$

where σ_1 is the standard deviation of the longitudinal wind speed at hub height and I_{15} is the turbulence intensity under the wind speed of 15 m/s.

4 THE PERFORMANCE OF TMD FOR VIBRATION AND LOAD REDUCTION

4.1 *The case of NWP*

Before the EOG cases, the normal wind profile (NWP) in Eq. (9) is first studied as the flow condition. The wind speed is set to be 10 m/s and the simulation time is 200 s. Fig. 2(a) shows the time traces of the displacement of the tower top in the y direction. Under the steady wind, only about 4% vibration reduction is obtained through TMD. Fig. 2(b) shows the time traces of the load on the tower top and only about 5% load reduction is achieved. Under the steady wind speed, the frequency of TMD is considerably different from the frequency of the wind turbine which is mainly determined by the rotational speed of the wind rotor. Hence the effect of TMD is lower than the expected result.

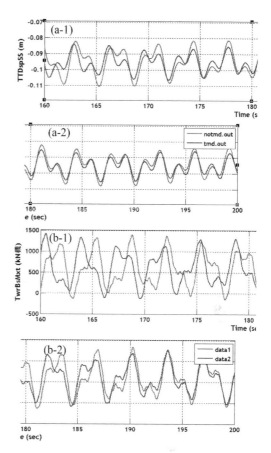

Figure 2. The effect of TMD under NWP with a wind speed 10 m/s. (a) time traces of the displacement of the tower top of the wind turbine in y direction; (b) time traces of the thrust force of the tower top. (Red: no TMD, Blue: with TMD).

4.2 *The case of EOG*

For a comprehensive analysis of the TMD system damping effect, two wind speeds were set at the hub height, with one at below the rated wind speed (8 m/s) and the other at higher than the rated wind speed (15 m/s). The EOG was added into the simulation during the 1950s. The operating time of the EOG is 10.5 s and the total simulation is set to be 100 s.

Fig. 3(a) shows the time traces of the displacement of the tower top in the x direction. Due to the applying of EOG, the displacement becomes very large. A TMD in the nacelle does change the response of the structure but no obvious vibration reduction is obtained. Fig. 3(b) show the time traces of the thrust force on the tower top, which is very similar to Fig. 3(a). Therefore, it can be concluded that under a small wind speed, the TMD only plays a small role in the structural control of floating wind turbines. Fortunately, under these conditions, both the vibration and the load are not very big.

Fig 4(a) shows the time traces of the displacement of the tower top in the y direction under the EOG of 15 m/s. As can be observed, TMD changes the response

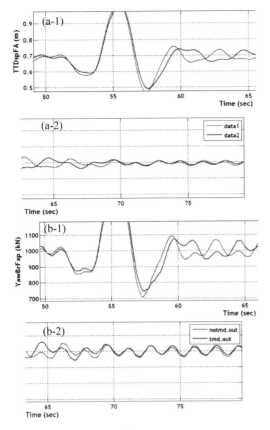
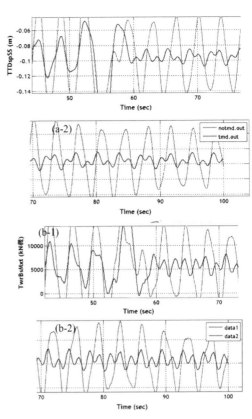

Figure 3. The effect of TMD under EOG with a wind speed 8 m/s. (a) time traces of the displacement of the tower top of the wind turbine; (b) time traces of the thrust force of the tower top. (Red: No TMD, Blue: with TMD).

Figure 4. The effect of TMD under EOG with a wind speed 8 m/s. (a) time traces of the displacement of the tower top in y direction; (b) time traces of the thrust force of the tower top. (Red: No TMD, Blue: with TMD).

of the structure greatly. Up to 15% reduction of vibration is achieved. Fig. 4(b) shows the time traces of the axis load of the tower top. Up to 30% load reduction is achieved by using the TMD.

5 CONCLUSIONS

In order to achieve vibration and load reduction in floating wind turbines, the Tuned Mass Damper (TMD) is introduced into the nacelle. Taking the 5 MW offshore floating wind turbines as a research object, the coupled simulation model, including aerodynamics, structure, hydrodynamics and control, is established and based on the FAST program. The effect of TMD on the structural control is studied. The conclusions are as follows:

(1) Under a normal wind profile, with a velocity of 10 m/s, the frequency of TMD is considerably different from the frequency of the wind turbine, which is mainly determined by the rotational speed of the wind rotor. Only about 5% vibration and load reduction is achieved under a steady wind.

(2) Under small wind speeds of EOG, the TMD only plays a small role in the structural control of the floating wind turbine. Fortunately, under these conditions, both the vibration and the load are not very big. Under stronger wind speeds of EOG, up to 15% vibration reduction and about 30% load reduction is achieved, which plays an important role in the protection of wind turbines from damage.

ACKNOWLEDGMENT

Thanks for the support of the Fundamental Research Funds for the Central Universities. JB2014080.

REFERENCES

Calderon, B. Design and optimization of a wind turbine tower by using a damper device. 2009. Master's thesis. Stuttgart University.
Enevoldsen, I., & Mørk, K. J. 1996. Effects of a Vibration Mass Damper in a Wind Turbine Tower. Journal of Structural Mechanics, 24(2), 155–187.

Henderson, A. R., Zaaijer, M. B., Bulder, B., Pierik, J., Huijsmans, R., et al. 2004. Floating windfarms for shallow offshore sites. In Proceedings of the 14th International Offshore and Polar Engineering Conference, Toulon, France.

Lackner, M, Rotea, M. Structural control of floating wind turbines. 2011(a). Mechatronics, 21(4):104–719.

Lackner, M., & Rotea, M. 2011(b). Passive structural control of offshore wind turbines. Wind energy, 14(3), 373–388.

Murtagh, P. J., Ghosh, A., Basu, B., & Broderick, B. M. 2008. Passive control of wind turbine vibrations including blade/tower interaction and rotationally sampled turbulence. Wind Energy, 11(4), 305–317.

Musial, W., Butterfield, S., & Boone, A. 2004. Feasibility of floating platform systems for wind turbines. In 23rd ASME Wind Energy Symposium, Reno, NV.

Rotea, M, Lackner M, Saheba, R. 2010. Active structural control of offshore wind turbines. In: 48th AIAA aerospace science meeting and exhibit. Orlando (FL).

Stewart, G. M., & Lackner, M. A. 2011. The effect of actuator dynamics on active structural control of offshore wind turbines. Engineering Structures, 33(5), 1807–1816.

Wayman, E. N., Sclavounos, P. D., Butterfield, S. et al. 2006. Coupled dynamic modeling of floating wind turbine systems. In Offshore Technology Conference (Vol. 139). Houston.

Wilmink, A. J., & Hengeveld, J. F. 2006. Application of tuned liquid column dampers in wind turbines. In Proceedings of the European Wind Energy Conference.

Environmental science and engineering

Controlled synthesis of manganese oxides with different morphologies and their performance for catalytic removal of gaseous benzene

WenXiang Tang, ShuangDe Li & YunFa Chen
Institute of Process Engineering, Chinese Academy of Sciences, Beijing, PR China

ABSTRACT: A series of manganese-related particles with different features, such as nanowires, and nanoparticles, urchin-spindle-like particles, were synthesized by reducing $KMnO_4$ with ascorbic acid (AA) under reflux condition. The dosage of AA in the reaction process had a great effect on the morphology and composition of products. Manganese oxide nanoparticles with a mesoporous structure were obtained by calcining the precursors at 450° in air and the morphologies mostly kept very well. The crystalline structure, morphology, specific surface area, porosity, and reducibility of the samples were investigated by XRD, SEM, BET and H_2-TPR. The performances of benzene combustion on these manganese oxides were carried out and the best-performing catalysts can effectively catalyse the total oxidation of benzene at lower temperatures (T90 = 240° at space velocity = 60,000 ml g^{-1} h^{-1}).

1 INTRODUCTION

Morphology is an important element in determining the physiochemical properties of nanomaterial, and it has become an essential point in the process of materials fabrication to obtain different kinds of structure [1, 2]. As a significant metal-based material, recently manganese compounds (such as Mn_3O_4, MnO_2, MnOOH, MnC_2O_4, MnS, etc.) have attracted a lot of interests due to their excellent properties in many fields such as magnetic application [3], catalysis [4], electronics [5] etc. In the last decade, the shapes of manganese compounds have been controlled by various techniques [6–8]. For an example, a variety of Mn-based materials including Mn_3O_4 octahedrons, MnOOH nanorods, MnO_2 nanowires and aggregated $MnCO_3$ nanoparticles in the form of spindles were prepared by a hydrothermal method [9], but the affected factors are difficult to control. Developing a facial method to synthesis various products with different morphology is still a great challenge for the materials' researchers.

Volatile Organic Compounds (VOCs) emitted from industrial process and fossil fuels' combustion are an important class of air pollutants. The emissions of VOCs can cause many environmental problems, such as ozone generation, and photochemical smog etc [10]. Low concentrations of VOCs will be also a great threat to human beings' health [11] and the abatement of VOCs is highly desirable. Catalytic combustion is one of the most important approaches for the abatement of VOCs at low temperature. Nobel metals, typically Au [12], Pd [13], and Pt [14], are usually used for catalysing the deep oxidation of benzene and exhibit perfect activity at very low temperatures. However, because of their high cost and some related problems, such as volatility, sintering, and a susceptibly poisoning tendency, they cannot be used in industrial process widely. Recently, many ordinary metal oxides (MnO_x, Co_2O_4, NiO, CuO, CeO_2) [15] are described as the active catalysts for the combustion of various VOCs. In these metal oxides catalysts, Mn-based materials have been proved to own high activity for VOCs oxidation [16,17] and their morphologies or microstructures play an important role. Dai et.al [18] compared the activity for the toluene oxidation of manganese oxides with different morphology and found the rod-like α-MnO_2 catalyst had the highest activity.

Herein, we report a facile reflux method to synthesize a series of manganese compounds with various features by only adjusting the mole ratio of the reactants. All the precursors were calcined at 450° in air and every morphology of the precursors was kept very well. BET analysis showed that all samples after calcination had a high surface area with mesoporous structure. The probable process of reaction has also been investigated. In addition, benzene was selected as a VOC model to test their catalytic activities.

2 EXPERIMENTAL

2.1 *Chemicals and Sample preparation*

All the chemical reagents used in this study were from Xilong Cop. (China) and were analytically graded without further purification. The water for preparing solutions was made from Millipore Milli-Q water (18 MΩ cm). A series of manganese compounds were synthesized by reducing KMnO₄ with ascorbic acid (AA). Typically, 10 mmol KMnO₄ was dissolved in 100 ml distilled water, and then another 100 ml solution containing AA was added to the above solution. The concentrations of solution containing AA were varying from 0.02 to 0.08 M. The mixed solution was transferred to a 250 ml flask and stirred for 24 h under a reflux condition at 100°C. The resulting precipitates were filtered, washed several times with distilled water, and then dried at 80°C in air. The products a ∼ d were prepared by different concentration of AA solution (a: 0.02M; b: 0.03M; c: 0.04M; d: 0.08M). All the products are calcined at 450°C for 4 h in air and marked as a1∼d1 for.

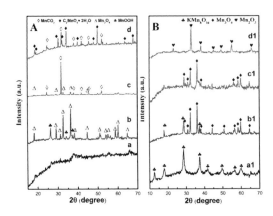

Figure 1. The XRD patterns of samples a∼d (refluxed at 100°C for 24 h) and products a1∼ d1 (calcined at 450°C in air).

2.2 Materials characterization

The crystal phases were identified by X-ray powder diffraction (XRD), using a Panalytical X'Pert PRO system, with Cu-Kα radiation in the diffraction angle (2θ) range 5°–90° at a sweep rate of 3° min⁻¹. The morphology of the products was characterized by a scanning electron microscopy (SEM, JEOL JSM-6700F). The specific surface areas and porosities of all catalysts were obtained with N₂ adsorption-desorption method on an automatic surface analyser (AS-1-C TCD, Quantachrome Cor., USA). In order to understand the reducibility of catalysts, H₂ temperature programmed reduction (H₂-TPR) was carried out in a U-shaped quartz reactor under a gas flow (5% H₂ balanced with Ar, 25 ml min⁻¹). 30 mg sample (40∼60 mesh) was used in each procedure and the temperature was raised to 1000 K from room temperature at a constant rate 10°C min⁻¹.

2.3 Catalytic activity measurement

Catalytic activities of varied manganese oxides for total oxidation of benzene were performed in a continuous-flow fixed-bed quartz microreactor (i.d. 6 mm) at a weight hourly space velocity (WHSV) of 60,000 mL g⁻¹ h⁻¹. In each procedure, 100 mg (40∼60 mesh) catalyst was loaded in the quartz reactor with quartz wool packed at both ends of the catalysts bed and a continuous flow (100 ml min⁻¹) with 1000 ppm of gaseous benzene in air was used for catalytic test. The concentration of benzene in the effluent gas was analysed by a gas chromatograph (Agilent 6890A) equipped with a flame ionization detector (FID). The conversion of benzene ($W_{Benzene}$, %) was determined as follows:

$$W_{Benzene} = (C_{Benzene\ in} - C_{Benzene\ out})/C_{Benzene\ in} \times 100\%$$

where $C_{Benzene\ in}$ (ppm) and $C_{Benzene\ out}$ (ppm) are the concentrations of benzene in the inlet and outlet gas, respectively.

3 RESULTS AND DISCUSSION

3.1 Crystal phase, morphology and BET analysis

Fig. 1 a and b show that the XRD patterns of the as-prepared samples a∼d and the samples a1∼d1 calcined at 450°C respectively. The sample a is an amorphous product because it has no obvious diffraction peak in Fig. 1a and it turns into cryptomelane KMn₈O₁₆(JCPDS 004-0603) as shown in Fig. 1b. The most of XRD peaks on sample b can be indexed to the tetragonal Mn₃O₄ phase (JCPDS card 24-0734) and the others can be indexed to the monoclinic MnOOH phase (JCPDS 74-1632). Fig. 1b shows that the sample b1 includes KMn₈O₁₆ and Mn₃O₄ phase. In Fig. 1a, the major peaks of sample c correspond to the reflection of the rhombohedral MnCO₃ phase (JCPDS 83-1763), and there is a little Mn₃O₄ phase, but the pure Mn₃O₄ phase is obtained after calcination as shown in Fig. 1b. All the peaks of sample d in Fig. 1a shows the pure MnCO₃ phase (JCPDS 83-1763). Fig. 1 shows that sample d has both MnCO₃ phase (JCPDS 83-1763) and C₂MnO₄2H₂O phase (JCPDS 025-O544). From the XRD patterns shown in Fig. 1b, the sample d1 is the pure Mn₂O₃ phase (JCPDS 041-1442) after calcined the sample at 450°C in air. By analysing the crystal phase of refluxed products, it is noted that the crystal structures of products were affected by the dosage of reducing agent (AA).

Fig. 2 displays SEM images of products obtained at different concentrations of AA solution and calcined at 450°C in air. It is observed that the sample a consisted of large quality of nanowires with diameters 30–100 nm and lengths in the range of 2–5 μm which were the typical feature of Cryptomelane KMn₈O₁₆. Besides nanowires in sample b, some nanoparticles with an average size of 100 nm can be viewed clearly. From the XRD patterns of b and b1, it is explained that the nanoparticles in sample b would be Mn₃O₄ and the nanowires would be mainly KMn₈O₁₆. The size of nanoparticles in sample c and c1 were about 30–60 nm

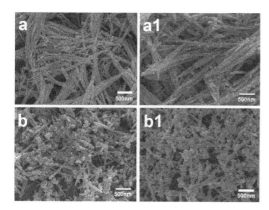

Figure 2. SEM images of as-prepared manganese compounds samples (a, a1, b, b1).

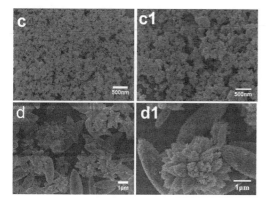

Figure 3. SEM images of as-prepared manganese compounds samples (c, c1, d, d1).

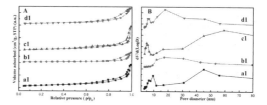

Figure 4. N$_2$ adsorption-desorption isotherms curves and pore size distribution calculated from desorption branch of as-calcined manganese oxides (a1-d1).

as shown in Fig. 3. Interestingly, the sample d consisted of not only spindle-like but also new urchin-like particles. The new urchin-like particles with diameters 3 μm were made up of some half spindles and this new feature of manganese compound has never been reported in other literature. In addition, the morphology of sample d1 was perfectly maintained after thermal decomposition at 450°C in air atmosphere and a porous structure was obtained.

Fig. 4 shows the nitrogen adsorption-desorption isotherms and the pore size distribution of the calcined products from the precursors. All the isotherms were similar to the type IV isotherm and had typical hysteresis loops which indicated the existence of mesopore structure on the samples. It is noticeable that most distributions of pore size were asymmetric and wide, which indicated that the mesoporous manganese oxides had highly irregular pore shapes. All the textural properties are listed in Table 1. The specific surface areas of sample d1 is 78.8 m^2·g^{-1} which is the highest value among the as-prepared samples. The porous structure was formed after the heat treatment which can be ascribed to the decomposition of MnCO$_3$ or MnC$_2$O$_4$ species. As reported in other publications [19, 20], such porous structure will create numerous active sites and this special structure will facilitate the adsorption and diffusion of organic molecules, thus avoiding limitations of interphase mass transfer and hence promoting their catalytic activities.

3.2 Catalytic performance of benzene combustion

The catalytic activity of benzene combustion over the sample a1~d1 is revealed in Fig. 5. For the purposes of comparison, the reaction temperatures T$_{10\%}$, T$_{50\%}$, T$_{90\%}$ (corresponding to the CO$_2$ yield = 10, 50, 90%) were used to evaluate the catalytic performances of the catalysts as summarized in Table 1. The complete combustion temperature of nanowires (sample a1) was about 350°C which indicated the poor performance for benzene oxidation on this structure. However, sample b1 including nanoparticles and nanowires had a better activity than sample a1 that only consisted of nanowires. Mn$_3$O$_4$ nanoparticles can apparently improve the catalytic performance due to the increase of the reactivity and mobility of lattice oxygen which is similar to another report [21]. The manganese oxide c1 calcined from MnCO$_3$ precursors had a higher activity for benzene oxidation and the reaction temperature of T$_{90\%}$ was 262°C. It can be seen that the sample d1 exhibited the best activity and the T$_{10\%}$, T$_{50\%}$, T$_{90\%}$ values were 161, 206 and 240°C for benzene oxidation, respectively. The outstanding performance of benzene combustion over sample d1 can be attributed to its high surface area, mesoporous structure and particular morphology. Moreover, the MnC$_2$O$_4$ species obtained in the sample d would have a spinodal transformation into oxides by the calcination in air which lead to generate many mesopores with high specific surface areas and Frey et al. [4] reported that these special structures were highly active for CO oxidation. Hence, the morphology and components of precursors had an important effect on their activity for benzene combustion.

3.3 H$_2$-TPR analysis

H$_2$-Temperature-programmed reduction analysis was used to investigate the reducibility of catalysts and the results are presented in Fig. 6. As we know, the reduction of manganese oxides can be described by the

Table 1. BET surface areas, Pore diameter, total pore volumes, micro-pore volumes and catalytic activities of as-prepared samples.

sample	BET surface area (m²·g⁻¹)	BJH pore diameter (nm)	Pore volume (cm³·g⁻¹)	$T_{10\%}$	$T_{50\%}$	$T_{90\%}$
a1	69.7	6.4	0.29	215	273	341
b1	22.6	11.7	0.13	206	258	296
c1	69.6	4.8	0.31	168	221	262
d1	78.8	5.5	0.21	161	206	240

BET analysis columns and Catalytic activity (°C) columns.

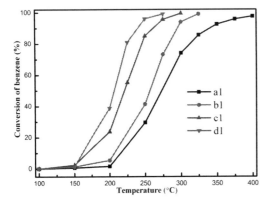

Figure 5. Benzene oxidation over as-calcined manganese oxides (a1~h1) (1060 ppm benzene in air, F = 100 ml min-1, F/W = 60,000 ml gcat-1 h-1.

successive processes: MnO₂ → Mn₂O₃ → Mn₃O₄ → MnO. Every reduction process appears at different temperature zones which gives us an effective method to determine the accurate composition of the catalysts. Fig. 6 represents the H₂-TPR profiles of the manganese oxides calcined from precursors. There are two or three clear peaks in each profile and it depends on the components of samples. The first peak appeared at a lower temperature zone (<320°C) and this was ascribed to reduction of Mn₂O₃ to Mn₃O₄ and another peak at a higher temperature zone (>400°C) can be corresponded to the reduction of Mn₃O₄ to MnO. This result is similar to the H₂-TPR results of manganese oxides reported in other literature [22, 23]. It is worth mentioning that the reducibility at lower temperature zones on manganese oxide catalysts has a great effect on the performance of VOCs combustion [22, 24]. As shown in Fig. 6, one can conclude that the general decreasing low temperature reducibility trend was: a1 < b1 < c1 < d1 which was in good agreement with the trend of catalytic performances.

3.4 Possible formation mechanism of Mn-based nanostructures

The growth of manganese precursors was affected by the dosage of ascorbic acid under reflux condition. In our research, the dosage of reducing agent (AA) not only played a vital role in the crystal structure of as-prepared products but also affected their final

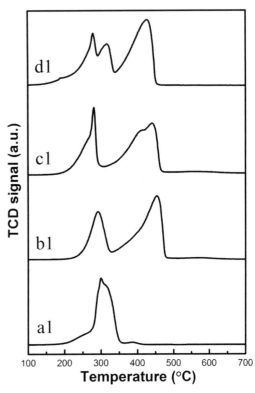

Figure 6. TPR profiles of as-calcined manganese oxides (a1-d1).

morphology. As we know, ascorbic acid is a biological reducing agent and is often used to prepare nanocrystals [25]. For the reducing ability of AA, KMnO₄ was first reduced to in the solution and AA was oxidized into dehydro-ascorbic acid (DHA) when the reactants were mixed together. Mn²⁺ generated at beginning of the reaction can also react to MnO₄⁻ while manganese oxides would be produced. Under reflux condition, DHA was hydrolyzed into 2, 3-diketogulonic acid and further oxidized into threonic acid and oxalate acid in a reaction solution [26], and the CO_3^{2-} will be generated from the deep oxidation of the threonic acid. Meanwhile, all MnO₄⁻ can be reduced to Mn²⁺ directly when the content of AA is enough. If the amount of AA is in excess in the solution, such as sample d, the Mn²⁺ will react with $C_2O_4^{2-}$ to form C₂MnO₄2H₂O. In addition, the products of AA decomposition were also served as a structure-directing agent to form a special structure.

4 CONCLUSIONS

In summary, a series of manganese compounds were synthesized by a facial reflux reaction between KMnO$_4$ and ascorbic acid (AA). Nanowires and nanoparticles, spindle-like and urchin-like particles, were obtained by adjusting the additional content of a reducing agent (AA). The morphology of all particles can be perfectly maintained after thermal decomposition. The mole ratio of KMnO$_4$ to AA was an important factor in influencing the crystal structure and the morphology of the prepared products. The calcined samples, especially those in which the precursors have MnC$_2$O$_4$ species, performed an excellent catalytic activity for benzene combustion. Moreover, the morphology of samples also has an important effect on the activity of benzene oxidation.

REFERENCES

C.K. King'ondu, N. Opembe, C.-h. Chen, K. Ngala, H. Huang, A. Iyer, H.F. Garcés, S.L. Suib, 2011, Manganese Oxide Octahedral Molecular Sieves (OMS-2) Multiple Framework Substitutions: A New Route to OMS-2 Particle Size and Morphology Control, Advanced Functional Materials, 21: 312–323.

J. Zhang, Sun, Yin, Su, Liao, Yan, 2002, Control of ZnO Morphology via a Simple Solution Route, Chemistry of Materials, 14: 4172–4177.

J. Luo, H.T. Zhu, J.K. Liang, G.H. Rao, J.B. Li, Z.M. Du, 2010, Tuning Magnetic Properties of α-MnO2 Nanotubes by K+ Doping, The Journal of Physical Chemistry C, 114: 8782–8786.

K. Frey, V. Iablokov, G. Sáfrán, J. Osán, I. Sajó, R. Szukiewicz, S. Chenakin, N. Kruse, 2012, Nanostructured MnOx as highly active catalyst for CO oxidation, Journal of Catalysis, 287: 30–36.

K.A.M. Ahmed, Q. Zeng, K. Wu, K. Huang, 2010, Mn3O4 nanoplates and nanoparticles: Synthesis, characterization, electrochemical and catalytic properties, Journal of Solid State Chemistry, 183: 744–751.

S. Lei, K. Tang, Z. Fang, Q. Liu, H. Zheng, 2006, Preparation of α-Mn2O3 and MnO from thermal decomposition of MnCO3 and control of morphology, Materials Letters, 60: 53–56.

T. Kokubu, Y. Oaki, E. Hosono, H. Zhou, H. Imai, 2011, Biomimetic Solid-Solution Precursors of Metal Carbonate for Nanostructured Metal Oxides: MnO/Co and MnO-CoO Nanostructures and Their Electrochemical Properties, Advanced Functional Materials, 21: 3673–3680.

N. Xu, Z.-H. Liu, X. Ma, S. Qiao, J. Yuan, 2008, Controlled synthesis and characterization of layered manganese oxide nanostructures with different morphologies, Journal of Nanoparticle Research, 11: 1107–1115.

J. Yin, F. Gao, Y. Wu, J. Wang, Q. Lu, 2010, Synthesis of Mn3O4 octahedrons and other manganese-based nanostructures through a simple and green route, CrystEngComm, 12: 3401–3403.

A.P. Altshuller, 1983, Review: Natural volatile organic substances and their effect on air quality in the United States, Atmospheric Environment (1967), 17: 2131–2165.

J.E. Cometto-Muñiz, W.S. Cain, M.H. Abraham, 2004, Detection of single and mixed VOCs by smell and by sensory irritation, Indoor Air, 14: 108–117.

D. Andreeva, P. Petrova, J.W. Sobczak, L. Ilieva, M. Abrashev, 2006, Gold supported on ceria and ceria–alumina promoted by molybdena for complete benzene oxidation, Applied Catalysis B: Environmental, 67: 237–245.

S. Zuo, Q. Huang, R. Zhou, 2008, Al/Ce pillared clays with high surface area and large pore: Synthesis, characterization and supported palladium catalysts for deep oxidation of benzene, Catalysis Today, 139: 88–93.

T. Garetto, M. Avila, C. Vignatti, V. Venkat Rao, K. Chary, C. Apesteguía, 2009, Deep Oxidation of Benzene on Pt/V2O5–TiO2 Catalysts, Catalysis Letters, 130: 476–480.

J.S. Yang, W.Y. Jung, G.D. Lee, S.S. Park, E.D. Jeong, H.G. Kim, S.-S. Hong, 2008, Catalytic combustion of benzene over metal oxides supported on SBA-15, Journal of Industrial and Engineering Chemistry, 14: 779–784.

D. Delimaris, T. Ioannides, 2008, VOC oxidation over MnOx–CeO2 catalysts prepared by a combustion method, Applied Catalysis B: Environmental, 84: 303–312.

A.R. Gandhe, J.S. Rebello, J.L. Figueiredo, J.B. Fernandes, 2007, Manganese oxide OMS-2 as an effective catalyst for total oxidation of ethyl acetate, Applied Catalysis B: Environmental, 72: 129–135.

F. Wang, H. Dai, J. Deng, G. Bai, K. Ji, Y. Liu, 2012, Manganese Oxides with Rod-, Wire-, Tube-, and Flower-Like Morphologies: Highly Effective Catalysts for the Removal of Toluene, Environmental Science & Technology, 46: 4034–4041.

B. Puertolas, B. Solsona, S. Agouram, R. Murillo, A.M. Mastral, A. Aranda, S.H. Taylor, T. Garcia, 2010, The catalytic performance of mesoporous cerium oxides prepared through a nanocasting route for the total oxidation of naphthalene, Applied Catalysis B: Environmental, 93: 395–405.

K. Ji, H. Dai, J. Deng, L. Song, B. Gao, Y. Wang, X. Li, 2013, Three-dimensionally ordered macroporous Eu0.6Sr0.4FeO3 supported cobalt oxides: Highly active nanocatalysts for the combustion of toluene, Applied Catalysis B: Environmental, 129: 539–548.

V.P. Santos, M.F.R. Pereira, J.J.M. Órfão, J.L. Figueiredo, 2010, The role of lattice oxygen on the activity of manganese oxides towards the oxidation of volatile organic compounds, Applied Catalysis B: Environmental, 99: 353–363.

S.C. Kim, W.G. Shim, 2010, Catalytic combustion of VOCs over a series of manganese oxide catalysts, Applied Catalysis B: Environmental, 98: 180–185.

K. Frey, V. Iablokov, G. Sáfrán, J. Osán, I. Sajó, R. Szukiewicz, S. Chenakin, N. Kruse, 2012, Nanostructured MnOx as highly active catalyst for CO oxidation, Journal of Catalysis, 287: 30–36.

S.C. Kim, W.G. Shim, 2008, Influence of physicochemical treatments on iron-based spent catalyst for catalytic oxidation of toluene, Journal of hazardous materials, 154: 310–316.

M.N. Nadagouda, R.S. Varma, 2007, A Greener Synthesis of Core (Fe, Cu)-Shell (Au, Pt, Pd, and Ag) Nanocrystals Using Aqueous Vitamin C, Crystal Growth & Design, 7: 2582–2587.

A.M. Rojas, L.N. Gerschenson, 2001, Ascorbic acid destruction in aqueous model systems: an additional discussion, Journal of the Science of Food and Agriculture, 81: 1433–1439.

The isolation, identification and biochemical reducing pathway of Cr (VI)-removal bacterium *Brevibacillus Parabrevis* from sludge biosystems

Yan Zhou, YanBin Xu, ShiHui Xu, XiaoHua Zhang & JiaXin Xu
School of Environmental Science and Engineering, Guangdong University of Technology, Guangzhou, People's Republic of China

JingSheng Luo, WuLong Gao & YaoJie Deng
Zhongneng Environmental protection Technology Limited Company of Zhongshan, Zhongshan, People's Republic of China

ABSTRACT: Strain 6# of Cr (VI) (hexavalent chromium) removal bacterium was isolated from sludge in Cr (VI)-contained wastewater treatment system and cultivated by UV mutagenesis to get a higher capability of Cr (VI) resistance and reduction. Based on morphology, biochemical results and 16S rRNA sequence, the strain was identified as *Brevibacillus parabrevis* CR1. Its plasmid (23 kb) was isolated and the plasmid missing tests indicated that the plasmid was closely related to the bacterial resistance to Cr (VI). According to the results of physiological-biochemical and basic biodegradation pathway of the main nutrients, the biochemical pathway of Cr (VI) reduction was summarized, which gives a guide to the application of B. parabrevis CR1 on the remediation of Cr (VI)-contaminated sites.

1 INTRODUCTION

The uncontrolled release of industrial wastes containing Cr (VI) has caused severe contamination of soil-water systems and subsequent chromium toxicity because of its carcinogenic, mutagenic, and teratogenic potential (Ackerley et al., 2006). Conventional remediation technologies for Cr (VI), contained in wastewater, include ion exchange, precipitation and adsorption on alum or kaolinite. However, this cannot be applied on a large scale because of high costs and secondary pollution. Therefore, bioremediation of the sites contaminated by toxic metals is drawing more and more attention because of its efficient, affordable and environmentally friendly advantages (He et al., 2010). The Cr (VI)-reducing capacity of chromate-resistant bacteria (CRB) has been widely reported (Camargo et al., 2005). Moreover, fungi, algae and other biomaterials are considered to possess a Cr(VI) adsorption capacity and beneficiation capacity (Han et al., 2007; Gupta and Babu, 2009; Poopal and Laxman, 2009). 'Adsorption-coupled reduction' is now widely accepted as the mechanism of Cr (VI) biosorption by natural biomaterials (Park et al., 2007a; 2007b; 2008). The main mechanism of Chromium (VI) biosorption on to Trapa dried powder was through the binding of Chromium ions with amide group of the biomass (Vankar et al. 2013). Additionally, the co-oxidoreductase (flavocytochrome b2) is considered as a potential candidate for chromate reduction by living cells in the presence of L-lactate (Smutok et al., 2011) and the co-existence of organic pollutants such as azo dye can offer electrons to reduce chromate (Wai et al., 2010). However, the most fundamental reason for Cr (VI) bio-removal has to be determined at gene level, and ChrA, B, C, F are generally considered as chromate resistant genes (Aguilera et al., 2004).

In this study, a predominant strain was isolated from the sludge acclimated by synthetic wastewater, containing Cr (VI), and its characteristics of morphology, biochemical tests and 16S rRNA sequence were analysed, and the mechanisms on Cr (VI) bioreduction and resistance were also studied. The results will be beneficial for the in-situ bioremediation of Cr (VI) contaminated sites.

2 MATERIALS AND METHODS

a) Sludge source

A small scale of bio-treatment system was acclimated by synthetic wastewater containing Cr (VI), and the Cr (VI) concentration of wastewater was 14.4 mg/L by adding $K_2Cr_2O_7$.

b) Medium

The isolation medium formula that was used was made up of 0.5 g of KH_2PO_4, 2.0 g of $(NH_4)_2SO_4$, 0.1 g of NH_4Cl, 0.5 g of Na_2SO_4, 0.1 g of $CaCl_2$, 0.5 g of $NaHCO_3$, 0.1 g of $MgSO_4$, 2.0 g of yeast extract, 3.0 g of sodium lactate, 0.02 g of Vitamin B$_2$, 0.02 g of Vitamin C 0.041 g of K$_2$Cr$_2$O$_7$, 15 g of agar, dissolved in 1 L of deionized water, with pH adjusted to 7.0–7.2, which was also used in the enrichment of isolates.

Broth nutrition medium and other media including dextrose peptone medium, nitrate liquid medium, starch medium, gelatine liquefaction medium, Hugh and Leifson's second semi-solid medium, phenylalanine medium from ammonia enzyme, citrate medium and yolk medium were used in physiological-biochemical tests and prepared according to the description in Bergey's Manual of Systematic Bacteriology (Holt et al. 1994).

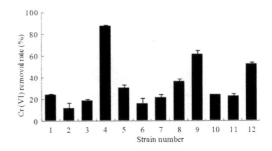

Figure 1. Cr (VI) removal rates of 12 predominant strains isolated from acclimated sludge.

c) Isolation of Cr (VI)-removal strains

Strains were isolated from the acclimated sludge by dilution plate method (Somasundaram et al., 2009). Isolated colonies on solid plates were picked off and inoculated onto fresh plates to obtain pure cultures. The pure cultures were transferred to a 250-mL flask with 100 mL of liquid medium and incubated at 30°C, 125 rpm for 48 h to get the bacterial suspension of each isolate, and the density of bacterial suspension was about 10^6 CFU/mL.

d) Ultraviolet ray mutagenesis of Cr (VI)-removal strains

1 mL of the above-mentioned bacterial suspension (106 CFU/mL) was irradiated by 15-W ultraviolet light (UV) with a 30-cm lighting distance for 10, 20, 30, 40 and 60 s, respectively. 0.5 mL of irradiated bacterial suspension was transferred to the solid isolation medium and incubated at 30° for 24 h under dark conditions; tri-duplicates were made with each UV irradiation time. In order to determine the variation of bacterial resistance to Cr (VI), mutated strains were inoculated on the medium with Cr (VI) of different concentrations (20, 40, 60, 80, 100, 120, 160, 200 mg/L) and incubated at 30° for 48 h, then a single colony of each mutated strain under different mutagenesis condition was enriched to get its bacterial suspension with a density of 106 CFU/mL or so.

10 mL of mutated bacterial suspension was introduced into a 150 mL beaker with 50 mL synthetic wastewater containing Cr (VI) of 14.4 mg/L, and the mixture was stirred rapidly for 1 min and stirred slowly for forty-five minutes and then was centrifuged (High Speed Refrigerated Centrifuge, 2-16K, SIGMA) at 3000 rpm for ten minutes. Cr (VI) and total Cr concentration in the supernatant indicated the Cr (VI) removal capability of every strain and its mutants. The predominant strain was selected for the following tests according to the results of Cr (VI) removal tests. The formula of Cr (VI) or total Cr removal rate was as follows:

$$\eta\% = [(C_1 - C_2)/C_1] \times 100\%$$

C$_1$ represents Cr (VI)/total Cr concentration of raw water (mg/L);

C$_2$ represents Cr (VI)/total Cr concentration of supernatant after adsorption (mg/L).

e) Characterization and identification

Cell morphology of the selected strains was observed using an electron microscope (FEI Tecnai 12, Holland). Conventional physiological and biochemical characterization tests were carried out as described in Bergey's Manual of Systematic Bacteriology (Holt et al., 1994). Identification of strains was carried out by 16S rRNA sequence analysis (Zheng et al., 2008).

f) Plasmid extraction and missing test

Plasmid was isolated by using the Plasmid DNA Extraction Kit (Dingguo Changsheng Biotechnology Co. Ltd., Beijing, China). The plasmid missing tests were used to judge the relativity between the plasmid and Cr (VI) removal capacity. The strain was subcultured twice by means of the medium and a similar medium with 30 mg/L of sodium dodecyl sulfonate (SDS), sequentially, and then the growth of strain was observed.

3 RESULTS

3.1 Isolation

12 strains with high capability of Cr (VI) removal were isolated from acclimated sludge and the Cr (VI) removal rates were assayed and compared. The removal rates of strains 4$^\#$, 9$^\#$ and 12$^\#$ were 87.62%, 61.11% and 52.62%, respectively, while those of strain 2$^\#$ and 6$^\#$ were only 12.03% and 15.71%, respectively (Fig. 1).

3.2 UV mutagenesis

Breeding by UV mutagenesis was used to improve the bacterial tolerance to Cr (VI) and their Cr (VI) removal capacity. Five strains including strains 2$^\#$, 4$^\#$, 6$^\#$, 9$^\#$ and 12$^\#$ were selected to be cultivated under UV irradiation. Mutated strains were incubated in

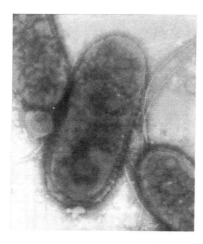

Figure 2. Electro-micro photo of strain 6#(×35000).

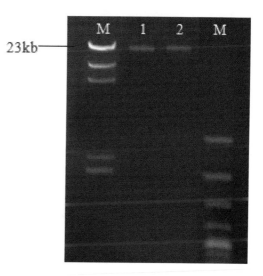

Figure 3. Electrophoresis pattern of plasmid DNA. (M represents for the DNA marker; 1 and 2 represent plasmids from strain 6#)

nutrient medium with Cr (VI) of different concentrations (20, 40, 60, 80, 100, 120, 160, 200 mg/L) at 30° for 24–48 h. Table 1 showed that Cr (VI) resistance levels of all these five strains increased greatly; all strains except strain 4 were able to grow on a medium containing as much as 200 mg/L of Cr (VI).

Single bacterial suspension of every strain, including the original strains and mutated strains, were prepared and used to treat synthetic wastewater containing 100 mg/L of Cr (VI). Table.1 showed that Cr (VI) removal rates of mutated strains increased by 6.1%–48.15% and Cr (VI) removal rates of strain 2# and strain 6# increased to 49.14% and 50.18% after 30-s UV irradiation, respectively. Mutated strain 6# with the highest Cr (VI) removal rate was selected for further study in this paper.

3.3 Characterization and identification

Strain 6# is an endospore-forming, gram-negative, flagellate, rod bacterium according to its micrograph (Fig. 2). Their common characteristics are round colonies, light yellow, neat edged, acicular, lustrous, smooth and opaque in surface.

Results of physiological and biochemical tests of strain 6# showed that catalase and nitrate reduction reaction was positive. However, oxidase reaction was negative, so were Voges-Proskauer test (V-P test), starch hydrolysis, phenylalanine deaminase reaction and liquefaction of gelatine. Strain 6# could utilize D-glucose but not L-arabinose, D-mannitol, D-xylose as the carbon source. Citrate and lecithin were dissimilated. These data indicated that strain 6# could be a member of *Brevibacillus genus*. Strain 6# was 99% identical with the sequence of *Brevibacillus parabrevis* from GenBank database (GenBank accession: AB112714) according to its 16S rRNA sequence. Therefore, strain 6# was identified as *Brevibacillus parabrevis CR1*.

3.4 Mechanism analysis of Cr (VI) bio-reduction

Plasmid from *B parabrevis CR1* had a size of 23 kb (Fig. 3). In the plasmid missing test, strain lacking plasmid will not grow, while the original strain 6# could grow on the medium with a Cr (VI) final concentration of 200 mg/L. This result showed that the bacterial tolerance to Cr (VI) was closely related to its plasmid and Cr (VI) reduction-adsorption genes could exist in plasmid.

According to the physiological biochemical characteristics of strain 6# and the degradation pathways of sugar, fat and protein, a potential biochemical pathway of Cr (VI) reduction was concluded (Fig. 4). *B. parabrevis CR1* can obtain acetyl-CoA by preliminary degradation and then release in the presence of reducing electron acceptors, nicotinamide-adenine dinucleotid (NADH)$_2$. Acetyl-CoA is then transported Krebs cycle (TCA), releasing a large number of electron acceptors NADH$_2$, which were possibly being oxidized or generating ATP simultaneously (channel II) through a respiratory chain. In the presence of Cr (VI), due to its strong oxidability, Cr (VI) could replace oxygen as electron acceptors to accept the electron carried by reducing electron carriers and was then possibly reduced (channel I). Due to low Cr (VI) catalase activity, the reaction in which NADH$_2$ transferred electrons into oxygen, generating H$_2$O$_2$, occurred only in aerobic conditions, while more reducing electron acceptors transferred electrons to other electron acceptors, such as Cr (VI).

Therefore, strain 6# transferred a small number of reducing electron acceptors caused by a dehydrogenation decomposition process to Cr (VI) and Cr (VI) was reduced to Cr (III). In aerobic conditions, strain 6# transferred a mass of electron acceptors mainly from organic matter degradation, especially by TCA, to Cr (VI) and Cr (VI) was reduced to Cr (III). However,

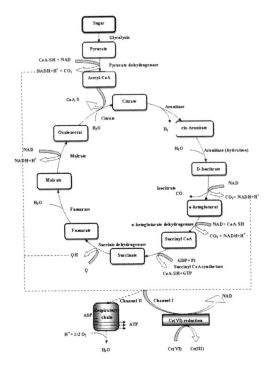

Figure 4. Biochemical pathway of Cr (VI) reduction by strain 6#.

the biochemical pathways could only be performed by strain 6# on the condition that the Cr (VI) ions could penetrate the cell wall and reach the plasma membrane or the possession of specific coenzyme by strain 6#.

So Cr (VI) reduction process was closely related to the biochemical processes of strain 6#. Due to its strong oxidability, Cr(VI) could compete with bacterial oxidative phosphorylation of electron transformation and strived for electrons of reducing electron acceptors NAD(P)H$_2$ after which Cr(VI) was reduced as electron acceptor.

4 DISCUSSION

In this study, 12 predominant strains were isolated from Cr (VI)-containing acclimated sludge. All the isolates were able to reduce Cr (VI), and the Cr (VI) removal rates range from 12.02% to 87.62%. The influence of UV mutagenesis treatment on strain 6# was the most significant and was selected for further study in the present paper.

Based on morphology, biochemical tests and 16S rRNA sequence, strain 6# was identified as *Brevibacillus parabrevis* CR1 and can tolerate Cr (VI) of up to 200 mg/L. *Brevibacillus parabrevis* and *Brevibacillus agri* from earthworm viscera, play an important role in the conversion of fish wastes to a liquid fertilizer, and they can tolerate low concentration of Cr (Kim et al., 2010). An autochthonous microorganism, *Brevibacillus brevis,* isolated from Cd amended soil can increase P and K acquisition, as well as immobilizing metals (Cd, Cr, Mn, Cu, Mo, Fe and Ni) to decrease their translocation to the plant shoot (Vivas et al., 2005). *Brevibacillus brevis* is a Gram positive and spore-forming bacterium. It can secrete large amounts of secondary metabolites, which are important for combating pathogens. Most of the research done on *Brevibacillus spp.* focus on the function of increasing plant growth and producing bacteriocin (Song et al., 2012), while the resistance to heavy metals of *Brevibacillus* sp. has not been studied extensively. Additionally, a large number of microorganisms have been reported as being able to tolerate high concentrations of Cr (VI) and reduce Cr (VI), such as *Bacillus sp.* (Cheng and Li, 2005), *Arthrobacter sp.* (Mishra and Doble, 2008). Cr (VI) bio-reduction is dependent on pH, temperature, Cr concentration, carbon and energy source. In addition, a low pH of about 2.0–3.0 and a general temperature of about 25° are found to be optimal for Cr (VI) removal (Şahin and Öztürk, 2005; Bhattacharya et al., 2008). Therefore, its ability to tolerate extreme heat (20 min at 120°C) and low pH (range of 3–10) makes *Brevibacillus* sp. a good candidate for remediation in sites heavily contaminated with heavy metals (Ghadbane et al., 2013).

It is known that the primary cause of Cr (VI) bio-reduction is dependent on the biochemical and genetic characteristics of microorganisms. In recent years, there have been several reports on chromate resistance genes and chromate reductases. The chrA gene of *Pseudomonas aeruginosa* plasmid pUM505 encodes hydrophobic protein ChrA, which confers resistance to chromate by energy-dependent efflux of chromate ions (Aguilera et al., 2004). The chrA gene of *Lysinibacillus fusiformis* ZC1 encoding a putative chromate transporter conferring chromate resistance has been identified, and a yieF gene and several genes encoding reductases have been found and they are possibly involved in chromate reduction (He et al., 2011). Tn5045, a novel antibiotic and chromate resistance transposon, was isolated from a permafrost strain of *Pseudomonas sp* and contains genes (chrB, A, C, F) of chromate resistance (Petrova et al., 2011). *Brevibacillus parabrevis* CR1 studied in the present paper carries a plasmid of 23 kb, which is larger than the circular plasmid (6,600 bp) of *Brevibacillus brevis* X23, an appropriate bio-control agent against bacterial wilt caused by *Ralstonia solanacearum*, as reported by Chen et al. (2012). Results of plasmid missing test show that the Cr (VI) tolerance of strain 6#was closely related to its plasmid and putative Cr (VI) reduction-adsorption genes should exist in the plasmid.

The rate of Cr (VI) reduction under anaerobic conditions decreases with increasing concentration of Cr (VI). Dissolved oxygen decreases the rate of Cr (VI) reduction, which is the result of oxygen molecule competing with Cr (VI) to accept electron by uncompetitive inhibition, as well as more energy generated for cells from thermodynamically oxygen reduction than from Cr (VI) reduction (Somasundaram et al.,

2009). Moreover, Cr (VI) reduction occurring in extracellular is different from that in intracellular. In the present study, the biochemical pathway of Cr(VI) reduction shows that Cr(VI) competes with bacterial oxidative phosphorylation for electrons NAD(P)H$_2$ and is reduced to the less soluble and less toxic Cr(III) by strain 6[#] because of its strong oxidizing property.

5 CONCLUSION

Microorganisms with their ability to tolerate and reduce Cr (VI) can be used to restore a Cr (VI)-contaminated environment. Twelve predominant strains isolated from sludge acclimated by Cr (VI)-containing wastewater were cultivated under UV mutagenesis. The Cr (VI) removal rate of strain 6[#] increased most significantly and was therefore selected for study on its characteristics and Cr (VI) reduction pathway. Strain 6# was identified as *Brevibacillus parabrevis* CR1 from its morphology, biochemical test results and 16S rRNA sequence. Plasmid of 23 kb from *B parabrevis* CR1 was found to be closely related to the bacterial resistance to Cr (VI). In addition, genes determining the bacterial capability of Cr (VI) reduction-adsorption was proposed to be located in the plasmid. According to the physiological and biochemical characteristics of *B parabrevis* CR1, as well as its degradation pathways of sugar, fat and protein, a potential biochemical pathway of Cr (VI) reduction was concluded Further investigations will be conducted for the evaluation of Cr (VI) reduction by applying *Brevibacillus parabrevis* CR1 in the bioremediation of Cr (VI)contaminated sites.

ACKNOWLEDGEMENTS

We would like to extend our thanks to Natural Science Fund of China for granting us the Project (No.40801194). Our thanks also go to both China Postdoctoral Science Foundation for their support by granting us the Project (No. 20100470921) and Science and Technology Plan Project of Zhongshan (2013A3FC0243).

REFERENCES

Ackerley DF, Barak Y, Lynch SV, Curtin J, Matin A. 2006. Effect of chromate stress on *Escherichia coli* K-12. J Bacteriol. 188:3371–3381.

Aguilera S, Aguilar ME, Chávez MP, López-Meza JE, Pedraza-Reyes M, Campos-García J, Cervantes C. 2004. Essential residues in the chromate transporter ChrA of *Pseudomonas aeruginosa*. FEMS Microbiol Lett. 232(1), 107–112.

Bhattachary AK, Naiya TK, Mandal SN, Das SK. 2008. Adsorption, kinetics and equilibrium studies on removal of Cr (VI) from aqueous solutions using different low-cost adsorbents. Chem Eng J. 137(3):529–541.

Camargo FAO, Okeke BC, Bento FM, Frankenber WT. 2005. Diversity of chromium-resistant bacteria isolated from soils contaminated with dichromate. Appl Soil Ecol. 29(2):193–202.

Chen W, Wang YS, Li DJ, Li L, Xiao QM, Zhou QM. 2012. Draft genome sequence of *Brevibacillus brevis* strain X23, a biocontrol agent against bacterial wilt. J Bacteriol. 194(23):6634.

Cheng GJ, Li XH. 2009. Bioreduction of chromium (VI) by *Bacillus sp.* isolated from soils of iron mineral area. Eur J Soil Biol. 45:483–487.

Freitas DB, Lima-Bittencourt CI, Reis MP, Costa PS, Assis PS, Chartone-Souza E, Nascimento AMA. 2008. Molecular characterization of early colonizer bacteria from wastes in a steel plant. Lett Appl Microbiol. 47(4): 241–249.

Ghadbane M, Harzallah D, Laribi AI, Jaouadi B, Belhadj H. 2013. Purification and Biochemical Characterization of a Highly Thermostable Bacteriocin Isolated from *Brevibacillus brevis* Strain GM100. Biosci, Biotechnol, and Biochem. 77 (1):151–160.

Gupta S, Babu BV. 2009. Removal of toxic metal Cr (VI) from aqueous solutions using sawdust as adsorbent: Equilibrium, kinetics and regeneration studies. Chem Eng J. 150(2–3):352–365.

Han X, Wong YS, Wong MH, Tam NF. 2007. Biosorption and bioreduction of Cr(VI) by a microalgal isolate, *Chlorella miniata*. J Hazard Mater. 146:65–72.

He MY, Li XY, Guo L, Miller SJ, Rensing C, Wang GJ. 2010. Characterization and genomic analysis of chromate resistant and reducing *Bacillus cereus* strain SJ1. BMC Microbiology. 221:1–10.

He MY, Li XY, Liu HL, Miller SJ, Wang GJ, Rensing C. 2011. Characterization and genomic analysis of a highly chromate resistant and reducing bacterial strain *Lysinibacillus fusiformis* ZC1. J Hazard Mater. 185:682–688.

Holt JG, Krieg NR, Sneath PHA, Staley JT, Williams ST. 1994. Bergey's manual of determinative bacteriology (9th Ed). Williams and Wilkins, Baltimore, MD.

Kim JK, Dao VT, Kong IS, Lee HH. 2010. Identification and characterization of microorganisms from earthworm viscera for the conversion of fish wastes into liquid fertilizer. Bioresource Technol. 101(14): 5131–5136

Petrova M, Gorlenko Z, Mindlin S. 2011. Tn5045, a novel integron-containing antibiotic and chromate resistance transposon isolated from a permafrost bacterium. Res. Microbiol. 162(3):337–345.

Mishra S, Doble M. 2008. Novel chromium tolerant microorganisms: Isolation, characterization and their biosorption capacity. Ecotoxicol Environ Saf. 71:874–879.

Poopal AC, Laxman SR. 2009. Studies on biological reduction of chromate by *Streptomyces griseus*. J Hazard Mater. 169:539–545.

Park D, Lim SR, Yun YS. 2007a. Reliable evidences that the removal mechanism of hexavalent chromium by natural biomaterials is adsorption-coupled reduction. Chemosphere. 70(2):298–305.

Park D, Yun YS, Ahn CK, Park JM. 2007b. Kinetics of the reduction of hexavalent chromium with the brown seaweed *Ecklonia* biomass. Chemosphere. 66(5):939–946.

Park D, Yun YS, Lee HW. 2008. Advanced kinetic model of the Cr(VI) removal by biomaterials at various pHs and temperatures. Bioresource Technol. 99(5): 1141–1147.

Smutok O, Broda D, Smutok H, Dmytruk K, Gonchar M. 2011. Chromate-reducing activity of *Hansenula polymorpha* recombinant cells over-producing flavocytochrome b2. Chemosphere. 83(4):449–454.

Somasundaram V, Philip L, Bhallamudi SM. 2009. Experimental and mathematical modeling studies on Cr (VI) reduction by CRB, SRB and IRB, individually and in combination. J Hazard Mater. 172: 606–617.

Song Z, Liu QX, Guo H, Ju RC, Zhao YH, Li JY, Liu XL. 2012. Tostadin, a novel antibacterial peptide from an antagonistic microorganism *Brevibacillus brevis* XDH. Bioresource Technol. 111:504–506.

Şahin Y, Öztürk A. 2005. Biosorption of chromium (VI) ions from aqueous solution by the bacterium *Bacillus thuringiensis*. Process Biochem. 40:1895–1901.

Vankar PS, Sarswat R,. Dwivedi AK, Sahu RS. 2013. An assessment and characterization for biosorption efficiency of natural dye waste. Journal of Cleaner Production. 60: 65–70.

Vivas A, Barea JM, Azcón R. 2005. Interactive effect of *Brevibacillus brevis* and *Glomus mosseae*, both isolated from Cd contaminated soil, on plant growth, physiological mycorrhizal fungal characteristics and soil enzymatic activities in Cd polluted soil. Environ Pollut. 134(2): 257–266.

Wai TN, Cai QH, Wong CK. 2010. Simultaneous chromate reduction and azo dye decolourization by *Brevibacterium casei*: Azo dye as electron donor for chromate reduction. J Hazard Mater. 182:792–800.

Zheng YG, Chen J, Liu ZQ, Wu MH, Xing LY, Shen YC. 2008. Isolation, identification and characterization of *Bacillus subtilis* ZJB-063, a versatile nitrile-converting bacterium. Appl Microbiol Biotechnol. 77:985–993.

Effect of operating parameters on the photocatalytic oxidation disinfection of swimming pool water

C.W. Kan & Y.L. Pan
Institute of Textiles and Clothing, The Hong Kong Polytechnic University, Hung Hom, Kowloon, Hong Kong

H. Chua
Faculty of Science and Technology, Technological and Higher Education Institute of Hong Kong, Tsing Yi Island, New Territories

ABSTRACT: In this study, works were carried out to develop a photocatalytic oxidation system using RuO_2 electrolytic and TiO_2 photocatalytic systems for a swimming pool water disinfection system. Chlorine will be used in this disinfection system. Thus, the objective of this study is to evaluate the formation and the disinfection efficiency of the free chlorine, generated by the photocatalytic oxidation system under different chloride concentrations, electric potentials, and the contents of organic substances.

1 INTRODUCTION

Photocatalytic Oxidation (PCO) disinfection is an electro-activation technology and capitalizes on a special coating technique of a specifically designed formula of ornamented titanium dioxide on D.C. electrodes (Zhao & Yang 2003; Zhang et al., 2009; Yu and Brouwers, 2009). The electrically activated coated surfaces transform naturally occurring chloride ions in water into activated chlorine with effective disinfecting power. This technology has been modified into a compact system that has a small footprint, low power consumption, no chemical addition requirement, fully automatic operation, and a stable and reliable water disinfection system.

Secondly, the PCO component in the system utilizes TiO_2 as a photocatalyst to generate oxidation/reduction reactions. When a photo catalyst medium is exposed to radiation of ultraviolet rays, energized electrons will break free from the TiO_2 coating (Zhao & Yang 2003; Yu and Brouwers, 2009). These electrons leave behind positively charged pockets called 'positive holes'. The positive holes vigorously attract hydroxide ions (OH^-) from ambient water. The positive holes then take an electron from an OH^- turning it into extremely unstable OH hydroxyl radicals (Ananpattarachai et al., 2009). To stabilize themselves, the OH hydroxyl radicals take electrons from nearby organic compounds and pollutants. This breaks up the water-borne organic compounds, including bacteria and viruses, thus decomposing them into harmless carbon and water that are released into the air.

Thus, in this study, PCO developed with RuO_2 electrolytic and a TiO_2 photocatalytic system will be used as a swimming pool water disinfection system. Chlorine will be used in this disinfection system. The formation and the disinfection efficiency of the free chlorine generated by the photocatalytic oxidation system under different chloride concentrations, electric potentials, and the contents of organic substances will be studied.

2 EXPERIMENTAL

2.1 PCO disinfection system

A bench-scale PCO disinfection system, using a RuO_2 electrolytic and TiO_2 photocatalytic system, was used and Figure 1 shows the schematic diagram of the system. A water tank was connected to the system.

2.2 Operation parameters

A total volume of 70 L tap water was filled in the tank for testing. The amount of chloride ion in the

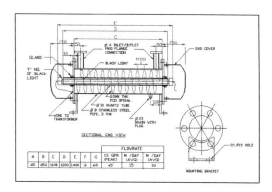

Figure 1. Bench-scale PCO disinfection system.

tap water was adjusted by adding sodium chloride (NaCl). In this study, chloride concentrations of 0, 10, 50, 75, 100 mg/l were added in the tap water, respectively. Concentrations of an organic substance (urea) that represents the common contaminants/pollutants in swimming pools, in the tank were controlled at 0, 5, 10, 20, 40 mg/l, respectively.

The electric voltage applied in water electrolysis processes was varied because of the influence of different amounts of NaCl in the water on the conductivity. The electric voltage was 10, 15, 20 and 25 V when the amount of added chloride ion was 0, 10, 50, 75, 100 mg/l, respectively.

The water samples in the tank were pumped to pass the electrolytic-photocatalytic device and then returned to the tank and recycled at the flow rate of 2 m^3/h. The flow rate was controlled by a series of control valves and internal recirculation circuits. Flow rate, pH and temperature were measured during the test. The concentration of free chlorine was monitored during the process of electrolysis at each five minute interval. The free chlorine concentration shown in the Results and Discussion section was regarded as the free chlorine in the water tank.

The effectiveness of this process was quantified by evaluating the total bacteria that were removed in the process. The method for bacterial enumeration was as follows: the microbial culture media, glassware, deionised water were sterilized in an autoclave at 121°C for fifteen minutes. Dilutions of microbial samples were carried out using sterilized deionised water. An appropriate sample volume was filtered through a sterile membrane filter. The filter papers were then placed on the surface of a plate containing agar medium for determining total bacteria content, and were incubated at 38°C for 48 hours. The colonies were counted in a standard microbial counter after the complete cycle of incubation.

3 RESULTS AND DISCUSSION

3.1 *Effect of chloride concentration on the formation of free chlorine*

The production of free chlorine over the period of a forty minute operation under different chloride concentrations is shown in Figures 2 and 3. The results indicated that free chlorine was not formed during the electrolysis process with neither NaCl nor 10 mg/l NaCl added. It was because of a low chloride concentration in the tap water. This was attributed to both low chloride concentration and low electric current (2.2A) set in this test. Low electrolyte concentration (the simulated swimming pool water) results in poor conductivity of the solution. In addition, the decline of free chlorine is due to the water motion in the system, which was somewhat like the effect of aeration in the water that purges out free chlorine.

Figure 3 shows the free chlorine concentration under operation with 50, 75 and 100 mg/l chloride ion added. The results showed that free chlorine was

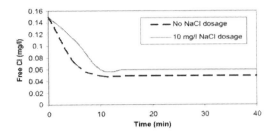

Figure 2. Free chlorine under the operation without NaCl addition and with 10 mg/l NaCl addition.

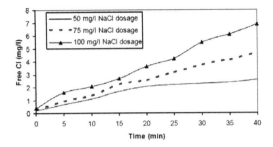

Figure 3. Free chlorine concentrations under the operation with 50, 75, and 100 mg/l NaCl addition.

formed during the electrolysis process depending on how many chloride ions existed in the water. Less than ten minutes of operation was required to generate more than 1 mg/l of free chlorine when more than fifty mg/l of NaCl was used. The concentration of free residual chlorine reached the requirement of the standard of public and private swimming pools in Hong Kong (>1 mg/l). The formation of free chlorine in the water was obvious and substantial during the electrolysis process when the chloride concentration reached a higher level of 75 and 100 mg/l. About 5 min after the beginning of the electrolysis process, free chlorine in the water reached 1 mg/l. In addition, the process period for sufficient free Cl generation could be shortened by increasing the NaCl dosage.

According to the test results, the chloride concentration in water should be maintained at around 50 mg/l, which is the optimum dosage under the electric voltage of 30 V, in order to generate sufficient free chlorine (1 to 3 mg/l) for legislative requirements and effective disinfection.

3.2 *Effect of electric potential on the formation of free chlorine*

The effect of electric potential on the generation of free chlorine by the electro-photo-disinfection system was determined by varying the voltage from 10 to 25 V at different NaCl dosage. The results of free chlorine generation at different voltage and 50 mg/l NaCl dosage are illustrated in Figure 4.

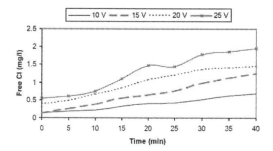

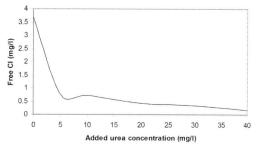

Figure 4. Free chlorine generation under different voltages and at 50 mg/l NaCl dosage.

Figure 5. Free chlorine generation at various urea concentrations.

Table 1. Formation of free chlorine (mg/l) under various conditions after 30 min operation.

Quantity of chloride concentration added in the water (mg/l)	Electric potential applied the electrode (V)			
	10	15	20	25
10	0.05	0.05	0.05	0.05
20	0.09	0.36	0.67	1.20
50	0.51	0.98	<u>1.40</u>	<u>1.78</u>
75	0.71	<u>1.44</u>	<u>2.12</u>	<u>2.88</u>
100	<u>1.01</u>	<u>1.98</u>	3.30	4.67

Under lower voltage (10V), the current was small in the system, therefore, the rate of free chlorine formation was also very slow. When the voltage was increased to 15 V, the rate of chlorine generation was significantly increased. The concentration of free chlorine in the water reached 1 mg/l after 30-minute operation.

As the electric voltage was further increased to 20 and 25 V, the formation of free chlorine was substantially increased due to the higher current applied. The reaction time required to generate 1 mg/l free chlorine was shortened to twenty minutes and fifteen minutes for 20 V and 25 V, respectively. As the electrolysis went on, the concentration of free chlorine accumulated. This indicated that the system is capable of generating a free chlorine production rate that reaches a dynamic equilibrium with the rate of chlorine consumption in the pool water disinfection process.

The concentrations of free chlorine formed under various electrode potentials and externally-dosed chloride concentrations are presented in Table 1. When chloride concentration was controlled above a certain value, the formation of free chlorine was positive and was proportional to the electric voltage applied across the electrodes. The higher the voltage applied in an electrode, the higher rate of free chlorine formation in the water. The electric potential should be applied to the system for effective disinfection is shown in Table 1 (the underline values), in order to acquire sufficient free chlorine (1 to 3 mg/l) in the pool water for effective disinfection. The optimum operational condition of this system is 50 mg/l NaCl dosage and 20 to 25 V.

3.3 Effect of organic substance content on the formation of free chlorine

The effect of organic substance on the formation of free chlorine was examined by varying the concentration of organic substance with 75 mg/l of external NaCl addition and at 30 V electric potential. Urea was selected as a simulated organic matter commonly introduced by swimmers.

According to the result shown in Figure 5, the effect of organic contents on the formation of free chlorine was significant. Higher organic concentration obviously reduced the quantity of free chlorine formed in the water solution. This is because the free chlorine was consumed in the oxidation-reduction reactions with oxidizable organics (e.g. urea). However, the concentration of organic substances is usually very low in swimming pools, while the effect of organic substances on the formation of free chlorine is not obvious under normal circumstances (less than a few ppm of organic matters) and normal operations.

3.4 Effect of free chlorine on disinfection

3.4.1 Formation and disinfection efficiency of free chlorine

The microbial inactivation efficacy of the electrolysis system at an electrode potential 30 V with 50 mg/l NaCl addition is shown in Figure 6. The removal of total bacteria was increased with increasing chlorine generation. After twenty-five minutes operation (free Cl >2 mg/l), which is well above the legislative requirements, seven logarithmic reductions of total bacteria were achieved. These results illustrated the effectiveness of electrochemically generated chlorine species for disinfection.

3.4.2 Effects of free chlorine concentration and contact time on inactivation efficiency

The effects of Cl concentration and contact time on the disinfection performance are assessed by varying the free chlorine concentration from 0.5 to 4.0 mg/l and the contact time from five to twenty minutes, respectively. According to the results, the germicidal efficiency of disinfection, as a measurement of bacteria survival, depends primarily on the concentration of free chlorine and contact time. The result of inactivation efficiency

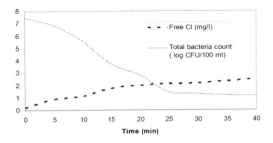

Figure 6. Microbial inactivation efficiency under the operation at 30 V and 50 mg/l NaCl addition (y-axis indicates the amount of bacteria).

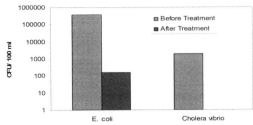

Figure 7. The bacteria count (CFU/100 ml) of samples collected before and after the disinfection.

Table 2. Ratio of total bacteria survival in a batch reactor.

Contacting time (min)	Concentration of free chlorine used (mg/l)				
	0.5	1.0	2.0	3.0	4.0
5	0.48	0.31	0.015	0.009	0.0056
10	0.33	0.02	0.0061	0.0005	0.0004
15	0.21	0.008	0.0007	0.0003	0.0002
20	0.015	0.007	0.0003	0.0002	0.0001

under various operating conditions is presented in the Table 2. The recommended contact time is between fifteen and twenty minutes, according to the desirable CT values and the ratio of total bacteria survival (<0.01). These superior results show that this novel electro-photo-disinfection system is suitable for application in swimming pools. The optimum contact time, which is about fifteen to twenty minutes, is long enough in the normal swimming pool operation with a close loop recirculation flow system. These results (desirable CT values) can establish a database which is important for the process scale up in full scale applications.

3.4.3 *Disinfection efficiency of the pathogenic microorganisms*

The disinfection efficiency of the pathogenic microorganisms of this system is evaluated by the removal of the two selected pathogens (*Escherichia coli* and *Cholera vibrio*). The removal efficiency of these two pathogenic microbes is determined by measuring the quantity of microbes in the samples collected from the tank before and after the disinfection process. The excellent performance (99.96% and 100% removal efficiencies on *E. coli* and *Cholera vibrio*, respectively) on the treatment of pathogenic microbes was observed by the system operated under the optimum conditions (Figure 7).

4 CONCLUSION

The PCO system used in this study was found to have good disinfection function on swimming pool water. Experimental results revealed that the operating parameters played important role in affecting the performance of the PCO disinfection performance. Therefore, the system should be carefully designed in order to obtain the optimum effect.

ACKNOWLEDGEMENT

Authors would like to thank the financial support from the Teaching Company Scheme under The Hong Kong Polytechnic University.

REFERENCES

Ananpattarachai, J., Kajitvichyanukul, P. & Seraphin, S. 2009. Visible light absorption ability and photocatalytic oxidation activity of various interstitial N-doped TiO$_2$ prepared from different nitrogen dopants. *Journal of Hazardous Materials.* 168(1): 253–261.

Yu, Q.L. & Brouwers, H.J.H. 2009. Indoor air purification using heterogeneous photocatalytic oxidation. Part I: Experimental study. *Applied Catalysis B: Environmental* 92 (3–4): 454–461.

Zhao, J. & Yang, X. 2003. Photocatalytic oxidation for indoor air purification: a literature review. *Building and Environment* 38(5): 645–654.

Zhang, M., Wang, Q., Chen, C.C., Zang, L., Ma, W. & Zhao, J. 2009. Oxygen atom transfer in the photocatalytic oxidation of alcohols by TiO$_2$: oxygen isotope studies. *Angewandte Chemie* 48(33): 6081–6084.

Study on photochemical degradation of sulfamethazine in an aqueous solution

HuaHua Xiao, GuoGuang Liu, ZhiMing Chen & RuiHui Wu
College of Environmental science and engineering, Guangdong University of technology, Guangzhou, P.R. China

ABSTRACT: The photochemical degradation of Sulfamethazine (SMT) under solar simulator irradiation was investigated. Results showed that the photolysis of SMT followed the first order kinetic equation and the photolytic rates of SMT was 0.0186 when the initial concentration was 10 mg/L. Under the same photolysis conditions, the photolytic rates were decreased while increasing the initial concentration. The highest photolytic rate was found at pH of 8.0, which was followed by pH 9.0, pH 7.0, pH 3.0, pH 4.0 and pH 5.0. After sixty minutes of irradiation, the photolytic rate of SMT in buffer aqueous solution of pH 8.0 was 75.12%. Decreasing the concentration of dissolved oxygen in water was beneficial to the photodegradation of SMT. Reactive oxygen species such as •OH, $^1O_2$ and O_2^- = were produced and their contributions to the autosensibilization photolysis reaction were 15.41%, 6.48% and 26.05%, respectively. All the intermediates produced in the photolysis reaction showed lower toxicity than the parent SMT.

Keywords: Sulfamethazine; photodegradation

1 INTRODUCTION

Contamination of wastewater is a long-term, and maybe a never-ending topic in the field of environmental protection, due to the development of industrial factories, including textile mills, pesticides and fertilizers, pharmaceuticals, and hospitals, etc. which discharge large amounts of wastewater, containing various chemical pollutants, year by year (Huang, Q.X. et al., 2011; Stuart et al., 2012). The elimination of these chemical pollutants from wastewater is one of the most important subjects in pollution control today. Two strategies are normally used in this respect, one is to seek assistance from an external force, such as catalysts, adsorbents, to remove the pollutant, and the other is by a self-degradation process under irradiation conditions (C.M. et al., 2011; D.F.K. et al., 2011; Domínguez, J. et al., 2011; Elmolla, E. et al., 2011; Fenoll, J. et al., 2012; Ignasi, S. et al., 2012; L.P.R. et al., 2012). The latter is most suitable from the viewpoint of cost and energy savings, especially when the pollutant is in tiny amount.

In clinical medicine Sulfamethazine (Figure 1), SMT, is one of the most important sulpha antibiotics and is extensively used in human bodies, animal husbandry, and aquaculture, because of its excellent ability to destroy bacteria and inflammation. However, about 80% of the used SMT will be released from the organism as excreta, in the form of metabolites or in its original form, into the surface water, contaminating the environment. The elimination of these SMT through sewage treatment is incomplete and residual SMT, with concentrations of ng/L-μg/L, could still exist in the outlet effluents (García-Galán et al., 2010; García-Galán et al., 2012; Yan, C.X. et al., 2013). Although such antibiotics have a short half-life period, they will never vanish due to their large and continuous use in curing diseases. These released SMT will not only kill some microorganisms in the environment, but also induce drug resistant bacteria, and thus their removal or degradation is required and has attracted wide attention recent years.

Photochemical degradation is an efficient way of removing pharmaceutical pollutants in aqueous solutions including SMT, but the degradation behaviour may be influenced under different environments. Thus the understanding of photodegradation pathways and kinetics is essential to predict the behaviour and the environmental impact of these pollutants in waters, which is also a criterion in evaluating the practicability of a medicine before its practical application. The influence of environmental conditions on the photodegradation behaviour of sulpha antibiotics has been observed by many authors. Someone (Niu et al., 2013) investigated systematically the photolysis of SMX in aqueous solutions and found that high SMX concentration, fulvic acid, suspended sediments, NTB and high pH values would suppress the photodegradation rate of SMX, whereas H_2O_2 facilitated the photolysis process. Someone (Trovó, A.G. et al., 2009) studied the photolysis behaviours of SMX in different water matrices and found that the photolysis in distilled water (DW) is

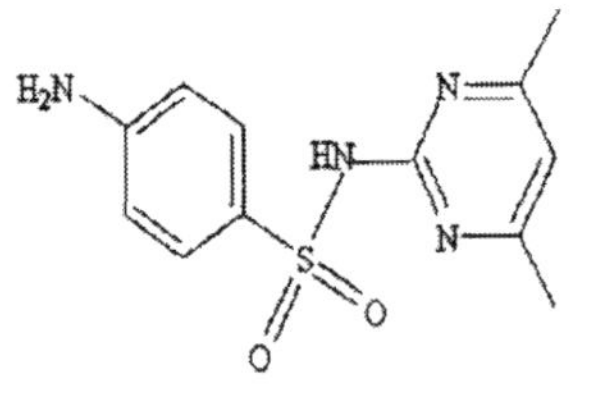

Figure 1. Chemical structure of sulfamethazine.

faster than that in seawater (SW). Scholars (Gao et al., 2013) tested the influence of gases, anions, alcohols, ferrous ion on the degradation rate of SMT, finding that different behaviours were exhibited at different conditions.

In this work we attempted to study the self-degradation behaviour of SMT under different environments, with solar simulator irradiation, to arrive at a more complete understanding of its kinetics, degradation behaviour and its degradation mechanism. It was shown that the degradation follows first-order kinetics and the behaviour was significantly influenced by the environmental parameter, such as the initial SMT concentration, the pH value, the soluble oxygen concentration, the type of quenching agent. Especially the degradation products showed less toxicity towards the environment relative to the mother SMT.

2 MATERIAL AND METHODS

2.1 Chemicals

Chemicals: sulfamethazine was purchased from J&K Chemical Co. Ltd (Beijing, China), HPLC-grade methanol was obtained from German Merk company (Germany), Isopropanol was obtained from Chengdu Kelong Chemical Company (Chengdu, China), Sodium Azide was obtained from Tianjin Fuchen Chemical Company (Tianjin, China), p-benzoquinone was obtained from Adamas Reagent Co. Ltd (Shanghai, China), Photobacterium phosphoreum was purchased from The Chinese Academy of Sciences, Nanjing Institute of Soil Microorganisms (Nanjing, China). All of the chemicals used were of analytical grade, without needing further purification. Ultra-pure water from a Smart2 Pure water process (TKA, Germany) was used for preparing all aqueous solutions.

2.2 Photodegradation experiments and analytical methods

Photolysis experiments were performed on the equipment that had hollow cylindrical quartz tubes photoreactor (SGY-II, Nanjing STO Co. Ltd.), which was put vertically outside a double-walled quartz cooling jacket, A 350 W xenon lamp was put in the double-walled quartz cooling jacket and temperature was controlled by the constant-temperature liquid-circulating apparatus. Additives were placed into the 250 mL volumetric flask and diluted with ultra-water,

Figure 2. Schematic of the experimental setup.

then the reaction liquid was put into nine quartz tubes surrounding the double-walled quartz cooling jacket, as shown in Figure 2. After the reactor had been started up for five minutes the instrument timer started. Before irradiation, aliquots of 1 mL of every spiked sample were taken and put in the dark (aluminium wrapped vials) at room temperature (26°C). During the whole experiment, one tube was taken periodically out at an interval of every ten minutes from the quartz tube and immediately analysed by a reversed-phase high-performance liquid chromatography system (Shimadzu LC-20AT). The analytical column was a 250 mm × 4.6 mm Waters C18 column. A Waters Guard column (C18, 4.6 mm × 20 mm) was used to protect the analytical column (both purchased from Waters). The injection volume was 20 L. The mobile phase was mixtures 55% HPLC-grade methanol and 45% ultra-water at a constant flow rate of 1.0 mL · min-, and the detection wavelength was set at 266 nm. When studying the effect of dissolved oxygen and the quencher on the photodegradation, the reaction mixture sparged with nitrogen or oxygen, or adding the corresponding quencher (isopropanol, sodium azide, p-benzoquinone), while keeping the other conditions constant.

2.3 Data processing

First-order kinetics is generally used to express the photodegradation of a micropollutant (Liu et al.). The rate constant k was calculated from the first-order equation:

$$dC/dt = -kC \qquad (1)$$

where C is the concentration of SMT; k is the rate constant; t is the reaction time. By integrating the equality, the following equation could be obtained:

$$C_t/C_0 = \ln kt \qquad (2)$$

where C_t is the SMT concentration at time t, C_0 is the initial concentration of SMT. Furthermore, when the concentration of SMT reduces to 50% of its initial

Table 1. Kinetics of SMT photodegradation at various initial concentrations.

Concentration/mg/L	k/min^{-1}	t$_{1/2}$/min	R^2
1	0.0262	26.46	0.990
5	0.0230	30.14	0.990
10	0.0186	37.27	0.998
15	0.0166	41.76	0.992
20	0.0133	52.12	0.994

Table 2. Kinetics of SMT photodegradation at various initial concentrations.

pH	k/min^{-1}	t$_{1/2}$ min	R^2
3	0.0175	39.61	0.998
4	0.0114	60.80	0.990
5	0.0129	53.73	0.996
7	0.0186	37.27	0.998
8	0.0245	28.29	0.998
9	0.0218	31.80	0.999

concentration, the half-life can be calculated from the rate constant as the equation:

$$t_{1/2} = \ln 2/k \quad (3)$$

3 RESULTS AND DISCUSSION

3.1 Effect of initial concentration

Before experiment we tested the degradation activity of SMT (10 mg/L) in dark condition, showing that only 1% of SMT was degraded. This indicated that the degradation by microorganisms and hydrolysis could be neglected. The result indicates that the rate decreased with the increase of initial concentration. This is in accordance to the phenomena observed by others (Huang et al., 2011), and is related to a free radical reaction mechanism of photolysis. That is, the SMT was first activated into an oxidative free radical by absorbing energy from the solar irradiation and then induced into the degradation process. The energy from the solar irradiation is responsible for the degradation process. Consequently, when the energy supplied by the solar irradiation is constant the average energy absorbed by each SMT molecule will be lowered if its concentration increases, leading to a lowered degradation rate. This is further supported by the kinetics calculated for the degradation process, Tab. 1. The kinetics were calculated by plotting the natural logarithm of degradation rate, $\ln(C_t/C_0)$, at different initial concentrations as a function of reaction time, which yielded straight lines with coefficient values (R^2), higher than 0.99, indicative of pseudo first-order kinetics. By comparison, it is seen that the rate constant for SMT at initial concentration of 1 and 20 mg/L is 2.62×10^{-2} and 1.33×10^{-2} min^{-1}, with the corresponding half-life ($t_{1/2}$) of 26.46 and 52.12 min. Namely, the degradation of SMT at initial concentration of 1 mg/L is almost two times faster than that at 20 mg/L.

3.2 Effect of pH value

As has been demonstrated in many previous works, the degradation of pharmaceuticals was largely influenced by the solution's pH value, (Yang, H. et al., 2010), thus the influence of pH value on the degradation rate of SMT was also tested in our work. The pH value was controlled by a desired buffer solution as indicated in the experimental section. The degradation rate was suppressed at low pH value (pH = 4) and reached the best at pH = 8, with half-life of 60.80 and 28.29 min. (Tab. 2) respectively, demonstrating that the pH value has a big influence on the degradation process and the reaction is more favoured in a basic, rather than in an acidic condition. This is in accordance with results reported by other scholars (Challis, J.K. et al., 2013) and could be explained by the structural change in different pH values.

It has been reported that sulfamide antibacterials (SAs) canexist in three forms of dissociation in solution, i.e., RH_2^+, RH and R^- (where R represents the parent sulphanilamide) depending on the solution pH and the dissociation constant of the chemicals (Boreen, A. et al., 2005). In the case of SMT, which has the dissociation constants of $pK_a = 2.6 \pm 0.2$ and $pK_b = 8 \pm 1$, someone (Gao, J. et al., 2005) demonstrated that SMT^0 is the major form at pH value between the two constants, and it trends to form SMT$^+$ or SMT$^-$ when the pH value is down- or up-towards to the pK_a and pK_b, respectively. Because of the different forms SMT exhibits and its different ability to light absorption thus the photochemical reactivity leads to different degradation rates at different pH values. Based on the results above, it was concluded that SMT$^-$ at near pK_b (pH = 8) has the strongest ability to light absorption. SMT$^+$ at near pK_a (pH = 3) shows a weaker ability to light absorption, and SMT^0, which is the major form at pH value between pK_a and pK_b (pH = 5), has the worst ability to light absorption. That is: the ability to light absorption is in order of SMT$^-$ > SMT$^+$ > SMT^0. It is thus suggested that the photodegradation of SMT should be done in a weak basic environment.

3.3 Effect of dissolved oxygen

It is known that in the natural environment molecular oxygen has certain amounts of solubility and can be dissolved in aqueous solutions. The dissolved oxygen, on one hand, can act as a quenching reagent in a photochemical reaction, decreasing the quantum yield, and on the other hand, can participate in the photochemical reaction as an active oxygen species, accelerating the reaction. These self-contradictory behaviours make it is hard to conclude whether the dissolved oxygen is disadvantageous or advantageous to the reaction. For this we investigated the effect of dissolved oxygen on the degradation rate of SMT, to specify if it plays a

Table 3. Photolysis of sulfamethazine at various gas.

Gas	k/min^{-1}	t$_{1/2}$/min	R^2
static air	0.0186	37.27	0.998
N$_2$	0.0363	19.09	0.991
O$_2$	0.0147	7.15	0.990

Table 4. Photolysis of sulfamethazine at different quenching agents.

Quenching agents	k/min^{-1}	t$_{1/2}$/min	R^2
Blank	0.0186	37.27	0.998
isopropanol	0.0130	53.32	0.992
Sodium azide	0.0102	67.96	0.993
p-quinone	0.0110	63.01	0.996

positive or negative role in the overall reaction. In controlling the amount of dissolved oxygen we passed through the solution with a) N$_2$, b) O$_2$ or c) no gas (i.e., in static air). Table 3 lists the degradation kinetics of SMT obtained at different atmospheres, showing that the half-life increases from 19.09 to 37.27 and further to 47.15 min when increasing the dissolved oxygen, i.e., the atmosphere changes from N$_2$ to static air and to O$_2$. This indicates that the dissolved oxygen plays a negative role in the overall photochemical degradation of SMT.

3.4 Mechanism of Photodegradation of SMT

As is known that photolysis can be conducted in three ways: direct, indirect and autosensibilization, which can be identified by comparing the calculated degradation rate (from the relationship between the concentration and rate constant k) with the real one, and/or by comparing the degradation rate measured with or without the addition of external substances. In our case as the calculated degradation rate is lower than the real one, see Table 1, and no external substrate was added to the reaction, it is therefore inferred that the autosensibilization reaction was occurred. That is, some reactive oxygen species (e.g., •OH, ^{1}O$_2$ and O$_2^-$) were produced during the SMT degradation process, accelerating the reaction rate. To support this we did three quenching experiments and monitored the degradation rate, using isopropanol (for quenching •OH species), sodium azide (for quenching •OH and ^{1}O$_2$ species) and p-quinone (for quenching O$_2^-$) as the quenching reagents. As expected, the photochemical degradation of SMT was suppressed when the quenching reagent was added, Table 3, and the half-life of SMT measured at different quenching reagents is in order of none reagent > isopropanol > p-quinone > sodium azide. This indicates that reactive oxygen species were produced and a photochemical autosensibilization reaction occurred during the reaction, and the O$_2^-$ species contributes more to the autosensibilization reaction than the •OH species as the suppressing effect resulting from p-quinone is stronger than that from isopropanol.

The contribution of reactive oxygen species •OH, ^{1}O$_2$ and O$_2^-$ to the autosensibilization reaction could be evaluated according to the formula (4) to (6), where R$_{•OH}$, R$_{1O2}$ and R$_{O2^-}$ represent the contribution of •OH, ^{1}O$_2$ and O$_2^-$ to the SMT autosensibilization reaction, K$_{•OH}$, K$_{1O2}$ and K$_{O2^-}$ represent the rate constant of •OH, ^{1}O$_2$ and O$_2^-$ in the reaction, K$_{isopropanol}$, K$_{NaN3}$, K$_{p-quinone}$ and K represent the rate constant of SMT photolysis in the presence of quenching reagents isopropanol, NaN$_3$, p-quinone and that without quenching reagent, respectively. Based on these equations and the data in Tab. 4, the R$_{•OH}$, R$_{1O2}$ and R$_{O2^-}$ was calculated to be 15.41%, 6.48% and 26.05%, respectively, supporting the above conclusions.

$$R_{•OH} = \frac{k_{•OH}}{k} \approx \frac{k - k_{isopropanol}}{k} \quad (4)$$

$$R_{^1O_2} = \frac{k_{^1O_2}}{k} \approx \frac{k_{isopropanol} - k_{NaN_3}}{k} \quad (5)$$

$$R_{O_2^-} = \frac{k_{O_2^-}}{k} \approx \frac{k - k_{p-quinone}}{k} \quad (6)$$

3.5 Photolysis toxicity of SMT

The toxicity was evaluated using the *Vibrio fischeri* luminescent bacteria assay and the evolution of toxicity was expressed as % inhibition as shown in Figure 3, which was calculated from the change in the HgCl$_2$ concentration recorded at different reaction time. The starting inhibition rate of SMT (0 min) was 33.1% and no appreciable change was observed during the initial ten minutes which could have been that only a small amount of SMT was photodegradated or the toxicity of the intermediate was similar to that of the parent SMT. With the increase of reaction time, the inhibition rate decreased significantly to 18.43% after twenty minutes, while a slight increase at reaction time of thirty minutes was observed, indicating a more toxic intermediate than that at after twenty minutes. The inhibition rate decreased almost straight to 8.81% after a reaction time of sixty minutes and no more toxic intermediate was observed thereafter, suggesting that the toxicity of SMT and its photolysis intermediates to Vibrio fischeri luminescent bacteria can be significantly lowered by a reaction time of sixty minutes. Overall, these results indicate that the toxic SMT can be degradated into less or no toxic intermediates after photolysis. More toxic evaluations of SMT and its intermediates are taking place in our group to give better understanding of its toxicity to the environment.

4 CONCLUSION

Photolysis of SMT under solar simulator irradiation follows the pseudo first-order kinetics, with rate constant of 0.0186. The rate constant of SMT photolysis

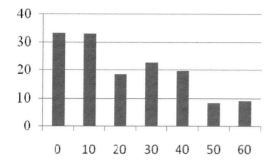

Figure 3. Bioluminescent inhibition of the photoldegradation Sulfamethazine to *Vibrio fischeri*.

decreases with the increase of SMT concentration, with otherwise identical conditions.

The solution pH value affects greatly the photolysis rate. The value at near pK_b that favours the formation of SMT^- is more facilitated by the reaction, while that in between the pK_a and pK_b that favours the formation of SMT^0 is the worst to the reaction. Dissolved oxygen plays an overall negative effect on the photolysis reaction. Reactive oxygen species such as •OH, $^1O_2$ and O_2^- were produced and their contributions to the autosensibilization photolysis reaction were 15.41%, 6.48% and 26.05%, respectively. All the intermediates produced in the photolysis reaction showed lower toxicity than the parent SMT.

REFERENCES

Boreen A. & Williama A. et al. 2005. Triplet-SensItized photodegradation of sulfa drugs containing six—membered heteroeyclic groups: Identification of an SO_2 extrusion photoproduct. *Environ.Sci.Technol.* 39(10): 3630–3638

Challis J.K., Carlson J.C. et al. 2013. Aquatic photochemistry of the sulfonamide antibiotic sulfapyridine. Journal of Photochemistry and Photobiology A: Chemistry 262: 14–21

C. M. & M. C.L. 2011. Aqueous degradation of diclofenac by heterogeneous photocatalysis using nanostructured materials. *Applied Catalysis B: Environmental* 107: 110–118

D.F.K. & M.I.V. 2011. Transformation products of pharmaceuticals in surface waters and wastewater formed during photolysis and advanced oxidation processes – Degradation, elucidation of byproducts and assessment of their biological potency. *Chemosphere* 85: 693–709

Domínguez J.R. & Teresa. G. 2011. Removal of common pharmaceuticals present in surface waters by Amberlite XAD-7 acrylic-ester-resin: Influence of pH and presence of other drugs. *Desalination* 269: 231–238

Elmolla E.S. & Chaudhuri M. et al. 2011. The feasibility of using combined TiO_2 photocatalysis-SBR process for antibiotic wastewater treatment. *Desalination* 272: 218–224

Fenoll, J. & Hellin. P. 2012. Photocatalytic degradation of five sulfonylurea herbicides in aqueous semiconductor suspensions under natural sunlight. *Chemosphere* 87: 954–961

Gao, J. & Pedersen, J. 2005. Adsorption of sulfonamide antmicrobial agents to clay minerals. *Environ. Sci. Techno l.* 39(24): 9509–9516

Gao, Y.Q. & Yang, N.Y. 2013. Factors affecting sonolytic degradation of sulfamethazine in water. *Ultrasonics Sonochemistry* 20: 1401–1407

García-Galán, M.J. & González, B.S. 2012. Ecotoxicity evaluation and removal of sulfonamides and their acetylated metabolites during conventional wastewater treatment. *Science of the Total Environment* 437: 403–412

García-Galán, M.J. & Garrido, T. 2010. Simultaneous occurrence of nitrates and sulfonamide antibiotics in two ground water bodies of Catalonia (Spain). *Journal of Hydrology* 383: 93–101

Huang, C.N. & Li, X.D. 2011. Photochemical degradation of sulfamethazine in aqueous solution. *Environmental pollution and control (China)* 33(12):59–64

Huang, Q.X. & Yu, Y.Y. 2011. Occurrence and behaviour of non-steroidal anti-inflammatory drugs and lipid regulators in wastewater and urban river water of the Pearl River Delta, South China. *J. Environ. Monit.* 13: 855

Ignasi, S. & Enric, B. 2012. Remediation of water pollution caused by pharmaceutical residues based on electrochemical separation and degradation technologies: A review. *Environment International* 40: 212–229

L. P.R. & S. M.C. 2012.Treatment of emerging contaminants in wastewater treatment plants (WWTP) effluents by solar photocatalysis using low TiO2 concentrations. *Journal of Hazardous Materials* 211–212: 131–137

Niu, J.F. & Zhang, L.L. et al. 2013. Effects of environmental factors on sulfamethoxazole photodegradation under simulated sunlight irradiation: Kinetics and mechanism. *Journal of Environmental Sciences* 25(6): 1098–1106

Stuart, M. & Lapworth, D. et al. 2012. Review of risk from potential emerging contaminants in UK groundwater. *Science of the Total Environment* 416: 1–21

Trovó, A.G. & Raquel F.P. 2009. Photodegradation of sulfamethoxazole in various aqueous media: Persistence, toxicity and photoproducts assessment. *Chemosphere* 77: 1292–1298

Yan, C.X. & Yang, Y. et al. 2013. Antibiotics in the surface water of the Yangtze Estuary Occurrence, distribution and risk assessment. *Environmental Pollution* 175: 22–29

Yang, H. & Li, G.Y. 2010. Photocatalytic degradation kinetics and mechanism of environmental pharmaceuticals in aqueous suspension of TiO2: A case of sulfa drugs. *Catalysis Today* 153: 200–207

Study on landscape diversity characteristic of Xilin Gol grassland, based on remote sensing technology

M.X. Huang, G. Bao & Y.H. Bao
Inner Mongolia Normal University, Hohhot, China

F.H. Zhang
Capital Normal University, Beijing, China

ABSTRACT: In this study, Xilin Gol, the study area, is based on Landsat ETM/TM satellite data and a grassland thematic survey data. A thematic map including sixteen types of grassland landscape was generated, and twelve kinds of landscape diversity indices were constructed. Landscape diversity indices were calculated using Fragstats 4.2, and the landscape status, landscape diversity characteristics, and its spatial pattern were analysed. The study results show that the bunch grass rhizome grass, a typical grassland is the dominant type covering the largest area in the study area. The average patch area is large, not rich in landscape diversity, high with landscape aggregation degree, while the landscape diversity indices in different spatial regions are quite different.

1 INTRODUCTION

Grassland resources are one of the important ecosystems of the earth's biosphere, and also comprise the largest ecosystem in China's land surface, and as a kind of natural resource which plays an important role and is of great value for livestock pasturage, providing fodder, water conservation, biodiversity maintenance, soil and water conservation, ecological balance, climate regulation, leisure and recreation, and nutrient cycling (Sha, 2008). Traditional grassland resources monitoring methods mainly rely on grass and costly, time-consuming, labour-intensive, field survey methods which are not suitable for large area monitoring. While in recent years, with high spatial and high spectral resolution remote sensing flourishing, remote sensing technology can effectively distinguish the nuances of different types of ground cover, and combined with the chemical and structural characteristics of plants, it has a greater potential in biological diversity research and conservation (Turner, 2003), and has become the preferred option in monitoring and evaluation studies of grassland diversity, and has been widely used (Chi, 2011; Wu, 2010; Mehner, 2004; Rocchini, 2007; Root, 2002).

Biodiversity refers to all living creatures and genetic material found on the planet, and the diversity and extent of ecosystems posed by the interaction between these living creatures and the environment. Currently, biodiversity can be studied on four levels, genetic diversity, species diversity, ecosystem diversity, and landscape diversity (Li, 2011), of which landscape diversity means that space, functional mechanisms, and temporal dynamics' diversity or variability of landscape, constituted by different types of landscape elements or ecosystems (Wang, 2007). Landscape diversity, as an aspect of biodiversity, starting from the mid-1980s, was studied by many researchers, using remote sensing methods (Forman, 1986; Gould, 1997; Wu, 2000), such as Hua Zhang et al., who studied the landscape diversity of Songshan Nature Reserve in Beijing' Yanqing county using satellite images of TM and the Beijing-1 (Zhang, 2013); Xing Fang and Yuan applied multiphase image of Landsat-5 TM and Landsat-7 ETM + to analysis the variation trend of landscape spatial structure since 1988 on both sides of the Pearl River (Fang, 2012).

In this study, we took Xilin Gol League as the study area, and using Landsat ETM / TM satellite data and grassland thematic survey data, a grassland landscape thematic map was generated. In the meanwhile, a landscape diversity index was constructed and analysed and a landscape diversity index was calculated, using Fragstats 4.2. We also researched and analysed the current landscape situation, landscape diversity characteristics, and its spatial pattern in the Xilin Gol League. This study provides technical support for grassland landscape mapping and a diversity characteristic analysis in Inner Mongolia and offers new ways and means for grassland diversity research.

2 STUDY AREA

Xilin Gol Grassland is located in central Inner Mongolia Autonomous Region of China, located

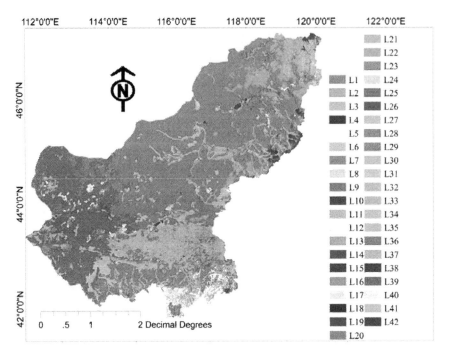

Figure 1. Landscape spatial distribution in the study area.
L1:Bunch grass-rhizome grass typical grassland; L2:Trees garden; L3:Trees green fields; L4:Transportation land; L5:Artificial vegetation; L6:Low wetland vegetation grassland; L7:Other grassland; L8:Residence land; L9:Industrial land; L10:Evergreen needle leaved forest; L11:Evergreen coniferous shrub; L12:Drought grassland; L13:Forb meadow at forest edge-grasses-weed meadow steppe; L14:Water area; L15:Deservoir/pits; L16:Desert/sand; L17:Sandy desert grassland; L18:River; L19:Lake; L20:Scrub; L21:Bush wetland; L22:Shrubs garden; L23:Shrubs green space; L24:Alkaline land; L25:Short grass-short subshrub desert grassland; L26:Grasses-subshrub grassland; L27:Grasses-subshrub desert grassland; L28:Sparse forest; L29:Sparse shrubbery; L30:Sparse grassland; L31:Grasscluster; L32:Desertification grassland; L33:Sandy vegetation grassland; L34:Herbal wetland; L35:Herbal green space; L36:Meadow steppe; L37:Deciduous coniferous forest; L38:Broadleaved deciduous forest; L39:Deciduous broad-leaved shrub; L40:Bare soil; L41:Bare rock; L42:Stope.

between 115° 13′–117° 06 ′E and 43° 02′–44° 52′N, which is an important livestock base for China and the western development frontier, and also is the nearest grassland from the Beijing-Tianjin-Tangshan region. Xilin Gol grassland is rich in vegetation resources, dominated by various grasses, and accompanied by sandy trees, low wetland vegetation and short shrubs. The largest vegetation type in Xilin Gol is typical steppe, and its edificators are Stipa and Chinensis, followed by drought grassland, and its dominant species are Stipa and Cleistogenes. At present, Xilin Gol Grassland is the relatively preserved temperate grasslands in northern China, and is of a natural, representativeness and typicality, and is the ecological barrier of northern China.

3 MATERIALS AND METHODS

3.1 Data preprocessing

We took the study area's medium-high-resolution Landsat TM/ETM of 2010 as the data source, to which after geometric precision correction, single-band extraction, false colour composite, image stitching, and image mask as follows. After this, a false colour composite image was acquired. We then interpreted the remote sensing image, using interactive visual interpretation methods to get a land use thematic map. Moreover, for the purpose of segmentation of grassland type, within the land use types, we then applied overlay analysis and processing methods to land use thematic map and recent grassland resources survey thematic map. Finally we achieved a thematic map containing forty two kinds of landscape types (Figure 1).

3.2 Landscape diversity index

A landscape diversity index is a concentrated landscape pattern with information, which uses quantitative indicators to reflect compositional structure and characteristics and the spatial configuration characteristics of the landscape. In order to analyse the grassland diversity characteristics of the study area, this study selected twelve indices, the Shannon diversity index, landscape dominance index, and the landscape richness index and so on, as the diversity indices, as shown in Table 1. Fragstats 4.2 was used to calculate the landscape diversity index. The spatial resolution of the

Table 1. Landscape diversity index (LDI) and their descriptions.

Landscape Index	Abbreviation	Description
Landscape Richness Index	PR	PR is equal to the total number of all landscape types, $PR \geq 1$.
Landscape Richness Density Index	PRD	$PRD = PR/A$, n/100 ha, PRD is equal to the number of patch types per 100 hectares.
Number of Patch	NP	$NP = N$, The total number of landscape patches, $NP \geq 1$.
Patch Density	PD	$PD = N/A$, The number of patch per 100 hectares, $PD \geq 0$, no limit, the greater the value, the greater the density
Class Area	CA	$CA = \sum_{i=1}^{n} a_i \left(\frac{1}{10000}\right)$ a_i is the area of the patch i (m^2), unit: ha, $CA > 0$, no limit, the greater the value, the greater the area.
Mean Patch Size	MPS	$MPS = (A/N) \times 10^6$ A: all the area of the patch (m^2), N: the total number of patches, $MPS > 0$, no limit, the greater the value, the greater the average patch size.
Largest Patch Index	LPI	LPI is a block type occupying proportion of the whole landscape area $0 < LPI < = 100$, unit: %, it helps determine the dominant landscape type, its value determine the size of the dominant species in the landscape and inside species' abundance of and other ecological characteristics.
Landscape Shape Index	LSI	$LSI = 0.25E/\sqrt{A}$ E: the total length of all the plaques boundary (m), A: total area of the landscape (m^2), 0.25 is square calibration constant. $LSI \geq 1$, no limit, the larger the value, the irregular the shape.
Shannon Diversity Index	SHDI	$SHDI = -\sum_{i=1}^{m} P_i \ln(P_i)$ p_i: each patch type of account for the proportion of the total area of the landscape, $SHDI \geq 0$, no limit, the bigger the value, the diversity increased.
Shannon Evenness Index	SHEI	$SHEI = SHDI/H_{max}$ H_{max}: the maximum possible diversity under given landscape abundance, $0 \leq SHEI \leq 1$, SHEI=0 show that landscape consists only of a plaque, no diversity, SHEI=1 show that each patch types were distributed evenly, having the largest diversity.
Landscape Dominance Index	DH	$DH = H_{max} + \sum_{i=1}^{m} P_i \ln(P_i)$ A big advantage degree includes one kind or several dominant landscape types; a small advantage degree shows that each type has a similar proportion.
Contagion Index	CONT	$CONT = \left[1 + \sum_{i=1}^{m} \sum_{j=1}^{n} \frac{P_{ij} \ln(P_{ij})}{2\ln(m)}\right]$ m: the total number of patch type, P_{ij} is the probability of randomly selected two adjacent grid cells belong to the type i and j. An aggregation index usually measures aggregation degrees of the same patch type, but its values are affected by the type, the number, and the type evenness, $0 < FN \leq 100$.

landscape diversity index for the entire study area is 500 m. and for each county is 30 m.

4 RESULT AND ANALYSIS

4.1 *Landscape characteristic of Xilin Gol grassland*

In order to further the quantitative analysis of each grassland landscape, this study also extracted an area and its percentage of study area for each grassland landscape, as shown in Figure 2. The largest grassland landscape is bunch grass-rhizome grass typical grassland (G8) with an area of approximately 912×10^4 ha, accounting for 45.37% of the study area and it is the dominant landscape of the study area, mainly distributed in the middle, northern, and south of the study area. The second one is short grass-short subshrub desert grassland (G5) with an area of approximately 338×10^4 ha, accounting for 16.79% of the study area, mainly distribute in the west of the study area. While the area of sandy vegetation grassland (G11), low wetland vegetation grassland (G6), Meadow steppe (G1) and forb meadow at forest edge-grasses-weed meadow steppe (G14) are quite balanced, their area being 119×10^4 ha, 118×10^4 ha, 113×10^4 ha, 99×10^4 ha, respectively, and accounting for 5.93%, 5.89%, 5.6%, 4.95% of the study area, respectively. Grass cluster (G15) and herbal green space (G16) occupy the smallest area in the study area, scattered in the research area.

4.2 *Landscape diversity characteristics*

For analysing landscape types in different regions and their diversity pattern in the study area, this study compiled statistics of the landscape types and a diversity index in twelve regions of the study area, with Figure 3 showing the landscape Pie Chart of each region in the study area. Figure 3 shows that

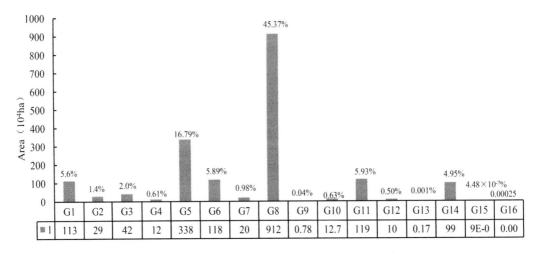

Figure 2. Area of each grassland landscape and its percentage:
G1: Meadow steppe; G2: Sparse grassland; G3: Drought grassland; G4: Other grassland; G5: Short grass-short subshrub desert grassland; G6: Low wetland vegetation grassland; G7: Grasses-subshrub desert grassland; G8: Bunch grass-rhizome grass typical grassland; G9: Herbal wetland; G10: Desertification grassland; G11: Sandy vegetation grassland; G12: Sandy desert grassland; G13: Grasses-subshrub desert grassland; G14: Forb meadow at forest edge-grasses-weed meadow steppe; G15: Grass cluster; G16: Herbal green space.

Table 2. Landscape diversity index in the study area.

LDI	A	NP	MPS	LPI	LSI	PD	PR	PRD	SHDI	SHEI	D_H	CONT
Value	20109325	26577	756.64	30.58	76.68	0.13	42	0.0002	2.01	0.54	1.7	63.22

ErLianHoTe and SuNiTeYouQi have the largest number of landscape types, having twelve species. ZhengXiangBaiQi and ZhengLanQi's grassland have fewer types i.e. five kinds of each, and the other counties' situation is an average of about ten kinds. In terms of the dominant grassland landscape type (i.e., the largest grassland types in each county), bunch grass-rhizome grass is the typical grassland, with obvious advantages, as it is the dominant grassland landscape type for six counties. The second one is short grass-short, subshrub desert grassland, occupying the area of ErLianHaoTe, SuNiTeYouQi, and SuNiTeZuoQi with 75%, 60% and 60% respectively. The following one is Drought grassland, occupying the area of DuoLunXian and TaiPuSiQi with 37% and 57% respectively. Besides, ZhengLanQi' dominant grassland type is Sandy desert grassland, with an area percentage of 42%.

4.3 *Spatial pattern of landscape diversity*

Twelve landscape diversity indices of the study area are shown in Table 2: Xilin Gol' area is about 20,109,325 ha with 26,577 patches, the mean average patch size is about 756.64 ha, and patch density index is 0.13, overall, patch density is not big; the largest patch index (LPI) is 30.58%. We analysed that the largest landscape patch is short grass-short subshrub desert grassland, not bunch grass-rhizome grass typical grassland which has the largest area in the study area. Landscape richness index (PR) is 42, that is a total of forty two grassland landscapes in the study area. Landscape richness density index (PRD) is 0.0002, which means that, on the average level, there is an index of 0.0002 landscape per 100 ha. The Shannon diversity index (SHDI), Shannon uniform index (SHEI), landscape dominance index (DH) and landscape aggregation index (CONT) were 2.01, 0.54, 1.7, 63.22, respectively. The overall patch areas of the study area and the number of landscape types per unit area were not large. The landscape type is not rich, but the advantages of the dominant landscape type is obvious, the advantage degree is high, and the aggregation of different types of ecological landscape is high.

Table 3 shows twelve landscape diversity indices for twelve regions in the study area. In Table 3, ZhengLanQi has the largest patch number with 40750, while the minimum number appears in ErLianHaoTeShi with 171 patches, which may be relevant to the area, ErLianHaoTeShi' area has 17612.4 ha, only about 1.73% of ZhengLanQi; the largest mean patch area is ABaGaQi with 271 ha, and the smallest one is ZhengLanQi with 25 ha. From the above two perspectives, ZhengLanQi's landscape fragmentation degree is relatively strong, the largest pieces existing within DONGWUZHUMUQINQI being 68 ha, the smallest pieces existing in Duolun, the size of 7 ha. Landscape shape index is the largest in Zhenglanqi, 170,

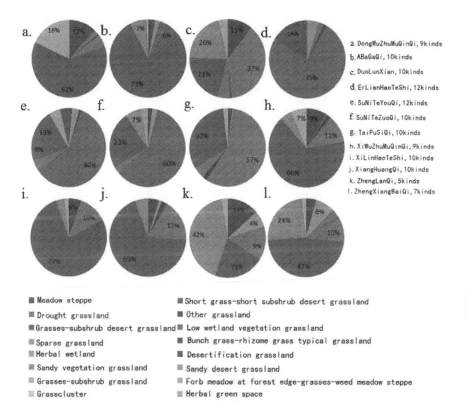

Figure 3. Landscape types of each Xilin Gol' county.

Table 3. A variety of grassland landscape diversity evaluation index (LDI) of each Xilin Gol' county.

COUNTY	LDI A	NP	MPS	LPI	LSI	PD	PR	PRD	SHDI	SHEI	DH	CONT
DongWuZhuMuQinQi	2737886	15149	181	68	52	0.5533	26	0.0009	1.0355	0.3178	2.2228	83
ABaGaQi	4612213	16995	271	46	74	0.3685	30	0.0007	1.3817	0.4062	2.0198	78
DuoLunXian	387948	6958	56	7	75	1.7007	27	0.0070	2.2702	0.6888	1.0257	60
ErLianHaoTeShi	17612.4	171	103	26	10	0.9709	18	0.1022	1.5607	0.5400	1.3295	69
SuNiTeYouQi	2584497	10715	241	48	50	0.4146	29	0.0011	1.6937	0.5030	1.6735	73
SuNiTeZuoQi	3417172	18588	184	54	59	0.544	29	0.0008	1.4994	0.4453	1.8678	76
TaiPuSiQi	344387	5619	61	12	66	1.6316	20	0.0058	1.6513	0.5512	1.3445	67
XiWuZhuMuQinQi	2362232	11373	208	15	69	0.4815	28	0.0012	1.7362	0.5210	1.5962	72
XiLinHaoTeShi	1486131	7747	192	41	44	0.5213	29	0.0020	1.1062	0.3285	2.2612	82
XiangHuangQi	510825	4623	110	54	32	0.905	22	0.0043	1.1744	0.3799	1.9169	79
ZhengLanQi	1016810	40750	25	14	170	4.0076	25	0.0025	2.1145	0.6569	1.1044	60
ZhengXiangBaiQi	630136	15715	40	32	81	2.4939	24	0.0038	1.8780	0.5909	1.3002	66

plaque strongest effect, a minimum of ErLianHaoTe; plaque densest also Zhenglanqi as 4.008, more than two once again confirms Zhenglanqi landscape fragmentation is strong counties; landscape richness of the strongest Abagaqi; landscape diversity index was the biggest in Duolun, Zhenglanqi second, respectively 2.2702, 2.1145; landscape evenness Duolun County is the largest, followed Zhenglanqi were 0.6888, 0.6509; landscape dominance of the strongest for Xilinhot, is 2.2612; landscape aggregation is DONGWUZHUMUQINQI strongest, the weakest of Abagaqi, respectively 83,60. Overall, different spaces within the study area (administrative divisions), landscape diversity index differences, and different regions show different landscape diversity.

5 CONCLUSION

The Xilin Gol grassland type, complete with its plant and animal species and other characteristics, is one of the four world famous prairies, a Eurasian steppe zone, and an important ecological boundary in northern China. This study, based on satellite remote sensing data and grassland thematic survey data, has produced a thematic map containing forty two kinds of landscape types, relying on the construction of the twelve landscape diversity indices. The use of Fragstats 4.2 calculated throughout the study area and sub-region landscape diversity index, and analysed the current situation of grassland landscape in Xilin Gol, and researched the characteristics and spatial pattern of landscape diversity. Through this study, it was found that bunchgrass – typical prairie grass roots type of the largest, with an area of 912×10^4 ha, accounting for 45.37%, had the advantage of an ecological landscape, mainly in central, northern and southern regions of the study area. Overall plaque larger study area, much per unit area on the landscape, the landscape type is not rich, but the obvious advantages of landscape types, greater advantage of the high degree of aggregation of different types of ecological landscape. Different spatial study areas (administrative divisions), show landscape diversity index differences, and different regions show different landscape diversity and evident spatial heterogeneity.

ACKNOWLEDGEMENT

The financial support of Natural Science Found of Inner Mongolia (2011BS0609), Inner Mongolia Science and Technology Project (20110524), and the National Natural Science Foundation of China (40901233) are gratefully acknowledged.

REFERENCES

Chi E. Y., Guo C. L., Yang X. T. 2011. Ejina poplar tree crown extraction based on Quickbird image. Science and Technology Innovation Herald. (5):13–15.

Fang X., Fang Y. 2012. Analysis on Spatial Pattern of the Coastal Area of Pearl River Estuary Based on GIS and Fragstats in Landscape Scale. Hubei Agricultural Sciences. 51(4):841–842.

Forman R., Gordon M. Landscape Ecology. New York, USA: John Wiley and Sons. 1986.

Gould W. A., Walker M. D., 1997. Landscape scale patterns in plant species richness along an arctic river. Canadian Journal of Botany. 75: 1748–1765.

Li Y. J., Zheng S. W., Gong G. T. et al. 2011. Research progress of biodiversity. Journal of Sichuan Forestry Science and Technology. 32(4):12–19.

Mehner H, Cutler M, Fairbairn D, et al. 2004. Remote sensing of upland vegetation: The potential of high spatial resolution satellite sensors. Global Ecology and Biogeography. 13(4):359–369.

Rocchini D. 2007. Effects of spatial and spectral resolution in estimating ecosystem-diversity by satellite imagery. Remote Sensing of Environment. 111(4): 423–434.

Root R, Ustin S, Zarco-Tejada P, et al. 2002. Comparison of AVIRIS and EO-1 Hyperion for classification and mapping of invasive leafy spurge in Theodore Roosevelt National Park. Proceedings of the Eleventh JPL Airborne Earth Science Workshop, NASA Jet Propulsion Laboratory.

Sha R.N. 2008. On the value grassland resource. Journal of Inner Mongolia Normal University (Philosophy & Social Science). 37(4):44–48.

Turner W, Spector S, Gardiner N. et al. 2003. Remote sensing for biodiversity science and conservation [J].Trends in Ecology and Evolution. 18(6): 306–314.

Wang B. S., Peng S. L., Guo L. et al. 2007. Diversity of tropical forest landscape type in Hainan Island, China. ACTA ECOLOGICA SINICA. 27(5):1690–1695.

Wu J., Peng D. L. 2010. Tree-Crown Information Extraction of Farmland Returned to Forests Using QuickBird Image Based on Object-Oriented Approach. Spectroscopy and spectral analysis. 30(9):2533–2536.

Wu Y. N., Li Z. H. 2000. Changing of Landscape Diversity with Time in Xilinguole Steppe. CJPE. 24(1): 58–63.

Zhang H. Li C., Zhang B., et al. 2013. Remote Sensing Monitoring and Evaluation of Landscape Diversity of Beijing Songshan National Nature Reserve. Journal of Green Science and Technology. (6):1–6.

Plant diversity and community stability in two subtropical karst forests in Southwest China

Z.H. Zhang & G. Hu
Key Laboratory of Beibu Gulf Environment Change and Resources Utilization, Guangxi Teachers Education University, Ministry of Education, Nanning, China
School of Chemistry and Life Science, Guangxi Teachers Education University, Nanning, China

B.Q. Hu
Key Laboratory of Beibu Gulf Environment Change and Resources Utilization, Guangxi Teachers Education University, Ministry of Education, Nanning, China

ABSTRACT: Karst forest in Southwest China has a specific habitat, complex community structure, and rich biodiversity, providing an ideal place for ecological study in the karst geological background. In this study, two 1 ha (100 m × 100 m) forest plots were established in the Maolan National Natural Reserve, Southwest China. All freestanding woody plants (including lianas and vines) in two plots, with Diameter at the Breast Height (DBH) ≥1 cm were investigated. Species diversity of forest community was investigated by calculating Shannon-Wiener Diversity Index, Simpson Diversity Index, and Pielou Evenness Index. The results showed that karst forest in Maolan has rich biodiversity and the two plots have a high diversity index. The values of the diversity and evenness index among different layers are listed in a decreasing order: shrub layer > arboreal layer > herb layer. Habitat heterogeneity may play an important role in regulating the species diversity in karst forest. Community stability of two karst forests was studied by using the M. Godron method. The cross points between the species proposition lines and accumulation relative frequency lines were located close to the stabilizing point (20/80), indicating a typical stability state of the two forest communities. Two plots with rich species diversity, to a certain extent, reflect also the stability of the communities in this subtropical karst forest.

1 INTRODUCTION

Plant diversity and community stability is a central issue in ecology (McCann 2000, Ives & Carpenter 2007). Stability is an integrated feature of structure and function of a plant community. Generally, the most important indicator of the stability is species composition and the structural diversity of plant community (Wang 2002). Species diversity characterizes the complexity of the community structure and function, reflecting the community structure, development stage, stability and habitat differences. The plant communities play an important role in the stability of the ecosystem. Thus, species diversity and stability are important indicators to describe the structure and function of plant communities and ecosystems (Wu et al. 2011). The analysis of species diversity and stability of plant community can reveal the structure and dynamics of communities, the structure and function of ecosystems, and the vegetation recovery of degraded ecosystems (McGrady-Steed et al. 1997).

The Karst forest in the Maolan National Natural Reserve in the Guizhou Province represents such vegetation in the mountainous area of Southwestern China. It is the only subtropical evergreen deciduous broad-leaved mixed forest at the same latitude in China or even in the world, and also a rare, original forest remnant in the mid-subtropics. This forest has the specific habitat, complex community structure and rich biodiversity, providing an ideal place for ecological study under a karst geological background (Zhang et al. 2012). Many studies have been carried out on plant species composition, the structure and dynamics of a plant community, the relationship between vegetation distribution and environmental factors and vegetation restoration in Maolan (Zhu 2002, Guo et al. 2011), but knowledge on the species diversity and community stability and their relationship was very limited. In this study, we analysed the species diversity and stability in two subtropical Karst forests in Maolan National Nature Reserve in Guizhou Province, China. The objectives of this study were: (1) to reveal the plant diversity in two subtropical forests, (2) to explore the stability of plant communities and then compare the relationships between the species diversity and stability of two forest communities.

2 MATERIALS AND METHODS

2.1 Study area

The study was conducted in the Maolan National Natural Reserve (MNNR, 25°09′20″–25°20′50″N, 107°52′10″-108°05′04″E). It is located in Libo County, Southern Guizhou Province of Southwestern China. In this ca. 20,000 ha reserve the elevation ranges from 430 to 1078.6 m with an average of 800 m. Topography is characterized by a typical karst fengcong depression. The carbonate rocks (mainly pure limestone and dolomite) are usually exposed on the surface. Soils are therefore thin and discontinuous in the karst terrain. However, the shallow black limestone soil is rich in organic matter, nutrients (N, P and K) and Ca. This region has a subtropical monsoon climate with a feature of plateau climate: mild temperature and plenty of rainfall through the year. Mean temperature ranges from 8.3°C in January to 26.4°C in July with an annual mean of 15.3°C. Mean annual frost-free periods last 315 days. Average annual precipitation is 1320.5 mm and mean annual relative humidity is 83%. The mean sunshine percentage is only ca. 20%.

2.2 Field sampling

In July 2008, two 1 ha (100 m × 100 m, named P1 and P2, respectively) permanent plots were established in the core zone of MNNR. Plot P1 is located in the Dongge site (25°18′25″N, 107°57′48″E; alt. 876m). This plot was established on a steep southeast-facing slope from valley bottom to hilltop covered by thin soil and more bare rocks. Plot P2 is located on the top of another low mountain at the Gengzheng site (25°18′07″N, 107°57′21″E; alt. 915m), with relative thick soil and less outcrop rock. Detailed information of the two plots can be found in Zhang et al. (2012). Forests in these two plots have less human disturbances and are representatives of subtropical karst forests in Maolan according to field surveys. Using the DQL-1 forest compass (Harbin Optical Instrument Factory, China), each plot was divided into a hundred contiguous 10 m × 10 m subplots as workable units. All freestanding woody plants in the plot, with diameters at the breast height (DBH) ≥1 cm, were investigated. The species' names, the relative location of each individual, DBH, and height within each subplots were recorded.

2.3 Calculation of species diversity

The trees in two plots were divided into the following three categories by height: shrub layer; tree height of more than 1 m and less than 5 m sub-tree layer; tree height of more than 5 m and less than 12 m; tree height of more than 12 m.

To determine species richness, we pooled woody species lists, registered for each subplot. The Shannon–Winner index (H') and Simpson index (D) for species diversity were calculated according to

Table 1. The diversity indices of two forest plots in Maolan National Natural Reserve.

Plot	Species richness index	Simpson index	Shannon-Wiener index	Pielou evenness index
P1	199	0.9726	4.114	0.777
P2	191	0.9736	4.153	0.791

$H' = -\sum p_i \log p_i$, $D = 1 - \sum p_i^2$, where p_i is the proportion of importance value of the ith species ($p_i = n_i/N$, n_i is the importance value index of ith species and N is the importance value index of all the species) (Magurran 1988). The Pielou evenness index was calculated according to the formula $E = H'/\ln S$, where S is the total number of species (Magurran 1988). Species importance values were computed as the average of the relative basal area, density, and frequency.

2.4 Calculation of community stability

The M. Godron method was applied to analyse community stability of the karst forest in this study (Godron 1972). According to the calculation of M. Godron, the ratio of the percentage of plant and accumulation of relative frequentness is 20/80, which is the steady point, and the more close to 20/80, the more stable is (Godron 1972, Zheng 2000).

3 RESULTS

3.1 Species diversity of forests

Of the two plots, plot P1 with 199 species was slightly higher than plot P2 with 191 species (Table 1). The Shannon-Weiner index and Pielou's evenness index values were just slightly, but insignificantly, greater for the P2 plot than the P1 one, whereas they both had the same Simpson index values, indicating that the woody plant species diversity among the two plots were almost the same.

The values of species richness index and Shannon-Wiener index among different layers in both two plots were in a decreasing order: shrub layer > sub-tree layer > tree layer (Figure 1). The value of Simpson index between sub-tree layer and shrub layer showed small differences with the lowest value being the tree layer. The values of Pielou evenness index among different layers of plot P1 were in a decreasing order: tree layer > sub-tree layer > shrub layer, while the lowest value occurred in tree layer for plot P2.

3.2 The stability of plant communities

The coordinate of crossing point in plot P1 and P2 were 21/77 and 23/78, respectively (Table 2 and Figure 2). The crossing points between the species proposition

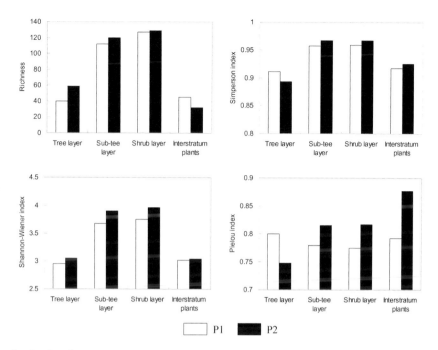

Figure 1. Species diversity of different tree layers of two forest plots in Maolan National Natural Reserve.

Table 2. Results of stability of two subtropical karst forests in Maolan National Natural Reserve.

Plot	Curve type	Coefficient	P value	Crossing point
P1	$-0.0110x^2 + 1.627x + 31.74$	0.9830	<0.01	21/77
P2	$-0.0121x^2 + 1.684x + 33.02$	0.9857	<0.01	23/78

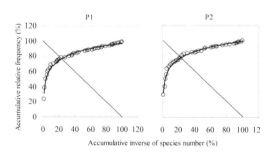

Figure 2. Stability graphs for plant communities of the two forest plots in the Maolan National Natural Reserve.

lines (x) and accumulation relative frequency lines (y) were located close to the stabilizing point (20/80), indicating a typical stability state of the two forest community.

4 DISCUSSION

The Maolan karst forest in Guizhou is a well conserved original forest. Because of the various edaphic condition and frequent rock outcrops topography. The forest has unique floristic composition. The habitat diversity is a major driving force for species distribution and diversity, because it determines ecological processes and thus triggers the level of adaptation and species selection (Wittmann et al. 2008). Our results showed that karst forest in MNNR has rich species diversity. The unique and combined geomorphologic configuration in Maolan forms a specific geological landscape of forest. The various microclimates, heterogeneitic environment, and diverse habitats (e.g. stone face, stone groove, stone gap, and soil surface) as well as their random distribution, provide rich regeneration niches for different plant species. So, this is the foundation of rich species in the Maolan karst forest (Long 2007). The slightly greater species richness (eight species more) obtained in P1, as compared to P2, may be attributed to site heterogeneity in P1, with more micro-environmental condition b (e.g., more irregular ground surface and rock outcrops) found within the site, with more soil cover, flat and with less rock outcrop terrain, while P2 has a more or less uniform terrain.

The values of richness and the Shannon-Wiener index of the two plots were in a decreasing order: shrub layer > sub-tree layer > tree layer. Compared to shrub layer in the karst forest, fewer species dominated in the tree layer, most species are distributed in the sub-tree tree and shrub layers. In addition, the shrub layer not only includes shrub species, but also tree saplings. Therefore, the species diversity of sub-tree tree and shrub layers is higher than the tree layer. The Simpson index also showed a similar trend.

The cross points between the species proposition lines (x) and accumulation relative frequency lines (y) were located close to the stabilizing point (20/80), indicating a typical stability state of the two forest community. Generally, the more complex the community structure, with higher species diversity and richness, the more stable plant communities were (Loreau 2000). Our study found that species diversity and stability in a karst forest community are positively correlated. Tilman (2000) suggested that the higher plant community diversity can increase the productivity and stability of ecosystems. However, a large number of studies showed that there were complex relationships between the stability and diversity of plant communities and plant diversity did not directly reflect the stability of a community (McCann 2000). Biodiversity can stabilize ecosystems by functional complementarity, with different species thriving under different conditions thereby buffering the effects of environmental change (MacDougall et al. 2013). Therefore, High species diversity ecosystems should be stable and complex. The forests in plot P1 and P2 in MNNR are mature forests. Meanwhile, the high species diversity of the two plots to some extent reflects the stability of the community.

ACKNOWLEDGMENTS

This study was supported by the National Natural Science Foundation of China (31300351) and the Guangxi Natural Science Foundation (2013GXNSFBA019085).

REFERENCES

Godron, M. 1972. Some aspects of hetenrgeneity in grasslands of Cantal. *Statistical Ecology* 3: 397–415.

Guo, K., Liu C.C. & Dong, M. 2011. Ecological adaptation of plants and control of rocky-desertification on karst region of Southwest China. *Chinese Journal of Plant Ecology* 35 (10): 991–999.

Ives, A.R. & Carpenter, S.R. 2007. Stability and diversity of ecosystems. *Science* 317: 58–62.

Long, C.L. 2007. Comparison of species diversity in karst forest among different topography sites—a case study in Maolan natural reserve, Guizhou Province. *Carsologica Sinica* 26: 55–60.

Loreau, M. 2000. Biodiversity and ecosystem functioning: recent theoretical advances. *Oikos* 91: 3–17.

Magurran, A.E. 1988. *Ecological Diversity and its Measurement*. Princeton: Princeton University Press.

McCann, K.S. 2000. The diversity–stability debate. *Nature* 405: 228–233.

McGrady-Steed, J., Harris, P.M. & Morin, P.J. 1997. Biodiversity regulates ecosystem predictability. *Nature* 390: 162–165.

MacDougall, S., McCann, K.S., Gellner, G. & Turkington, R. 2013. Diversity loss with persistent human disturbance increases vulnerability to ecosystem collapse. *Nature* 494: 86–89.

Tilman, D. 2000. Causes, consequences and ethics of biodiversity. *Nature* 405: 208–211.

Wang, G.H. 2002. Further thoughts on diversity and stability in ecosystems. *Biodiversity Science* 10 (1): 126–134.

Wittmann, F., Zorzi, B.T., Tizianel, F.A.T., Urquiza, M.V.S., Faria, R.R., Sousa, N.M., Módena, É.S., Gamarra, R.M. & Rosa, A.L.M., 2008. Tree species composition, structure, and aboveground wood biomass of a riparian forest of the lower Miranda River, southern Pantanal, Brazil. *Folia Geobotanica* 43: 397–411.

Wu, Z.L., Zhou, X.N., Zheng, L.F., Hu, X.S. & Zhou, C.J. 2011.Species diversity and stability of natural secondary communities with different cutting intensities after ten years. *Journal of Forestry Research* 22(2): 205–208.

Zhang, Z.H., Hu, G., Zhu, J.D. & Ni, J. 2012. Stand structure, woody species richness and composition of subtropical karst forests in Maolan, south-west China. *Journal of Tropical Forest Science* 24(4): 498–506.

Zheng, Y.R. 2000. Comparison of methods for studying stability of forest community. *Scientia Silvae Sinicae* 36 (5): 28–32.

Zhu, S.Q. 2002. *Ecological Research on Karst Forest III*. Guiyang: Guizhou Science and Technology Press.

Evaluating the smart land use of small towns in the context of Chinese new-type urbanization: A case study of Shanglin County in South China

B.Q. Hu, J.M. Wei, G. Hu & Z.Q. Zhang
Key Laboratory of Beibu Gulf Environment Change and Resources Utilization, Guangxi Teachers Education University, Ministry of Education, Nanning, China

ABSTRACT: Smart use of land is considered one of the most effective means to promote urbanization, transformation, and upgrading. In the future, development of new-type urbanization finds its potential in small urban areas. The study, using the example of eleven towns in Shanglin County as units, has built an evaluation index system of smart land use in small urban areas from the four perspectives of economic intensification, resource conservation, urban and rural coordination, and environment protection. A variation coefficient method and a matter-element extension model are adopted to analyse the level of smart land use. The study has shown that only the land use of Dafeng Town, where the town of Shanglin proper lies, reaches smartness level while the Xiangxian Town and Qiaoxian Town featuring grain producing areas stay at the basic-smartness level and most towns disappointingly remain at the level of non-smartness in land use. The study therefore is expected to provide a scientific and theoretical foundation for the path selection of regional smart land use as well as urbanization, transformation, and upgrading.

1 INTRODUCTION

Laying emphasis on development of urban areas but overlooking that of rural areas and agriculture, traditional urbanization focuses on extensive expansion and development at the cost of the environment and worsens the unbalance of the urban-rural dual structure. New-type urbanization, characterized by urban-rural coordination, resource conservation, environmental protection, and production development proves to be a more efficient, fairer, and greener low-carbon path for urbanization development. Propelled by economic intensification, new-type urbanization is oriented towards 'urban-rural coordination and environmental protection' through resource conservation. History of both domestic and overseas urbanization sufficiently tells us that land use methods bear closely on urbanization which will inevitably lead to the diversification of land use methods and fundamental changes in land use structures (Zhu & Qin 2012). Thus, exploration into the smart use of land, in the process of new-type urbanization, will on the one hand avert a 'great leap forward' and 'horse-riding for land enclosure', but on the other hand help grasp the development trend of new-type urbanization (Ma et al. 2013), optimize the land use methods and more importantly provide a reference for the path selection of smart land use.

Covering vast territory, China witnesses a big gap between the development of urban and rural areas. Metropolises can hardly accommodate the continuous influx of population while small rural areas have the potential to support urban-rural coordination. According to scholars like Zhang Shengze, for the benefit of new-type urbanization, big cities should be held in leash and prominence given to the development of medium and small-sized urban areas, especially the small towns (Feng et al. 2013). With relevance to studies on the smart use of land, most scholars carry out research on aspects that include land use structural optimization (Ren et al. 2008, Gabriel et al. 2006, Moglen et al. 2003), land use in urban areas (Ma 2012, Li & Pan 2006, Bao et al. 2009, Fu et al. 2007), and intensive land use (Li 2008, Wang 2011), guided by the concept of Smart Growth. Some scholars have probed into the comprehensive measurement index and methods of urban smart growth (Tan et al. 2012) as well as the measurement index system and methods of smart use of both urban and rural areas (Cao 2013). From an overall perspective, studies on measurement indexes are limited to cities and research on measuring small towns is almost next to none, still less the research on new-type urbanization promoting smart land use.

In recent years, tourism, together with the entire tertiary industry driven by it, has achieved rapid development. Tourism projects and auxiliary programs have put an urgent demand on land resources. Under such circumstances, selection of land use methods in Shanglin County has a direct bearing on the success of new-type urbanization. Studies on the smart land use of county-administered towns are of both practical and theoretical significance. Based on the eleven towns of Shanglin County as evaluation units, this paper adopts a variation coefficient method and matter-element extension model to analyse the level of smart land use in small towns in the process of stimulating a new-type of urbanization, in the hope of providing a scientific and theoretical foundation for accelerating the transformation of land use methods and formulating reasonable strategies for the new-type urbanization.

2 STUDY SITES

Located in the South Asia tropical monsoon climate zone, Shanglin County (23°12′-23°48′N, 108°22′-108°52′E) lies at the foot of the northeast Daming Mountains in the middle south of Guangxi Province. The west part of the county is mountainous and the whole terrain leans to the southeast. The county's total area is 1,871 km^2. The soil is fertile in the southeast and middle plains and low hills of the county. There are 11 towns and 129 administrative villages in the county. Shanglin County is a typical impoverished county in the karst region of Guangxi with large-scale agricultural production as the major commercial activity.

3 STUDY METHODS

3.1 Building of the evaluation index system of smart land use

New-type urbanization is expected to promote sound and rapid economic development. Resource conservation keeps a tight rein on the way of utilizing resources and urban-rural coordination and environment protection are set as new targets in order to resolve prominent contradictions in current urbanization (Ma et al. 2013). In line with the principles and basis of index system selection, an evaluation index system of smart land use in small towns against the backdrop of new-type urbanization has been built into the paper (Table 1).

3.2 Building of the data analysis model of smart land use

3.2.1 Index weight calculation based on variation coefficient method

As the evaluation of smart land use involves a multi-goal, multi-level, and multi-index process, a variation coefficient method which turns out to be more objective in weight-definition is adopted so as to avert human-empowered subjectivity and take the small-sample size of the study into account. When defining the evaluation index weight, the approach overcomes the shortcomings of equalization of the index weight in entropy weight method. The mathematical evaluative model built based on the new method can yield more objective and rational results (Xi et al. 2010, Ma et al. 2010). The variation coefficient method is known to be relatively mature in defining the weight and detailed calculation procedures which can be referred to in the documents (Lin & Yan 2011).

3.2.2 Smart land use measurement based on the matter-element extension model

Considering the requirements that the new-type urbanization put upon smart land use and its characteristics, the comprehensive evaluation method, featuring an organic qualitative and quantitative combination, is suggested for use. This paper applies the matter-element extension model to evaluation of smart land use. The detailed calculation procedures can be referred to in the documents (Cai 1994, Xue et al. 2004, Zhang 1998, Jin et al. 2012).

4 RESULTS AND ANALYSIS

4.1 Evaluation index grade designation and dimensionless results of classic domain and joint domain

In the light of the real situation of land use in the towns of Shanglin County, in combination with the research framework of smart land use against the backdrop of new-type urbanization, evaluation grades of smart land use in eleven towns of the county are standardized into three levels: Level I (Smart), Level II (Basic Smart), Level III (Non-Smart). See more details in Table 2.

4.2 Dimensionless results of the evaluation index state value

For the sake of comparison of values of the same index and evaluation of smart use levels, the study makes use of the maximum difference normalization method taking the state value of each evaluation index and its property into account and carries out a dimensionless operation (Table 3).

4.3 Calculation results of comprehensive correlation functions

According to Formula (5), correlation functions of each index in each evaluation unit regarding the three smart use levels are calculated. The results can be seen in Table 4 in which the correlation functions of each evaluation index of the eleven towns, regarding Level I, are clearly listed while the others are not because of the limit of space. Matter-element weight refers to the weight of each evaluation index. $W = \{0.074, 0.085, 0.033, 0.120, 0.179, 0.184, 0.068, 0.052, 0.028, 0.078, 0.064, 0.036\}$. A variation coefficient method is adopted in this study to yield the weight vector of the evaluation index. Once the above-mentioned index weight values and the correlation functions of each evaluation unit index regarding smart use levels are placed in Formula (7), the result of comprehensive correlation functions of each town is then yielded (Table 5). Through comparison, the level corresponding with

Table 1. Evaluation index system of smart land use in small towns against the backdrop of new-type urbanization.

Target level	Principle level	Index level	Calculation formula	Units	Index
Evaluating the smart land use of small towns in the context of Chinese new-type urbanization	Economic intensification (A)	Expansion resilience coefficient of land for economic use and construction land A1	Regional GDP growth rate/Construction land growth rate	—	+
		Land economic density A2	Regional GDP/Total land coverage	Ten thousands Yuan /km^2	+
		Agricultural output value of agricultural land per unit A3	Regional agricultural output value/Coverage of agricultural land	Ten thousands Yuan/hm^2	+
		The second and tertiary industrial output value of construction land per unit A4	Regional gross value of the second and tertiary industrial output/Scope of construction land	Ten thousands Yuan/hm^2	+
	Resource conservation (B)	Coordination between arable land consumption and economic growth B1	Arable land reduction rate/Regional GDP growth rate	—	–
		Amount of newly-added consumed construction land for fixed assets investment per unit B2	Amount of newly-added construction land/Amount of fixed assets investment	hm^2/Billion Yuan	–
		Amount of consumed arable land for newly-added construction land per unit B3	Amount of arable land decrease/Amount of newly-added construction land	hm^2	–
	Urban-rural coordination (C)	Coordination between land and population urbanization C1	Land urbanization rate/Population urbanization rate	—	–
		Income ratio of urban and rural residents C2	Per capita disposable income of urban residents/Per capita net income of rural residents	—	–
		Growth factor of urban population C3	Increase rate of urban population/Increase rate of total population	—	+
	Environment protection (D)	Coordination of ecological economy D1	Change rate of ecological service value per unit area/Change rate of GDP per unit area	—	+
		Ecological service value per land D2	Ecological service value/total land area	Ten thousands Yuan /hm^2	+

Each index value is directly calculated or deduced based on the 2010/2012 Shanglin County *Statistical Yearbook* and relevant survey data concerning land use change.

the biggest comprehensive correlation function is the smart use level of the land administered by each town.

4.4 Evaluation result expression and analysis

A spatial distribution map showing the smart use levels of each town may fold out in line with the specific situation of the eleven towns (Figure 2).

Level I is the smart use level, found distributed in the mid-western Shanglin County, namely, the Dafeng Town where the county proper lies. As one of the main urban areas, the town boasts higher levels than its counterparts do in terms economic and social development. The prominence in land economic density, output value of agricultural land per unit and second, and tertiary industrial output value of construction land per unit, have significantly contributed to the reasonable circulation of crucial elements like population, technology, and the economy. Besides, Guangxi Damingshan National Natural Reserve is located in the west of the Dafeng Town, speaking volumes for the town's primitive advantage in ecological environment and natural conditions. Thus, ecological service value per land is generally speaking ideal.

Level II is basic smart use level, finding distribution in the southeast of the county. Xiangxian Town and Baixu Town, which are in major grain producing areas, lie there. The economic development of the two towns remains at a medium level. Coordination between their arable land consumption and economic growth and ecological economy is more favourable than that in Dafeng Town, but the land use methods are relatively

Table 2. Evaluation Index Grade Designation and Dimensionless Results of Classic Domain and Joint Domain.

Evaluation	Evaluation index	Classification index I	II	III	Classic domain I	Extensive domain II	III	Joint domain
Economic intensification (A)	Expansion resilience coefficient of land for economic use and construction land A1	6.5~10	2.5~6.5	0~2.5	0.65~1	0.25~0.65	0~0.25	0~1
	Land economic density A2	4~7	2~4	0~2	0.57~1	0.29~0.57	0~0.29	0~1
	Agricultural output value of agricultural land per unit A3	2.4~3	1.8~2.4	1~1.8	0.7~1	0.4~0.7	0~0.4	0~1
	The second and tertiary industrial output value of construction land per unit A4	55~80	25~55	0~25	0.69~1	0.31~0.69	0~0.31	0~1
Resource conservation (B)	Coordination between arable land consumption and economic growth B1	0~0.2	0.2~0.4	0.4~0.8	0~0.25	0.25~0.5	0.5~1	0~1
	Amount of newly-added consumed construction land for fixed assets investment per unit B2	0~60	60~120	120~200	0~0.3	0.3~0.6	0.6~1	0~1
	Amount of consumed arable land for newly-added construction land per unit B3	0~0.25	0.25~0.5	0.5~0.85	0~0.29	0.29~0.59	0.59~1	0~1
Urban-rural coordination (C)	Coordination between land and population urbanization C1	0~0.06	0.06~0.12	0.12~0.24	0~0.25	0.25~0.5	0.5~1	0~1
	Income ratio of urban and rural residents C2	1~3	3~4	4~5	0~0.5	0.5~0.75	0.75~1	0~1
	Growth factor of urban population C3	0.12~0.2	0.04~0.12	0~0.04	0.6~1	0.2~0.6	0~0.2	0~1
Environment protection (D)	Coordination of ecological economy D1	17~25	9~17	1~9	0.67~1	0.33~0.67	0~0.33	0~1
	Ecological service value per land D2	1.5~2	1~1.5	0.5~1	0.67~1	0.33~0.67	0~0.33	0~1

Table 3. Dimensionless Results of the Evaluation Index State Value.

Evaluation index	Dafeng town	Mingliang town	Xiangxian town	Baixu town	Chengtai country	Sanli town	Qiaoxian town	Mushan country	Tanghong country	Zhenxu country	Xiyan town
A_1	0.775	0.093	0.681	0.533	1.000	0.000	0.175	0.836	0.725	0.429	0.028
A_2	0.520	0.175	0.521	0.283	1.000	0.164	0.044	0.000	0.064	0.020	0.013
A_3	0.536	0.441	0.500	0.868	1.000	0.538	0.083	0.321	0.835	0.000	0.072
A_4	0.656	0.035	0.738	0.214	1.000	0.160	0.027	0.000	0.043	0.305	0.022
B_1	0.945	0.756	0.957	0.966	0.991	0.000	0.964	1.000	0.969	1.000	0.270
B_2	0.965	0.682	0.994	0.982	0.957	0.122	0.999	0.999	0.985	1.000	0.000
B_3	0.262	0.279	0.000	0.257	0.729	0.324	0.769	1.000	0.318	1.000	0.469
C_1	0.898	0.782	0.948	0.555	0.000	0.850	0.584	0.996	0.793	0.962	1.000
C_2	0.270	0.479	0.614	0.484	0.000	0.362	0.901	0.983	0.755	1.000	0.338
C_3	0.839	0.259	0.272	0.340	0.066	0.000	1.000	0.390	0.101	0.589	0.383
D_1	0.045	0.644	0.131	0.651	0.000	0.487	0.796	0.590	0.899	1.000	0.192
D_2	0.741	0.641	1.000	0.365	0.146	0.094	0.540	0.007	0.000	0.129	0.474

extensive. The amount of arable land used for newly-added construction land per unit is larger than that in Dafeng Town. Furthermore, being mainly involved in agricultural production, the towns play an unobvious role in gathering such elements as population, economy and technology. Population growth factor remains at a relatively low level.

Level III is the non-smart use level, found distributed in the north, northwest, northeast and in the middle of the county. Eight towns: Zhenxu, Tanghong, Qiaoxian, Mushan, Xiyantown, Sanli, Chengtai and Mingliang are located there. Except for the town of Chengtai, all the others are known as backward areas in economic development among their counterparts. Their land use investment, technology, and scope fall behind other areas. Recent years have seen intensification in land use strength and expansion in exploration scope in the towns, including Xiyan, Sanli,

Table 4. Correlation function values of each evaluation index of the 11 towns regarding Level I.

Evaluation index	Dafeng town	Mingliang town	Xiangxian town	Baixu town	Chengtai country	Sanli town	Qiaoxian town	Mushan country	Tanghong country	Zhenxu country	Xiyan town
A_1	0.358	−0.857	0.089	−0.200	−1.000	−1.000	−0.731	0.468	0.213	−0.340	−0.958
A_2	−0.097	−0.694	−0.096	−0.505	−1.000	−0.713	−0.924	−1.000	−0.889	−0.965	−0.977
A_3	−0.261	−0.369	−0.286	0.442	−1.000	−0.259	−0.882	−0.541	0.451	−1.000	−0.898
A_4	−0.083	−0.949	0.162	−0.689	−1.000	−0.767	−0.960	−1.000	−0.937	−0.557	−0.968
B_1	−0.945	−0.756	−0.957	−0.966	−0.991	−1.000	−0.964	−1.000	−0.969	−1.000	−0.500
B_2	−0.965	−0.682	−0.994	−0.982	−0.957	−0.593	−0.999	−0.999	−0.985	−1.000	−1.000
B_3	−0.892	−0.948	−1.000	−0.873	−0.729	−0.500	−0.769	−1.000	−0.500	−1.000	−0.500
C_1	−0.898	−0.782	−0.948	−0.555	−1.000	−0.850	−0.584	−0.996	−0.793	−0.962	−1.000
C_2	−0.540	−0.957	−0.614	−0.968	−1.000	−0.725	−0.901	−0.983	−0.755	−1.000	−0.676
C_3	0.403	−0.852	−0.819	−0.649	−1.334	−1.500	−1.000	−0.351	−0.832	−0.027	−0.362
D_1	−0.932	−0.059	−0.804	−0.044	−1.000	−0.270	0.387	−0.158	0.304	−1.000	−0.712
D_2	0.222	−0.066	−1.000	−0.452	−0.781	−0.859	−0.216	−0.990	−1.000	−0.806	−0.288

Table 5. Comprehensive correlation function values of each town.

Evaluation index	Dafeng town	Mingliang town	Xiangxian town	Baixu town	Chengtai country	Sanli town	Qiaoxian town	Mushan country	Tanghong country	Zhenxu country	Xiyan town
$K_1(N)$ (Level I)	−0.488	−0.713	−0.631	−0.664	−0.990	−0.787	−0.801	−0.770	−0.690	−0.810	−0.761
$K_2(N)$ (Level II)	−0.537	−0.466	−0.543	−0.420	−0.919	−0.670	−0.763	−0.743	−0.784	−0.673	−0.717
$K_3(N)$ (Level III)	−0.661	−0.320	−0.651	−0.524	−0.849	−0.511	−0.528	−0.728	−0.536	−0.649	−0.401
max $\{K_i(N)\}$	−0.488	−0.320	−0.543	−0.420	−0.849	−0.511	−0.528	−0.728	−0.536	−0.649	−0.401
Grade	I	III	II	II	III	III	III	III	III	III	III

and Chengtai, stimulated by their abundant natural and tourism resources. Due to the extensive land use methods, ecological service value per land stays at a low level. Although the towns of Zhenwei, Tanghong and Qiaoxian, which all lie in the karst areas, feature small land use scope, their ecological environment is vulnerable and ecological service value per land is likewise low. The economic development of Chengtai, as town known for its industrial concentration, is better than that of other towns, with the biggest land use scope and intensity. Its disturbance toward nature, however, is relatively greater than its counterparts. Therefore, its coordination between ecology and economy has reached a low level with a backward ecological service value per land.

5 DISCUSSION AND CONCLUSION

Setting Shanglin County as an example and giving the example of eleven towns under its administration as evaluation units, the study adopts variation coefficient method and a matter-element extension model in the process of building and evaluation the index system of smart land use. The study has shown that only the land use of Dafeng Town where the Shanglin proper lies reaches smartness level while the Xiangxian Town and Qiaoxian Town featuring grain producing areas

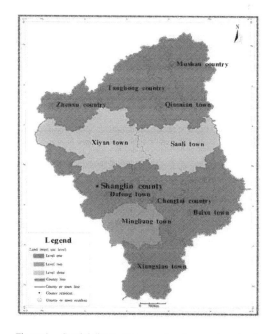

Figure 1. Spatial distribution map showing smart use levels of each town.

stay at the basic-smartness level and most towns disappointingly remain at the level of non-smartness in land use. From the perspective of the spatial distribution of smart land use levels, smart use levels are found in towns featuring a higher economic development level and a better ecological environment while those non-smart use levels are found in towns featuring a backward economy, extensive land use, and a fragile environment. As a result, there exist mutual influences and restrictions between smart land use levels and the economic development levels, land use methods, and the ecological environment (Xia 2013).

In the process of promoting new-type urbanization, besides focusing on economic growth, Shanglin County is also expected to switch to intensive and efficient land use from its original extensive land use, integrate the idea of ecological progress in land use, and gradually narrow the gap between urban and rural areas so that the Project of 'Four Guarantees' (i.e., Guarantees for development, resources, fairness and ecology) achieves a balanced development.

ACKNOWLEDGEMENTS

This study was financial supported by the Beibu Gulf Major Project of the Guangxi Natural Science Foundation (2011GXNSFE018003 & 2012GXNSFEA053001).

REFERENCES

Bao H.J, Feng K. & Wu C.F. 2009. Rational choice of urban spatial expansion in Zhejiang province based on smart growth. *China Population, Resources and Environment* 1: 53–58.

Cai W. 1994. Matter-element model and its application. Beijing: Science and Technology Literature Press.

Cao W., Zhou S.L. & Wu S.H. 2013. Design and application of the integrated measure indicator system on urban-rural land smart use. *Resources and Environment in the Yangtze Basin* 221: 1–7.

Feng G.J., Jiang R.K. & Zhang B.S. 2013. The development of New-type Urbanization requires further ameliorating the land policies for macro-level control. *China Land Science* 277: 94–95.

Fu H.Y., Hao J.M. & An P.L. 2007. Urban space expanding directions based on smart growth theory: a case in Taian city, Shandong province. *Resources Science* 1: 63–69.

Gabriel S.A., Faria J.A. & Moglen G.E. 2006. A multiobjective optimization approach to smart growth in land development. *Socio-Economic Planning Sciences* 3: 212–248.

Jin H.B, Zhang S.W. & Huang Y.F. 2012. Application of extension theory to evaluating land destruction extent in mining area. *Rock and Soil Mechanics* 319: 2705–2709.

Li D.D., Chen L.Q. & Zhao K.K. 2008. The enlightenment from theory of smart growth on land intensive utilization in China. *Shanxi Architecture* 348: 26–27.

Li W.M. & Pan R. 2006. The enlightenment of smart growth towards town space development in Zhejiang province. *Economic Geography* 2: 230–232.

Lin Y.L. & Yan H.W. 2011. Benefit evaluation of rural land comprehensive consolidation in theory. *Ludong University Journal* 272: 166–167.

Liu S.J. 2013. How to use the land in the process of New Urbanization. *Earth* 6: 27–28.

Ma Y.H., Zhang L.J. & Xu W.H. 2013. Scientific understanding of New-type Urbanization and advance development of realizing integration urban and rural area. *Urban Development Studies* 7: 98–102.

Ma Y.Q. 2012. Research of urban land use based on smart growth. *Modern Business Trade Industry* 16: 30–31.

Ma Y.X., Liu X.L. & Huang J.Z. 2010. The impact of land evaluation method to index weights- a case study of Yongdeng and Gaolan county in Lanzhou city. *Hunan Agricultural Sciences* 8: 32–33.

Moglen G.E., Gabriel S.A. & Faria J.A. 2003. A framework for quantitative smart growth in land development. *Journal of the American Water Resources Association* 4: 947–959.

Ren K., Zhou S.L & Zhang H.F. 2008. Optimization of regional land use based on smart growth. *Resources Science* 306: 912–918.

Tan J., Tao X.M. & Chen X. 2012. Comprehensive evaluation of urban smart growth based on improved entropy method: a case of 16 cities of the Yangtze River delta. *Resources and Environment in the Yangtze Basin* 212: 129–136.

Wang C.R. & Jing C.M. 2013. New Urbanization should focus on intensive use of land. *China Economic & Trade Herald*, 6, 53.

Wang R. 2011. A new inspiration to land intensive utilization in urban fringe area based on smart growth: a case in Fuzhou city. *China Urban Economy* 27: 194–195.

Xi R.B., Huang P. & Lai X.M. 2010. Discussion of combination assigning method to determine the weights. *China Collective Economy* 7: 75–76.

Xia F. 2013. Scale Effect, Population quality and the strategic considerations of New Urbanization. *Reform* 3: 25–27.

Xue J.B., Wu C.F. & Xu B.G... 2004. Research on index system and method of Project evaluation for decision-making in land consolidation. *Journal of Natural Resources* 193: 395–399.

Yang S.J. 2013. Consideration to issues of coordinated development with New-model Urbanization leading the "Three-modernizations". *Journal of Luoyang Institute of Science and Technology* 281: 37–38.

Zhang G.Y. 1998. Matter-element model on optimal allocation of land resource. Systems Engineering -theory & Practice, 1, 108–112.

Zhou X.Y. 2013. Land issues New Urbanization will be faced. *China Real Estate Market* 7: 16–17.

Zhu F.E. & Liu D. 2013. The meaning and development mode of New Urbanization. *Heilongjiang Social Sciences* 4: 60–62.

Zhu T.S. & Qin X.W. 2012. Urbanization path: the fundamental problem of changes in land use patterns. *Scientia Geographica Sinica* 32(11): 1348–1352.

Energy and Environmental Engineering – Wu (Ed.)
© 2015 Taylor & Francis Group, London, ISBN 978-1-138-02665-0

Assessing how rainy or snowy weather conditions change residents' home based and return-related travel behaviour

JunLong Li, XuHong Li & DaWei Chen
College of Transportation, Southeast University, Nanjing, China

ABSTRACT: To explore the relationship between rainy and snowy weather conditions and residents' home based and return-related travel behaviour, the hypotheses of the dependence of residents' travel behaviour on types of weather condition were formally tested. Behaviour data were collected by means of a stated adaptation experiment, which was administered by traditional paper-and-pencil questionnaires in Nanjing. To address the main research questions of this paper, the statistical technique of Pearson chi-square independence tests was adopted, following descriptive analyses of travel behaviour changes. Results underscore the significant dependency between weather conditions and home based and return-related travel behaviour. On the one hand, rainy and snowy weather conditions change home based travel behaviour of different purpose by the impact on residents' trip cancellation choices, departure time choices, and home based trip mode choices in different degrees. On the other hand, return trip mode choices and changing home based private modes and public transport contexts are highly depend on weather conditions. This paper contributes to the international literature on travel behaviour, and leads to a deeper understanding about residents' home based and return-related travel behaviour changes under different rainy and snowy weather conditions.

1 INTRODUCTION

Adverse weather, e.g. rain, snow, and temperature fluctuations, make a considerable influence on travel behaviour by increasing the risk of travel. (Kalkstein, Kuby et al. 2009) reported that ridership of transit and traffic volume have a significant change under inclement weather conditions. (Murray, Di Muro et al. 2010), (Sun Guanlin 2007), (Cools, Moons et al. 2010) published their findings that weather conditions determined travellers' behaviour in the choice of trip or not, choice of destination and transport mode, as well route choice. They also pointed out that weather had more impact on obligatory trips than elastic trips. Therefore, a deeper understanding of how weather conditions change residents' travel behaviour is essential for policy makers and transit agencies, since it provides insights that might help to alleviate negative effects associated with adverse weather.

The published literatures regarding the impact of weather conditions on residents' travel behaviour is mainly about home based trips and their results are varied (Saneinejad, Roorda et al. 2012). An important issue in the transferability of findings is the fact that travel behaviour varies across spatial and temporal contexts (Khattak and DePalma 1997). In addition, most weather-related studies make no differentiation based on a particular activity and rarely consider about return trip related travel behaviour (Cools, Moons et al. 2010). These are shortcomings in the literature because travellers have different modes of behaviour for different purposes and return trip related travel behaviour varies from home based behaviour. A fundamental question is how and how much adverse weather conditions triggers changes in return trip related travel behaviour.

In the light of the prior results and the main adverse weather conditions in China (In the northern area of China, the most frequent adverse weather conditions are mainly concerned with snow and frost conditions. While in southern area, it is primary concerned with rainy and fog conditions and there is a need to deepen the study on the impact of rainy and snowy weather conditions on home based and return trip related travel behaviour. The main objective of this present study is to accurately assess how weather conditions change Nanjing residents' home based and return trip related travel behaviour. To examine how weather conditions make an influence on residents' travel behaviour, the hypothesis that weather condition types determine the likelihood of travel behaviour changes (e.g., assessing whether people are more likely to cancel trip under inclement weather conditions) are tested, using statistical analyses on the basis of a stated adaptation study conducted in Nanjing, China.

Before addressing how the rainy and snowy weather conditions change residents' home based and return trip related behaviour, a short description of the fieldwork of stated adaptation survey is presented in the paper. The following presents the methodology used in this study, which is followed by illustrating the analyses results. The last part summarizes the most important conclusions and identifies directions for future research.

2 DATA AND METHODOLOGY

2.1 Data collection

In transport research, stated choice experiments are considered as an alternative method to explore travel choices under different situations. (D'Arcier, Andan et al. 1998) have presented the idea of stated choice experiments the first time. Now the method is widely used to indicate if and how respondents would change their behaviour considering experimentally varied attribute profiles, typically representing scenarios. Therefore, data about residents' travel behaviour under different weather conditions were collected by a stated adaptation experiment.

In total, eighty-four behavioural adaptations in response to different weather conditions were queried. The frequencies of four travel behaviour changes (making trip or cancellation choices, home based departure time choices, home based trip mode choices and return trip mode choices), in response to eight weather conditions (clear, light rain, moderate rain, heavy rain, rain storm, light snow, moderate snow and heavy snow), were determined. Moreover, the first three travel behaviour changes were repeated for two types of trips (working/study and shopping/entertainment/leisure). While the forth travel behaviour changes were repeated for six home based trip modes.

The stated adaptation questionnaire was split into three parts, corresponding to the object of this research. In the first part, respondents are asked to state their trip choices (whether they will cancel their trip plan or continue their trip plan under different weather conditions. If they continue their trip then how do they adapt their departure time). The second part is about mode choices for different trip purposes under inclement weather conditions. Residents are guided to make their choices based on some wildly known situations (e.g., working/studying trip from Andemen to Xinjiekou and shopping/leisure trip from Zhongyanmen to Gulouwith with detailed trip time and cost on a clear day). The last part is mainly about return trip mode shifting. Six scenarios (respondents made their home based trip by walking, cycling, driving, taxi, bus or metro with clear weather) are presented and respondents are asked to state their return trip mode shifting under different adverse weather conditions. As an illustration of the questionnaire style, Figure 1 shows questions on residents' travel behaviour under rainy or snowy weather conditions.

Three hundred questionnaires were distributed at altogether four businesses or service centres (e.g. library, city centre, book store and business centre) located inside the urban area of Nanjing. The data were collected between December of 2012 and February of

Figure 1. Stated adaptation questions on residents' travel behaviour under adverse weather conditions.

2013, during which adverse conditions were expected, by the traditional paper-pencil means.

Altogether 238 residents filled out the questionnaires, of which twenty-five cases were rejected because they were incomplete or had been meaninglessly completed.

2.2 Assessing methods for travel behaviour change

To test the hypothesis that the likelihoods of residents changing their different travel behaviour do in some degrees depend on rainy and snowy weather conditions, and to explore how the rainy and snowy weather conditions affect travel behaviour, chi-square independence tests were performed.

In the tests, four residents' travel behavioural changes and seven rainy/snowy weather conditions, as defined by the weather forecast, are combined. The Pearson statistic Q_P, which is defined in the following equation, is employed to test the null hypothesis between the multinomial variables.

$$Q_P = \sum_{i=1}^{k} \sum_{j=1}^{l} \frac{(n_{ij} - \mu_{ij})^2}{\mu_{ij}}$$

where n_{ij} is the observed frequency in cell (i, j), calculated by multiplying the observed chance by the sample size, and μ_{ij} is the expected frequency for table cell (i, j). When the row and column variables are independent, Q_P has an asymptotic chi-square distribution with $(k-1)(l-1)$ degrees of freedom.

In order to solve the problem there are more than 20% cells expected counts less than five in the two way contingency table and the Pearson statistic may not fit well with chi-square distribution. A contingency adjust is employed in hypothesis test.

$$Q_P = \sum \frac{(|f_o - f_e| - 0.5)^2}{f_e}$$

A criticism of the Pearson statistic is that it does not give a meaningful description of the degree of dependence (or strength of association). Cramer's contingency coefficient (Cramer's value, R_V) and Pearson's coefficient of mean square contingency (R_P) are methods for interpreting the strength of association. They are calculated by the following formulas:

$$R_V = \sqrt{\frac{Q_P}{N(MIN(R,C)-1)}} \qquad R_P = \sqrt{\frac{Q_P}{N+Q_P}}$$

where, N is the sample size and R, C are the row and column numbers of the two-way contingency table. The degrees of freedom are calculated by $(R-1)(C-1)$. Basically, Cramer's value and Pearson's coefficient of mean square contingency scales the chi-square statistic Q_P to a value between 0 (no association) and 1 (maximum association). The independence tests are conducted based on the hypothesis: residents' home based and return related travel behaviour are independent of rainy and snowy weather conditions (none hypothesis).

3 RESULTS

3.1 Aggregated analysis of travel behaviour change

The aggregated analysis was conducted in five travel behaviour changes. The trip mode shifting is defined by residents' trip mode changes: for example one resident's return trip mode choice B is different from his/her home based trip mode choice A and his/her trip mode shifting is from A to B.

Table 1 displays a brief understanding on how and how much the rainy and snowy weather types affect travel behaviour changes. It indicates significant influences on travellers' trip cancellation choice, home based departure time choices, home based trip mode choices (detailed and combined) and return mode shifting patterns with p-values less than 0.001.

On the other hand, Cramer's values and Pearson's coefficients, the difference of which is whether the degrees of freedom are considered, shows rainy and snowy weather types have the greatest impacts on trip cancellation choices and departure time choices with the Cramer's values of 0.41 and 0.23, while fewer impacts can be found on home based mode choices and return trip mode shifting patterns. The reasons may be that travellers tend to change their choices most likely in the early stages of the trip schedules and they may not want to change their regular habits such as

Table 1. Dependence test of travel behaviours on aggregate level.

Dependency on weather conditions	χ^2	DF	P-Value	C's v	P's Coe.	Sig.
all purposes, cancellation choices	501.7	6	0.00000	0.41	0.38	***
all purposes, departure time choices	203.2	12	0.00000	0.23	0.31	***
all purposes, modes choice (detailed)	331.6	35	0.00000	0.15	0.32	***
transit or private mode choice	27.7	7	0.00025	0.10	0.10	***
return trip mode shifting patterns	914.2	18	0.00000	0.19	0.32	***

***p-value < 0.001, **0.001 < p-value < 0.01, *0.01 < p-value < 0.05, NS = not significant p-value > 0.05.

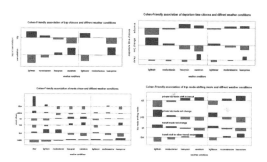

Figure 2. Cohen-Friendly association plots for different travel behaviour and weather conditions.

their regular trip mode for both home based trip and return trip.

The Cohen-Friendly association plots can give more details about the relationship between weather conditions and travel behaviour. Figure 2 shows (1) residents tended to make much more trips than expected in light rain and light snow, while they are more likely to cancel their trip when there are rainstorms and heavy snow conditions. (2) In the section of departure time choices, the difference between observed and expected frequencies are significant when there is light rain, heavy rain, rain storm and heavy snow weather conditions. (3) When referred to mode choices, we find observed trips by bike in Nanjing, no matter what the weather condition is, are very high. While for snowy context, trips by bike, taxi, and bus and metro display big differences between observed and expected frequencies. (4) In section of return trip, almost all contexts are significant. Especially return trip mode shifting for home-based private mode trip in light rain and

Table 2. Dependence test of home based trip decision.

Dependency on weather conditions	χ^2	DF	P-Value	C's v	P's Coe.	Sig.
working, trip cancellation or not	352.2	7	0.00000	0.46	0.41	***
shopping, trip cancellation or not	726.2	7	0.00000	0.66	0.55	***
different purposes, cancelling trip	94.6	6	0.00000	0.29	0.28	***
different purposes, keeping trip	134.3	6	0.00000	0.27	0.26	***

***p-value < 0.001, **0.001 < p-value < 0.01, *0.01 < p-value < 0.05, NS = not significant p-value > 0.05.

Table 3. Dependence test of home based departure time choices.

Dependency on weather conditions	χ^2	DF	P-Value	C's v	P's Coe.	Sig.
working, departure time choices	588.0	14	0.00000	0.44	0.53	***
shopping, departure time choices	150.9	14	0.00000	0.30	0.39	***
different purposes, not change departure time	9.7	6	0.13699	0.11	0.11	NS
different purposes, change departure time	103.1	6	0.00000	0.30	0.29	***

***p-value < 0.001, **0.001 < p-value < 0.01, *0.01 < p-value < 0.05, NS = not significant p-value > 0.05.

heavy snow weather conditions, provide the greatest contributions to Pearson statistics.

3.2 Disaggregated analysis of behaviour change

To further investigate the dependence of residents' travel behaviour changes on weather types, more-detailed analyses are performed: (a) For trip cancellation choices, departure time choices and home based trip mode choices, trip purpose was also considered as an influencing factor. (b) For return trip mode choices, all six home based trip contexts were respectively tested for whether or not travellers will change their return trip mode. Various conclusions could be drawn from these disaggregate analyses.

(1) Home based trip decision (trip cancellation?)

A descriptive analysis of the likelihoods of home based trip decision changes, resulting from rainy and snowy weather conditions, is conducted in disaggregated level. The results indicate that weather conditions types and trip purposes both can affect residents' travel decision. A first conclusion can be Drawn is the fact that rainy and snowy weather conditions play an important role in travel decisions of whether to make a trip or to cancel it. Impacts from snowy weather conditions are very similar to rainy weather conditions. Moreover, the influences on commuting trips are larger than on non-commuting trips. The reason is that a commuting trip has less elasticity than a non-commuting trip and residents have to make the trip no matter how inclement the weather condition is.

Results from the dependence test are showed in table 2. From the items of corresponding significance levels of these tests, it is immediately clear that weather types have significant influences on both working and shopping trip decisions (all p-values are smaller than 0.001). On the other hand, Cramer's values and Pearson's coefficients both indicate that:

(a) For trip purposes, weather conditions impact non-commuting (e.g., shopping) trips much greater than working/study based trip (Cramer's values are respectively 0.66 and 0.46, Pearson's coefficients are 0.55 and 0.41). This is in line with the preliminary conclusions drawn from the descriptive analysis.

(b) For trip decisions, the dependency of keeping trip appears to be smaller (lower chi-square values and same number of degrees of freedom), yet still high significant.

(2) Home based departure time choice

The statics results in table 3 confirm both for working and shopping based trip, the dependencies are highly significant with the p-values smaller than 0.001. Moreover, conclusion that rainy and snowy weather conditions produce many more changes for working/study based trip than non-commuting trip can be made (the Cramer's values are 0.44 and 0.30 and the Pearson's coefficients are 0.53 and 0.39). When focus on choices set, it is clear that residents' decision of not changing departure time is rarely affected by rainy and snowy weather types with a p-value much bigger than 0.05. While decisions of changing departure time (advance or delay) are significantly dependent on weather types with the same degrees of freedom. This is easy to be understood that travellers who do not change the home based trip departure time may do not care about the weather conditions and those tend to make changes are much more sensitive to rainy and snowy weather conditions.

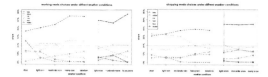
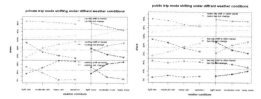

Figure 3. Home based trip mode choices under different weather conditions.

Figure 4. Return trip mode shifting under different weather conditions.

Table 4. Dependence test of home based mode choices.

Dependency on weather conditions	χ^2	DF	P-Value	C's v	P's Coe.	Sig.
working, different mode choices	342.6	35	0.00000	0.21	0.43	***
shopping, different mode choices	106.7	35	0.00000	0.12	0.26	***
different purposes, taking private mode	32.8	7	0.00003	0.19	0.18	***
different purposes, taking public transit	1.7	7	0.97399	0.03	0.03	NS

***p-value < 0.001, **0.001 < p-value < 0.01, *0.01 < p-value < 0.05, NS = not significant p-value > 0.05.

(3) Home based trip mode choice

Residents' home based mode choice is an important issue for exploring travel behaviour. In our adaptive experiments, respondents are asked to make decision referring to some wildly known trip situations with special trip originals and destinations. Figure 3 shows the descriptive analysis. It clearly indicates that for home based working or studying trip, metro accounts for the biggest share and the probabilities of taking a metro or a taxi, or driving a private car increase as the rain or snow is heaviest, while those of taking a bicycle, an e-bicycle or a bus decrease. The reasons are that safety and comfort of these two are affected by inclement weather conditions significantly, and due to the impact of weather conditions on road traffic, the speed of a bus is slower. However, residents have to arrive at their office of school on time. And therefore in order to resist the risk from rainy and snowy weather conditions travellers tend to take faster or more comfortable modes such as metro, taxi or driving. On the other hand, residents change their trip mode less for non-commuting purpose and they tend to take more comfortable or flexible modes, because their non-commuting trips are more flexible and unconstrained. E-bicycle is a flexible and fast mode and the trip mode share increases with the severity of weather conditions.

Deeper analyses are conducted and the results are shown in Table 4. They demonstrate that dependencies of commuting and non-commuting trips on weather conditions are definitely significant with p-values being much smaller than 0.001. In line with descriptive results, conclusions can be drawn that mode choices of home based working or study trips are more dependent on rainy and snowy weather conditions than non-commuting (e.g., shopping, entertainment and leisure) trips with the Cramer's value of 0.21 to 0.12 and the Pearson's coefficient of 0.43 to 0.26. When focused on mode choice set, it is clear that weather conditions make much more influences on the preference of private mode than on preference of public transit which does not significantly depend on weather conditions. The reason may relate to the fact that public transport shows better risk resistance capacity and residents do not have to considered things about having a car, owning bicycles, or parking problems, and so on.

(4) Return trip mode shifting

Return trip mode shifting is a complex issue. Figure 4 displays residents' responses of mode changes to weather conditions when they return. For home based private mode travellers, drivers change return trip mode slightly under various rainy conditions while they shift much more to (public?) transport when they return under snowy weather conditions. Compared to home based car trip makers, home based cyclists and walkers make many changes when returning. Especially under heavy rain, rainstorm moderate snow and heavy snow conditions, more than 70% tend to taking transport returning. A possible explanation for this contrast is the fact that a private car has a high capacity to resist rainy and snowy weather conditions and drivers do not want to leave their cars away from their living area and they may also not want to pay extra parking fees. While for home based cyclists and walkers, it is wise to take transport home under inclement rainy and snowy weather conditions.

The most marked finding from the comparison of the different home based transit takers under various weather conditions is the fact that rainy and snowy weather conditions have large impacts on return trip mode changes. Firstly, home based bus takers have a high likelihood of shifting to metro or taxi when returning. Because their experiments tell them clearly that rainy and snowy weather conditions are likely to cause poor efficiency of the bus system. Home based taxi takers care little about rainy conditions but pay more attention to snowy weather conditions, since snow can cause more road traffic jams and accidents

Table 5. Dependence test of return trip mode shifting.

Dependency on weather conditions	χ^2	DF	P-Value	C's v	P's Coe.	Sig.
Return trip mode choice for						
Home based walkers	509.7	18	0.00000	0.35	0.52	***
Home based cyclists	436.2	18	0.00000	0.33	0.49	***
Home based drivers	154.9	18	0.00000	0.20	0.32	***
Home based taxi takers	114.5	12	0.00000	0.21	0.28	***
Home based bus takers	369.5	12	0.00000	0.37	0.46	***
Home based metro takers	125.0	12	0.00000	0.22	0.29	***
Return trip mode shifting from						
Home based walkers	445.5	6	0.00000	0.57	0.50	***
Home based cyclists	358.1	6	0.00000	0.52	0.46	***
Home based drivers	156.6	6	0.00000	0.34	0.33	***
Home based taxi takers	72.9	6	0.00000	0.23	0.23	***
Home based bus takers	275.5	6	0.00000	0.45	0.41	***
Home based metro takers	87.8	6	0.00000	0.25	0.25	***

***p-value < 0.001,   **0.001 < p-value < 0.01,   *0.01 < p-value < 0.05, NS = not significant p-value > 0.05.

than rainy condition. On the other hand, home based metro takers tend to shift slightly to other public transit based on their experience that inclement weather conditions may cause more crowds in the metro system. Because the walking distance to the nearest metro station takes longer than to a bus or taxi station, home based metro takers may change more their return trio mode in rainstorm, moderate snow and heavy snow weather conditions.

A thorough look at how the home based trip makers of public transit and private mode adapt their return trip mode is conducted by analysing return trip mode choices and trip mode shifting as showed in table 5. From the degrees of freedom of these two analyses we can distinguish return trip mode choice and return trip mode shifting as: return trip mode choice analyses cover 3 or 4 choices sets (home based trip mode and public transit modes such as bus, taxi and metro); while return trip mode shifting analyses only involves two choices: keeping to home based trip mode or shifting to public transport. The dependence tests indicate:

(a) For return trip mode choices, weather conditions definitely have significant influences on six home-based travellers with all p-values less than 0.001. Compared to the indicators of degree of dependence, it provides, for private mode travellers, their return trip mode choices have the closest association with rainy and snowy weather conditions with the biggest Cramer's value of 0.35 and Pearson's coefficient of 0.52. In addition, home based cyclists have a more significant dependency on weather conditions types than home based drivers. On the other hand, return trip mode choices of home-based bus travellers are the most influenced by weather conditions in home based public transport contexts.

(b) For return trip shifting aspects, the main conclusions can be drawn are that for six categories of travellers' return trip mode, changing behaviour significantly depends on rainy and snowy weather conditions, and the influences of weather types on home based private trip mode contexts are bigger than home based public transit contexts with all p-values smaller than 0.001. As well as return trip mode choices, home based walkers and bus takers respectively share the highest dependencies of return trip mode shifting in home based private mode and public transit contexts.

4 CONCLUSIONS

From disaggregated analyses, the degree of influences from rainy and snowy weather conditions on home based trip choices (whether making a trip or cancelling it), home based departure time choices and trip mode choices vary between commuting and non-commuting purposes. Moreover, the behaviour of home based commuting trips indicates that change is most likely to adapt to inclement weather conditions. On the other hand, return trip mode choices and shifting are demonstrated to depend highly on weather conditions. Overall, weather conditions produce the greatest influence on home based private mode contexts. It means that home based walkers and cyclists prefer to make changes to their style of travel if there are inclement weather conditions.

This paper contributes to the international literature by not only looking at the underlying relationship between home based travel behavioural changes and weather conditions by means of a formal multifaceted adaptation approach, but also explores how and how much the rainy and snowy weather conditions impact on residents' return trip decisions.

The clear dependence of the likelihood of travel behavioural changes on weather conditions provides a deeper understanding of how and how much rainy and snowy weather conditions affect residents' home based and return related travel behaviour. The findings in this paper provide a solid basis for further analysis of weather related travel behaviour and weather-related transport preferences. For further research, data collection methods should attempt to survey both weather conditions and associated travel behaviour in as much detail as possible. On the other hand modelling the residents' return related travel behaviour and estimating travel preferences in inclement weather conditions are key challenges.

REFERENCES

Cools, M. and L. Creemers (2013). "The dual role of weather forecasts on changes in activity-travel behaviour." Journal of Transport Geography 28(0): 167–175.

Cools, M., E. Moons, et al. (2010). "Changes in Travel Behaviour in Response to Weather Conditions." Transportation Research Record: Journal of the Transportation Research Board 2157(-1): 22–28.

D'Arcier, B. F., O. Andan, et al. (1998). "Stated adaptation surveys and choice process: Some methodological issues." Transportation 25: 169–185.

Kalkstein, A. J., M. Kuby, et al. (2009). "An analysis of air mass effects on rail ridership in three US cities." Journal of Transport Geography 17(3): 198–207.

Khattak, A. J. and A. DePalma (1997). "The impact of adverse weather conditions on the propensity to change travel decisions: A survey of Brussels commuters." Transportation Research Part a-Policy and Practice 31(3): 181–203.

Murray, K. B., F. Di Muro, et al. (2010). "The effect of weather on consumer spending." Journal of Retailing and Consumer Services 17(6): 512–520.

Saneinejad, S., M. J. Roorda, et al. (2012). "Modelling the impact of weather conditions on active transportation travel behaviour." Transportation Research Part D: Transport and Environment 17(2): 129–137.

Sun Guanlin (2007). "Study of the urban network reliability under adverse weather." Master Thesis of Harbin institute of technology.

… Energy and Environmental Engineering – Wu (Ed.)
© 2015 Taylor & Francis Group, London, ISBN 978-1-138-02665-0

Establish evaluation system of haze pollution

Yue Shen, LiQiao Li, HanDi Ma & ZhiYu Wang
College of Civil Construction and Engineering, Beijing JiaoTong University, Beijing, China

ABSTRACT: Analysing the difference of the air quality standards at home and abroad after the standard has been revised in China. The conclusion is that human health should be included in the evaluation index. According to the characters of haze pollutions, the Meta, being an analysis method, could be used in the risk assessment to establish the system of risk evaluation. The EPA's Models-3 Community Multiscale Air Quality (CMAQ), being an Air-Quality model, provides one of the most reliable tools to the dynamic simulation. A set of comprehensive system of evaluation indicators is established by this model and its risk assessment.

Keywords: haze; mixing law; comprehensive system; risk assessment; evaluation system; CMAQ; Meta analysis

1 INTRODUCTION

The haze is composed of the ultrafine particles with the diameter less than or equal to 1 μm (i.e. PM_1) which could absorb and scatter light. PM_1 only occupies a very small part of the composition of the atmosphere, but it could impose stronger effect on the light absorption as its wavelength is closer to that of the visible light (0.39–0.77 μm). According to *the Haze Pilot Monitoring Report in 2010*, in the haze weather, the concentration of $PM_{2.5}$ is significantly higher than usual, which indicates the lower visibility[1]. Therefore, the particles with smaller diameter than $PM_{2.5}$ particles have greater impact on the atmospheric visibility. According to the observation results of PM_1 observation station in Guangzhou, PM_1 accounts for the vast majority of particles of $PM_{2.5}$.

As the size of haze particles is very small, they could stay longer in the air, which may affect the visibility and result in the inversion phenomenon due to their light scattering function. The cumulative impurity may reduce the pressure and increase the inhalable particles, thus resulting in poor air flow and the slower diffusion velocity of harmful bacteria and viruses. Finally, the virus concentration in the air increases, coupled with greater risk of disease transmission. When the diameter of particles is less than 2.5 micron ($PM_{2.5}$), they get into lungs and accumulate in the human body. When the particles are less than 1 micron (PM_1), they can get into the alveolar blood and do great harm compared with $PM_{2.5}$.

Therefore, the evaluation of the haze is of great significance to the analysis and control of environment quality as well as human health protection.

2 THE ESTABLISHMENT OF THE EVALUATION STANDARD

China developed the *Ambient Air Quality Standard GB-3095-82* in 1982 and has amended it twice, up until 2008. The pollution level of PM2.5 in some cities has been noted since 1998; however, the standard never regards the aerosol without fine particles as an evaluation indicator in the evaluation criteria. As early as in 1997, the United States proposed new standard which includes the PM2.5 (fine particles with the diameter less than or equal to 2.5 μm).

In recent years, China's realised that the prevention of environmental pollution should be improved from the decision-making, thus the evaluation system should regard the PM2.5 monitoring as the required work content[2]. Accordingly, in the new *Ambient Air Quality Standards GB3095-2012* released on 29 February 2012, the PM2.5 monitoring indicator has been added. This new standard will be fully implemented in 2016. In 2013, PM2.5 was officially named as fine particles.

2.1 *The revision process of evaluation standards*

In the three revision processes of ambient air quality standards, the emphasis is on the pollutant index, concentration limits, data validation rules and particles control, shown as follows:

(1) Adjust the pollutants and the classification of functional areas; in 2000, the nitrogen oxides were removed from the 10 pollutants of GB3095-96, a total of nine now; whilst according to the new

Table 1. The concentration of new items.

Item	Average time	Primary standard	Secondary standard
PM$_{2.5}$	Annual average concentration ($\mu g \cdot m^{-3}$)	15	35
	Daily average concentration ($\mu g \cdot m^{-3}$)	35	75
O$_3$/8h	Annual average concentration ($\mu g \cdot m^{-3}$)	100	160

Table 2. The list of variation items.

Variation item	Average time	Original standard	New standard	Percentage
PM10 Secondary standard	Annual average concentration ($\mu g \cdot m^{-3}$)	100	70	30%
NO2 Secondary standard	Annual average concentration ($\mu g \cdot m^{-3}$)	80	40	50%
	Daily average concentration ($\mu g \cdot m^{-3}$)	120	80	33%
	Average concentration per hour ($\mu g \cdot m^{-3}$)	240	200	16.7%

Table 3. The list of variation items.

Name	Average time	Original standard	New standard
Atmospheric pollutants	Daily average concentration	At least 12 (particles) or 18 (SO$_2$, NO$_2$, etc.) h	At least 20 average values per hour
	Average concentration per month (month)	At least 5 (particles) or 12 (SO$_2$, NO$_2$, etc.) daily average value	At least 27 daily average values
	Annual average concentration (year)	At least 60 (particles) or 144 (SO$_2$, NO$_2$, etc.) daily average values	At least 324 daily average values

Table 4. The list of changed analytical methods.

Pollutant items	Manual analytical method	Automatic analytical method
SO$_2$	Formaldehyde adoption pararosaniline spectrophotometric method; Tetrachloromercurate (TCM)-pararosaniline method	Ultraviolet fluorescence method, Differential absorption spectral analysis
O$_3$	Indigo Carmine spectrophotometric method; Ultraviolet photometric method	
NO$_2$	Hydrochloric acid naphthalene spectrophotometric method	Chemiluminescence, Differential absorption spectral analysis
PM (particulate matter)	Gravimetric analysis, GB6921-86 is updated to HJ618	Tapered Element Oscillating Microbalanee, βray method

standard, nitrogen oxides are added, coupled with PM$_{2.5}$, 8-hour ozone, cadmium and other pollutants, the total of 15; simultaneously, eliminate three types of functional areas and the third level of various pollutants.

(2) Set the concentration limit for the items; according to the new standard, the average concentration limits of PM$_{2.5}$ and 8-hour ozone, are as shown in Table 1.

(3) Reduce the concentration limits of PM$_{10}$ and other pollutants, as shown in Table 2.

(4) Reformulate the effectiveness rules of the predetermined data statistics; the percentage of available data has been improved from 30%–75% to 75%–90%, as shown in Table 3.

(5) Update the analytical methods of pollutant-free items and increase the analytical method of automatic monitoring, as shown in Table 4.

2.2 *International emission standards of PM$_{2.5}$*

(1) In October 6, 2006, WHO developed a standard value and target value for PM$_{2.5}$, as shown in Table 5[3].

(2) At the same time, in 2006 the United States issued a more stringent standard value of PM$_{2.5}$, Air Quality Index (AQI), the relationship between air quality and health warnings compared to the EPA in 1997[4], as shown in Table 6.

Table 5. The standard value and target value for PM$_{2.5}$ developed by WHO.

Item	Statistical pattern		PM$_{10}$/(μg·m^{-3})	PM$_{2.5}$/(μg·m^{-3})	Basis of concentration selection
Target value	IT-1	annual average concentration	70	35	For the standard value, the long-term exposure would increase mortality by around 50%
		daily average concentration	150	75	Regarding the various studies which have been published and the risk factors of Meta-analysis as the basis (short-term exposure would increase mortality by around 5 %). increase about 5% mortality
	IT-2	annual average concentration	50	25	Apart from other health benefits, compared to IT-1, exposure at this level would reduce mortality by 6%.
		daily average concentration	100	50	Regarding the various studies which have been published and the risk factors of Meta-analysis as the basis (short-term exposure would increase mortality by around 2.5%) y
	IT-3	annual average concentration	30	15	Apart from other health benefits, compared to IT-1, exposure at this level would reduce mortality by around 6%
		daily average concentration	75	37.5	Regarding the various studies which have been published and the risk factors of Meta-analysis as the basis (short-term exposure will increase mortality by around 1.2%)
New target		annual average concentration	20	10	The long-term exposure under PM$_{2.5}$ is a minimum safety level, where for the total mortality, mortality from lung cancer, and heart disease, the confidence is above 95%
		daily average concentration	50	25	Based on 24h and evaluation of security of exposure

Table 6. The US environmental air quality in 2006.

PM2.5/(μ·gm^{-3})	AQI	Air quality	Health warnings
≪15	0~50	Fine	nothing
16~40	51~100	Moderate	People who are physically sensitive should reduce the long-term or strenuous exercise
41~65	101~150	Not conducive to the health of sensitive people	People with heart or lung disease, the elderly and children should reduce long-term or strenuous exercise
66~150	151~200	Unhealthy	The elderly and children who have heart or lung disease should long-term or strenuous exercise; other people should reduce long-term or strenuous exercise
151~250	201~300	Extremely unhealthy	People with heart or lung disease, the elderly and children should reduce outdoor exercise; others should reduce long-term or strenuous exercise
≫251	301~500	Harmful	People with heart or lung disease, the elderly and children should stay indoors and reduce the standard of living; others should reduce long-term or strenuous exercise

2.3 *Diffusion and forecast model of haze*

Guangzhou, who planned to build PM1 Observatories, discovered that PM$_1$ accounted for the vast majority of particles of PM$_{2.5}$. In addition, the domestic and foreign researches indicate that cold air will aggravate the accumulation of air pollutants, which is seasonal. Meanwhile, they[5] also show that variations of the average daily mass concentration of PM$_{10}$, PM$_{2.5}$ and PM$_1$ caused by emissions of the same pollution sources tend to be similar, but the proportion of fine particles in PM$_{10}$ slightly increases. Measurement results[6] reflect the substantial increase of atmospheric particles, OC

Table 7. The characters of risk evaluation.

Number	Item	Comprehensive risk assessment
1	Types of risk	Sudden accident/poisonous and harmful
2	The main vulnerable groups	Human, ecological
3	Effect type	Sudden
4	Determination of source item	Great uncertainty of substance or reason
5	Physical effects that should be calculated	Light pollution
6	Nature of damage	Acute or chronic disease, sudden accident
7	Dispersal pattern	Multi-puff mode
8	Evaluation method	Probability/determination
9	Methods and measures	Necessary
10	Contingency strategy	Necessary

and EC in accumulation state is an important reason for haze formation. Experiment analyses[7] show that under the haze, acidic particles can often be observed in fine particulate matter. With the aggravation of haze, the percentage of fine particles grows, that is, the number of acidic particles increases. The severity of haze in long duration of ash haze is found to be associated with the amount of acidic particles. The comparison to morphological structure of fine particles of the aerosol, before and after the formation process of haze, demonstrates the significant differences.

The establishment of numerical prediction model[8] of urban haze based on the diffusion law of it and the use of a mathematical expression to reflect the pollution characteristics of the major pollutants (concentration distribution) will play a key role in the forecast and evaluation of the impact of the haze. The US Environment and Protection Association developed the third generation air quality forecasting and assessment system[9] Models-3/CMAQ, based on the existing air quality model CMAQ (U.S. EPA's CAMx model, mesoscale meteorological model; NAQPMS model of Chinese Academy of Atmospheric Physics), and the simulation model which has been widely used in air quality management at home and abroad. The system is composed of mesoscale meteorological mode, emissions mode and multi-scale air quality mode. The emission data required for the simulation of the entire system are provided by spatial and temporal distribution and plume rise calculation. The air quality mode combines meteorological elements and emission data to conduct air quality simulation whilst considering the atmospheric physical and chemical processes[10].

At the core of the system is CMAQ, that is, the multiscale air quality model. Based on the coordination of CCTM gasification transmission and conversion, MM5 mesoscale meteorological model, and SMOKE pollution discharge model, the comprehensive treatment of a variety of factors is realised, thus achieving air quality forecasts of the multiple pollutants.

Currently, CMAQ model is mostly used for $PM_{2.5}$ forecast at home and abroad, based on the fact that PM_1 accounts for the vast majority of the number of particles in $PM_{2.5}$. Therefore, the model can be used for simulating the pollution concentration distribution of haze, which will provide a theoretical basis for the haze analog computing.

3 COMPREHENSIVE ASSESSMENT

The chemical composition of haze is very complex, mainly including organic matter, sulfate, nitrate, ammonium, black carbon, heavy metals, and some other elements. The most harmful parts are the heavy metals and some organic matter. Due to their extremely small particle size, they would have a great impact on humans after being inhaled and accumulated.

Based on the environmental quality standards issued by WHO and US in 2006, human health is also regarded as an important indicator and it has been refined according to different concentrations of $PM_{2.5}$. Therefore, in addition to using the efficient forecasting system and analog systems to predict the pollution concentration of haze, it is necessary to assess its impact on public health, transportation, etc., and regard it as an important evaluation indicator.

3.1 Risk identification and feature analysis of haze

The damage caused by haze includes: (1) its complex components may affect health; (2) it may cause pessimistic mood and affect mental health; (3) its low visibility may impact traffic safety; (4) it may cause meteorological disasters and have an impact on the regional climate; (5) it may aggravate the ultraviolet light pollution of the city.

Based on the above analysis, haze may cause damage to public health, mental health, transportation safety and social unrest which have strong uncertainties. Therefore, the risk assessment is a kind of comprehensive risk assessment between the two concepts: "health risk assessment" and "social stability risk assessment".

Therefore, based on the characteristics of the risk assessment and with reference to the general problems, risk identification, source term analysis, consequence calculations and risk calculations, this paper carries out comprehensive risk evaluation for the haze, as shown in Table 7.

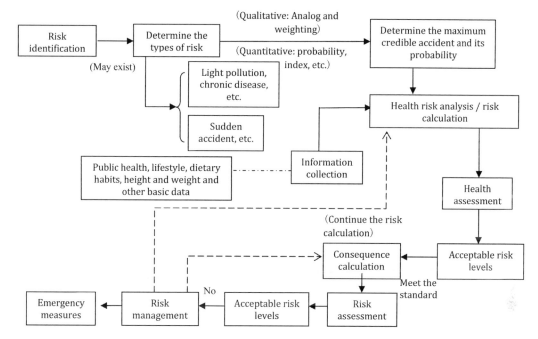

Figure 1. The haze risk evaluation.

3.2 The establishment of risk evaluation of haze

The risk evaluation system is people-oriented, and gives priority to health and safety issues of the public. It conducts evaluation and makes policy on the premise that public health is not threatened. Therefore, health evaluation needs to be added to the haze risk evaluation.

The design of the haze risk evaluation system is shown in Figure 1.

Because of uncertainty of the influence of haze pollution, it is impossible to directly analyse its effects on human health. Therefore, the comprehensive analysis of multi factors is needed to analyse and evaluate the effects.

The Meta analysis is a kind of multi-factor comprehensive analysis method. In 2005, Qian Xiaolin[11] et al. analysed the relationship between the atmospheric fine particles and the death of residents based on Meta-analysis, and mentioned the feasibility of the method. In 2010, Wang Ying[12] from Central South University introduced fuzzy mathematics to Meta-analysis in the demonstration research process of large scale location prediction technology in crisis copper mine in Tongling area, and constructed the Meta analysis model based on fuzzy mathematics, which was used to study the effectiveness of exploration technology and the effectiveness of prediction line, and obtained the results with statistical significance. Meanwhile, in this study, Meta-analysis is used for the first time in quantitative analysis and comparison of the importance of metallogenic index, and the expected results are obtained.

In fact, this method has been widely used in the medical field, displaying the superiority of the multi-attribute fuzzy decision-making performance.

Therefore, it is possible to use the Meta analysis model based on fuzzy mathematics to analyse the factors that influence the evaluation.

4 THE ESTABLISHMENT OF AN EVALUATION SYSTEM

According to the monitoring of atmospheric environment, achieve the concentration of pollutants. Then collect and observe the terrain and meteorological data of the evaluated zones, and predict the concentration of haze according to the selected prediction mode of haze. Meanwhile, carry out risk evaluation for the haze pollution characteristics, and propose some countermeasures.

The design process of evaluation system is as shown in Figure 2.

5 CONCLUSION AND PROSPECT

For the comprehensive haze evaluation system, the comprehensive evaluation with multi-attribute factors is one of the necessary technical focal points; however, the Meta-analysis technology is hardly applied in China. The optimised performance of this technology and expansion is based on the performance, such as the establishment of fuzzy mathematics-based model; In previous studies, Wu Changqian et al. established an

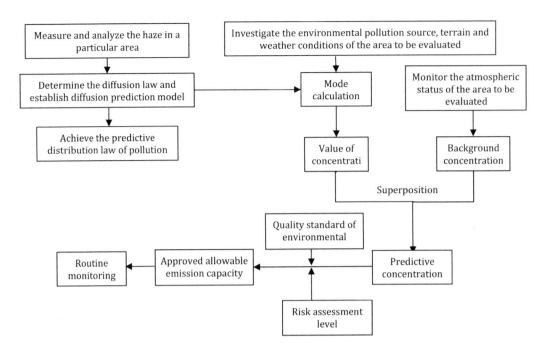

Figure 2. The haze evaluation system.

evaluation system with multi-level and multi-indicator based on the fuzzy comprehensive evaluation algorithm of information weight (entropy). However, business, society, or environment, etc., are all related to the case of multiple factors; therefore, to develop this technology and apply it to establishing a comprehensive evaluation system is the future direction. The software development and popularisation will bring convenience and improve our evaluation system.

In addition, for the air quality model, CMAQ model is selected as the predictive model of haze based on the previous research data and experience at home and abroad. Perfect simulations results play an important role in the atmospheric environmental quality management and decision-making[13]. However, the dynamic simulation technology still has some deficiencies in flexibility and interactivity, so the focus point of this technology lies in the interaction of each module. To sum up, weather simulation technology and various information-gathering techniques, or monitoring techniques are still challenging.

It is necessary to cultivate comprehensive talents with creative thinking in line with the situation in China. Flexible thinking and solid professional knowledge base are the basis for innovation. However, in order to promote the coordinated and sustainable development of China, achieve the harmony of economics, society and environment, it is necessary to enhance the awareness of environmental protection, energy saving and take some effective measures. In one word, to control and reduce pollutant emissions is the most effective way of environment protection.

REFERENCES

Cheng Dan-dan, Pan Xue-mei, Wei Wei. Su Jie. Study of the source of haze pollution base on CMAQ model in Handan [J]. Journal of Hebei University of Engineering (Natural Science Edition), 2012, 29(2), 45–48

Chow J. C., Watson J. G., Edgerton S. A, Vega E. Chemical composition of PM2.5 and PM10 in Mexico City during winter 1997 [J]. The Science of the Total Environment, 2002, 287(3), 177–201

Duan Yu-sen. Study on the haze pollution identification index system in Shanghai [J]. Environmental Pollution and Control, 2012, 34(3):1020–1023

Establishment of urban air quality evaluation system contains haze parameters [J]. Environmental Monitoring Management and Technology, 2010, 22(2):70

Geng Hai-qing, Ren Jing-ming. Control strategies for PM2.5 pollution in China and the corresponding proposals in environmental impact assessment system [J]. Environmental Protection, 2012, 38(6)

Harbusch Andreas. Schafer Klaus. Emeis Stefan. Sarigiannis Dimosthenis. Gotti Alberto. PM10, PM2.5 and PM1 spatial distribuition in the region of Munich determined by satellite images on the ICAROS NET platform [J]. Proceedings of SPIE – The International Society for Optical Engineering 2005.

Kuhlbusch T. A. J.,Neumann S. H Fissan. Number Size Distribution, Mass Concentration, and Particle Composition of PM_1, $PM_{2.5}$, and PM_{10} in Bag Filling Areas of Carbon Black Production [J]. Journal of Occupational and Environmental Hygiene, 2004, 1(10)

Li Yue-Yu, Shi Jian-Hua. SO2 diffusion study the distribution of Taiyuan City [J]. Journal of Shanxi University. 2011, 34(1), 153–157

Liu Hong-nian. Hu Rong-zhang, Zhang Mei-gen. Development and application of urban haze numerical forecast model [J]. Research of Environmental Science. 2009, 22(6):631–636

Lu Cheng-wei, Liu Sheng-yu. Zou Chang-wu, Yao Lin. Introduction of a CMAQ-based air quality simulating and forecasting system [J]. Journal of Anhui Agriculture Science, 2012, 40(10):6290–6291.6338

Miao Qin, Zhang Zhi-Qiang, Qu Jian-Sheng. Comparative analysis of international PM2.5 emission standards and their implementation [J] Environmental Pollution and Control, 2012, (10):96

Nie Bang-sheng. Introduction of common air quality modal in domestic and international [J]. Jiangsu Environmental Science and Technology, 2008, 22(1):119–121

Qian Gong-wang, Zhao Ling-xia. Correlation between particles with satellite droplet structure and haze [J]. Journal of South China University of Technology (Natural Science Edition), 2006, 34(5):5–10

Qian Xiao-lin, Kan Hai-dong, Song Wei-min et al. Meta-Analysis of association between air fine particulate matter and daily mortality [J]. Journal of Environmental Health, 2005, 22(4):246–248

Tan Ji-hua, Duan Jing-chun, Zhao Jin-ping, Bi Xin-hui, Sheng Guo-ying, Fu Jia-mo, He Ke-bin. The size distribution of organic carbon and element carbon during haze period in Guangzhou [J]. Environmental Chemistry, 2009, 28(02):67–271

Vecchi, G.R., Marcazzan G., Valli M., Ceriani C. Antoniazzi. The role of atmospheric dispersion in the seasonal variation of PM1 and PM2.5 concentration and composition in the urban area of Milan (Italy) [J]. Atmospheric Environment, 2004, 38(27): 4437–4446

Wang Kang, Zhu Yun, Zhou Qin, Chen Chu-Yi, JANG Carey. Air quality model based on a dynamic display technique Silverlight [J]. Application Research of Computers, 2001, 28 (3).

Wang Yin. Meta-analysis of diverse information mineralization and multi-attribute decision model of fuzzy optimal prediction based on. The University of Zhongnan, 2010

Wu Guo-Ping, Hu Wei. China four urban air pollution levels of PM2.5 and PM10, 1999, 19(2)

Xu En. Characteristics of ceramic production area air PM10, PM2.5 and PM1 mass concentration [J].Environment, 2006, (z1):132–133

Yu S. C., Rohit M., Kenneth S. et al. Evaluation of real-time PM2.5 forecasts and process analysis for PM2.5 formation over the eastern United States using the Eta-CMAQ forecast model during the 2004 ICARTT study [J]. Journal of Geophysical Research: Atmospheres (1984–2012), 2008, 113 (D6)

Performance and real-time control of a novel intermittent SBR based on simulating photovoltaic aeration for organics removal

Fang Shu Ma, Bei Hai Zhou, Yan Qin, Xin Du & Yong Jie Yuan
Department of Environmental Engineering, School of Civil and Environmental Engineering, University of Science and Technology Beijing, Beijing, China

ABSTRACT: A novel intermittent SBR based on simulating photovoltaic aeration for organics removal was established in order to save operating costs and reduce the organic matter inhibition on subsequent nitrification. The results showed that the organic removal process can be divided into the rapid degradation stage and the recalcitrant degradation stage. In the organic matter rapid degradation stage, when the temperature was constant (25°C), the increase in aeration rate resulted in a positive impact upon the SCOD decrease rate and the amount of NH_4^+-N assimilated by heterotrophic bacteria. The prolonged IDEL stage had no appreciable effect on both of the removal performances and the processes of SCOD and NH_4^+-N, however, it could enhance the biological phosphorous removal process by promoting the phosphorous release. The inflection point associated with the completion of organic degradation was positively identified in DO profiles under all operation conditions. Furthermore, slope changes in DO profiles (dDO/dt) were found to better represent the organic degradation process, which could be used for real-time control of the organics removal process after smoothing.

Keywords: PV intermittent aeration; SBR; organic removal process; real-time control; rural sewage

1 INTRODUCTION

With the improvement of living standards and the popularity of water flushing toilets in rural areas, the sewage production also increases gradually, however, its treatment is not optimistic. Taking China for instance, the proportion of domestic sewage treated was as low as 18.1% for county towns and 4.9% for rural villages by 2009 (Gong et al., 2012). The direct discharge of those poorly treated or untreated sewage containing large amounts of nitrogen, phosphorus and organic matter is responsible for many water bodies pollution (Ye and Li 2009, Li et al., 2009) and is a potential threat to public health (Starkl et al., 2013, Montgomery and Elimelech, 2007).

The high energy consumption of biological wastewater treatment technologies is often cited as the main barrier to their wide application in rural areas (Han et al., 2013). In order to overcome this limitation, some researchers have proposed to use the hybrid energy system combining conventional and renewable energy resources, in order to meet the energy requirement of the rural sector wastewater treatment plants (Devi et al., 2007). Others have attempted to solely use the novel photovoltaic technology without batteries for rural sewage treatment, thus further reducing the dependence of sewage treatment facilities on grid system and the construction costs. The promising performance proved it is feasible for rural sewage treatment (Han et al., 2013, Ma et al., 2012).

SBR has been widely used in domestic wastewater and industrial wastewater treatment, however, little attention is paid to photovoltaic intermittent aeration SBR for rural wastewater treatment. The characteristics of rural wastewater discharge, SBR operation and solar energy intensity have common ground, such as intermittence and periodicity. In other words, there is often no effluent discharge at night, at the same time, the light intensity reaches its minimum, and the SBR can stay at IDLE stage without influent and energy consumption. Based on the above-mentioned, utilising intermittent SBR based on photovoltaic aeration for wastewater treatment in rural areas can not only share the flexible operation and high performance with SBR, but also can achieve high, efficient use of energy and low infrastructure investment by eliminating the batteries and inverters.

Real-time control is believed to be useful in optimising the energy requirements and guaranteeing the treatment performance. Many researchers have considered the oxidation-reduction potential (ORP), the pH, and the dissolved oxygen (DO) levels for biological process monitoring and controlling, which are based on the characteristic bending points in these physics and chemistry indexes profiles. The ORP and DO levels can successfully indicate the oxidative and biological state of the wastewater, and pH is a good indicator of ongoing biological reactions (Ga and Ra, 2009, Ra et al., 1999). However, the usefulness of the information provided by those bending points is

Table 1. The running mode of SBR.

	FEED	REACTION	SETTLE	DRAW
Cycle 1	8:15–8:30	8:30–12:30	12:30–12:40	12:40–12:45
Cycle 2	12:45–13:00	13:00–17:30	17:30–17:40	17:40–17:45
IDLE	17:45 – next day 8:15			

site-dependent (Martín de la Vega et al., 2012). For different COD/TKN ratios of raw wastewater, the time points when the bending points in ORP and DO profiles appear vary largely, and there is no clear bending points in some conditions (Chang et al., 2002, Chang et al., 2004, Han et al., 2008).

The objective of this work is to remove organic matter in rural domestic wastewater treatment by a novel SBR based on simulating photovoltaic intermittent aeration, in order to reduce operational cost and avoid inhibition on subsequent nitrification, and investigate the effects of temperature and aeration rate on organic matter and nutrients removal processes. In addition, the correlation between the COD removal process and the ORP, DO and pH profiles was identified for potential real-time on-line application.

2 MATERIALS AND METHODS

2.1 Experimental set-up

The SBR reactor used in this study was a 28.3 L plexiglass vessel with the working volume of 25 L and 190 mm in diameter and 1000 mm in height. The reactor was wrapped with aluminum foil to prevent algae growth and nitrification inhibition by light. Oxygen was introduced into the reactor by an air compressor and distributed through fine-bubble diffuser stones with an air flow meter to control the aeration intensity. A heater was used to control the water temperature. Both the influent flow and the effluent flow were controlled by pre-calibrated peristaltic pumps (Chuangrui Co., Ltd., China).

Since the solar energy is abundant in the southwest of China (Liu et al., 2011), offering significant opportunity for the application of photovoltaic aeration for wastewater treatment, the operation mode of the reactor (Table 1) is designed according to the solar resource of these areas. The SBR was working in daytime (8:15–17:25) and idling at night (17:25 – next day 8:15). During the first 5 min of the cycle, 12.5 L wastewater was delivered into the reactor and the same amount of liquid was discharged at the end of the cycle, giving a volumetric replacement ratio of 50%.

2.2 Seed sludge and wastewater composition

The reactor was seeded with the return sludge of a local municipal wastewater treatment plant in Beijing, China. The mixed liquor suspended solids (MLSS) concentration was about 3000 mg/L.

Wastewater was taken from the septic tanks of the University of Science and Technology Beijing, which were adjusted appropriately according to sewage quality in rural areas by adding the glucose, urea and di-potassium hydrogen phosphate. The influent characteristics were as follows: COD of 269.4 ± 56.5 mg/L, TN of 47.8 ± 10.2 mg/L, TP of 5.5 ± 0.8 mg/L, NH_4^+-N of 36.2 ± 7.1 mg/L, NO_3^--N of 0.4 ± 0.2 mg/L.

2.3 Analytical methods

The pH was detected on-line using Hach HQ30d meter with PHC10103 probe, with the data acquisition interval of 5 min. ORP and DO were monitored by Hach HQ40d meter with MTC10103 probe and LDO10103 probe, with the data acquisition interval of 1 min. The SBR system performance was monitored by a variety of parameters including soluble chemical oxygen demand (SCOD), ammonium (NH_4^+-N), nitrate (NO_3^--N), nitrite (NO_2^--N) and dissolved total phosphorus (DTP). All chemical analyses were performed in accordance with standard methods given in APHA (APHA 1998). Samples were collected at 10–30 min intervals. All samples were analysed after filtration through 0.45 μm filter paper.

3 RESULTS AND DISCUSSION

3.1 The effect of aeration intensity on pollutants removal processes

Aeration intensity, playing a key role in wastewater treatment plants performance, determines the energy consumption of the conditional wastewater treatment plants, or the amount of solar panels and pumps installed in wastewater treatment plants driven by solar energy. Fig. 1 shows the variations of SCOD, NH_4^+-N, NO_3^--N, NO_2^--N and DTP during the reaction stage of the first cycle under the aeration rate of 1.0 L/min, 0.5 L/min, 0.4 L/min and 0.2 L/min, respectively, whilst the water temperature was maintained at 25°C. The organic removal process in all conditions can be divided into the rapid degradation stage and the recalcitrant degradation stage. In the rapid degradation stage, the higher the aeration rate, the faster SCOD decreases and the greater the amount of NH_4^+-N which was assimilated by heterotrophic bacteria. This result is in agreement with the observation of Tang et al., who observed that the NH_4^+-N assimilated by heterotrophic bacteria under the aeration rate of 100 L/h was 2 times higher than the result obtained under the aeration rate of 60 L/h (Tang et al., 2012).

Nitrate and nitrite were less than 0.5 mg/L during the organic matter rapid degradation stage under the aeration rate of 1.0 L/min and 0.5 L/min. Both nitrate and nitrite exhibited an increase due to nitrification during the organic matter recalcitrant degradation stage, however, the nitrification rate was much lower than reported conventional SBR used for organic matter removal (Tang et al., 2012, Zeng et al., 2004), while

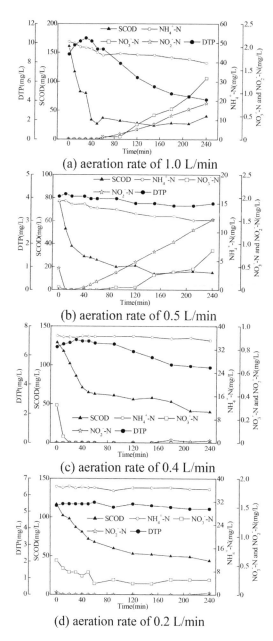

Figure 1. Effects of aeration rate on pollutants removal processes.

the influent COD was lower than the inhibition level (Zeng et al., 2004). The main reason for the low nitrification rate in this study could be that nitrification was hampered by long-term anoxic or anaerobic condition, since the aeration was cut off for 14.5 h during night in the simulating photovoltaic intermittent aeration SBR, the residual DO was gradually consumed by the respiration activity of the microorganisms, after the depletion of DO, the anoxic or anaerobic environment would prevail. Hu et al. (2007) showed that the average specific ammonia oxidation rate and average specific nitrite oxidation rate exhibited a decrease of 46.8% and 36.3%, respectively, after 12 h anoxic treatment. With the aeration rate of 0.4 L/min and 0.2 L/min, the nitrate and nitrite were negligible throughout the reaction period; at the same time, NH_4^+-N was essentially unchanged, indicating the nitrifiers were washed out of the system.

Biological phosphorus removal process is regulated by multiple operational factors; whilst it is indeed capable of efficient phosphorus removal performance, the stability and reliability can be a problem (Oehmen et al., 2007). In this study, the DTP increased during the earlier stage of reaction due to phosphorus release. Subsequently, a continuous decrease was found. However, there was a remarkable difference in the change extent under different operational conditions. Both the phosphorus release and uptake were most remarkable under the aeration rate of 1.0 L/min, giving the best phosphorus removal performance (Fig. 1a), even though the DO was near to 1 mg/L during the organic matter rapid degradation stage (Fig. 4a), which seemed undesirable to phosphorus release. There was anaerobic environment inside the microbial flocs as a result of DO gradients due to air diffusion limitations, where poly-P accumulating organisms (PAOs) can take up volatile fatty acids (VFAs) and store them in the form of polyhydroxyalkanoates (PHA) accompanied by degradation of poly-P and consequent orthophosphate release from the cell (Mino et al., 1998). In addition, the dissolved organic matter was sufficient for PHA formation under this condition, leading to phosphorus release more completely, so the subsequent phosphorus uptake was more obvious.

Under the aeration of 0.5 L/min, phosphorus removal was not obvious, since the insufficient SCOD resulted in inadequate phosphorus release (Fig. 1b). The phosphorous removal performance did not degenerate with the aeration rate decreased to 0.4 L/min. On the contrary, it was better than the result obtained under the aeration of 0.5 L/min mainly due to the the fact that SCOD was relatively abundant (Fig. 1c). When the aeration rate further reduced to 0.2 L/min, the biological phosphorous removal performance deteriorated, even though the organic load was comparable with the aeration rate of 0.4 L/min (Fig. 1d). That was because the liquid was poorly mixed under the aeration rate of 0.2 L/min, and therefore the mass transfer rate decreased.

3.2 *The effect of temperature on pollutants removal processes*

Under the aeration rate of 0.4 L/min, the variations of SCOD, NH_4^+-N, NO_3^--N, NO_2^--N and DTP during the reaction stage of the first cycle at 20°C and 30°C are shown in Figure 3. The results obtained under 25°C are shown in Figure 1c. Increasing the temperature from 20°C to 30°C had nearly no effects to the nitrification for nitrate and nitrite maintained insignificant (<0.5 mg/L) during the whole REACTION stage.

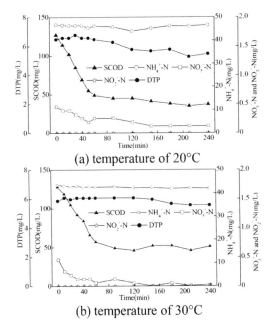

(a) temperature of 20°C

(b) temperature of 30°C

Figure 2. Effects of temperature on pollutants removal processes.

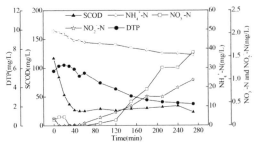

Figure 3. Parameter concentrations during the REACTION stage of the second cycle (aeration of 1 L/min, temperature of 25°C).

Table 2. Parameter concentrations in the supernatant during the IDEL stage (mg/L).

Duration of IDEAL (min)	SCOD	NH_4^+-N	NO_3^--N	NO_2^--N
0	23	36.8	1.6	1.0
120	24	37.3	1.5	1.0
240	20	37.7	0.8	0.6
880	25	39.2	undetected	undetected

However, the SCOD, NH_4^+-N and DTP removal performance deteriorated as the temperature rose from 20°C to 25°C, and to 30°C. The SCOD measured in the end of the REACTION stage were 35 mg/L, 40 mg/L and 51 mg/L, at 20°C, 25°C and 30°C, respectively. Furthermore, typical phosphorus release and uptake characteristics of the system were evidently noticed at 20°C, but started diminishing as the temperature increased.

Glycogen-accumulating organisms (GAOs) without phosphorous removal capacity are thought to compete with PAOs in biological phosphorus removal wastewater treatment systems, and be a cause of phosphorous removal failure because they can out-compete PAOs under some conditions (Tu and Schuler, 2013). Temperature appears to play an important role on the interaction between these two types of microorganisms. It is reported that PAOs have important advantages over GAOs at low and moderate temperatures (below 20°C), while higher temperatures (higher than 20°C) are more beneficial to GAO (Panswad et al., 2003, Whang and Park 2006, Lopez-Vazquez et al., 2009), so the PAOs may be gradually eliminated from the system as the temperature rose from 20°C to 30°C associated with the phosphorous removal efficiency decrease.

3.3 The impact of prolonged IDEL stage on pollutants removal processes

Figure 3 presents the SCOD and nutrients variations during the REACTION stage of the second cycle under the aeration of 1.0 L/min, temperature of 25°C, in order to investigate the extended IDEL stage on pollutants removal processes (For the purpose of comparing with the first cycle, the results obtained during the 0–240 min of the REACTION stage of the second cycle were used for drawing). Basically, SCOD and nutrients exhibited the same pattern similar to the first cycle (Fig. 1a). However, the DTP concentration (8.8 mg/L) at the beginning of REACTION stage of the first cycle was much higher than the result (5.7 mg/L) obtained in the second cycle, indicating apparent phosphorous release occurred during the IDEL stage (clearly demonstrated in Table 2). Though the SCOD in the supernatant remained at low level during the IDEL stage, there were still some undegraded organics adsorbed on the surface of the activated sludge, moreover, during substrate metabolism and biomass decay, soluble microbial products (SMP) excreted by microorganisms would increase the organic compounds pool (Barker and Stuckey 1999), the fermentation and hydrolysis of these organic matters during the prolonged IDEL stage would produce volatile fatty acids (VFAs) for PAOs to fuel phosphorous release.

The DTP reached as high as 10.5 mg/L after the phosphorous release and subsequently decreased to 4.7 mg/L at the end of the REACTION stage in the first cycle, while the highest DTP during the phosphorous release and the final DTP (240 min) during the second cycle were 6.3 mg/L and 2.3 mg/L, respectively. Thus, the extended IDEL stage could enhance the biological phosphorous removal process by promoting the phosphorous release. Wang et al. reported an aerobic/extended-idle (AEI) SBR process for phosphorous removal (Wang et al. 2009, Wang et al. 2012a).

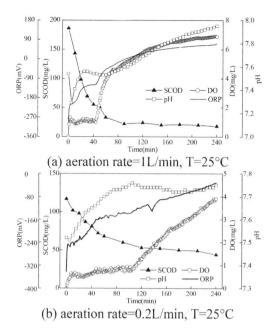

(a) aeration rate=1L/min, T=25°C

(b) aeration rate=0.2L/min, T=25°C

Figure 4. The correspondence between organics removal process and DO, pH and ORP.

In the AEI-SBR process a strict anaerobic stage is not required, whereas an extended-idle period is operated (210–450 min) between the DRAW and the next FEED phases. Efficient phosphorous removal could be achieved in the AEI-SBR and the phosphorous removal efficiency linearly increased with the increase of idle phosphorous release (Wang et al. 2012b).

3.4 Implication of DO profiles

Figure 4 shows the changes of SCOD with the ORP, pH and DO profiles in the first cycle of the SBR under different operation conditions. Some researchers reported that ORP is a robust indirect real-time control parameter for nutrients removal in SBRs (Ra et al., 1999, Martín de la Vega et al., 2012, Chang et al., 2002, Chang et al., 2004, Han et al., 2008). However, in this study, the ORP profiles show poor repeatability and weak correspondence with COD degradation process (Fig. 4). The undesirable results may be caused by the probe fouling after being immersed in wastewater for a certain period (Li and Irvin, 2007).

Also, there are some differences in pH profiles, which may be caused by the variation in the loading rate due to the influent strength fluctuation and the extend of biological reactions, e.g. ammonification, nitrification/denitrification and phosphorus release/uptake under different operation conditions. Therefore, if the pH profile were used for real-time control of the reaction stage, failure of the organics removal process control would have resulted during the operation.

In contrast to the OPR and pH profiles, the DO profiles were able to duplicate the feature points (DO

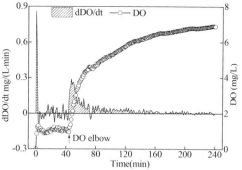

(a) the original DO profile and its derivatives

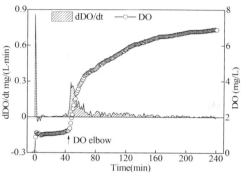

(b) the smoothed DO profile and its derivatives

Figure 5. The original and smoothed DO profiles and their derivatives in the reaction stage (aeration rate of 1.0 L/min).

elbow), corresponding to the completion of organics degradation under all the operation conditions. Using absolute values of DO to control the end of organic degradation process might not be completely reliable, as the absolute values of DO elbow mentioned above were not constant under different operation conditions. Detecting relative changes in sensor profiles (calculated by first derivative or second derivative) provided a more reliable control strategy for nutrient removal in SBRs (Martín de la Vega et al., 2012, Casellas et al., 2006, Ruano et al., 2012). The original DO signals and their derivatives, however, are contaminated by a substantial amount of high-frequency noise (Fig. 5a). Therefore, they must be pre-processed with digital filters to obtain good signal smoothing result, whilst at the same time, the feature points are fairly preserved (Fig. 5b). With the aid of computer and data acquisition system, these points can be readily recognised and timely applied to the process control.

4 CONCLUSIONS

The intermittent SBR based on photovoltaic aeration for organic removal, was feasible for rural sewage treatment in those area with abundant solar energy to reduce the energy consumption, or even make the wastewater treatment facilities self-sufficient. Efficient removal of SCOD can always be achieved under different operational conditions. Moreover, the extended IDEL stage had no appreciable effect on the removal performances and processes of SCOD and NH_4^+-N. However, it could enhance the biological phosphorous removal process by promoting phosphorous release.

Significant points for real-time control were identified in DO profiles. The pH and ORP profiles, however, failed to provide reliable evidence that corresponded to organic degradation process. Slope changes in DO profiles (dDO/dt) were found to better represent the organic degradation process, which could be used for real-time control of the organics removal process after smoothing.

REFERENCES

APHA. 1998. Standard methods for the examination of water and wastewater. American Public Health Association, Washington, DC 1268.

Barker, D.J. and Stuckey, D.C. 1999. A review of soluble microbial products (SMP) in wastewater treatment systems. Water Research 33(14), 3063–3082.

Casellas, M., Dagot, C. and Baudu, M., 2006. Set up and assessment of a control strategy in a SBR in order to enhance nitrogen and phosphorus removal. Process Biochemistry 41(9), 1994–2001.

Chang, C.H. and Hao, O.J., 1996. Sequencing batch reactor system for nutrient removal: ORP and pH Profiles. Journal of Chemical Technology & Biotechnology 67(1), 27–38.

Chang, C.-N., Ma, Y.-S. and Lo, C.-W., 2002. Application of oxidation–reduction potential as a controlling parameter in waste activated sludge hydrolysis. Chemical Engineering Journal 90(3), 273–281.

Chang, C.N., Cheng, H.B. and Chao, A.C., 2004. Applying the Nernst equation to simulate redox potential variations for biological nitrification and denitrification processes. Environ Sci Technol 38(6), 1807–1812.

Devi, R., Dahiya, R.P., Kumar, A. and Singh, V., 2007. Meeting energy requirement of wastewater treatment in rural sector. Energy Policy 35(7), 3891–3897.

Ga, C.H. and Ra, C.S., 2009. Real-time control of oxic phase using pH (mV)-time profile in swine wastewater treatment. Journal of Hazardous Materials 172(1), 61–67.

Gong, L., Jun, L., Yang, Q., Wang, S., Ma, B. and Peng, Y., 2012. Biomass characteristics and simultaneous nitrification-denitrification under long sludge retention time in an integrated reactor treating rural domestic sewage. Bioresour Technol 119, 277–284.

Han, C., Liu, J., Liang, H., Guo, X. and Li, L., 2013. An innovative integrated system utilizing solar energy as power for the treatment of decentralized wastewater. Journal of Environmental Sciences 25(2), 274–279.

Han, Z., Wu, W., Zhu, J. and Chen, Y., 2008. Oxidization-reduction potential and pH for optimization of nitrogen removal in a twice-fed sequencing batch reactor treating pig slurry. Biosystems Engineering 99(2), 273–281.

Hu, A.H., Zheng, P. and Jin, R.C., 2007. Mechanism of effect of anoxic stress on nitrification. CIESC Journal 58(10), 2587–2594.

Li, B. and Irvin, S., 2007. The comparison of alkalinity and ORP as indicators for nitrification and denitrification in a sequencing batch reactor (SBR). Biochemical Engineering Journal 34(3), 248–255.

Li, S., Li, H., Liang, X., Chen, Y., Cao, Z. and Xu, Z., 2009. Rural wastewater irrigation and nitrogen removal by the paddy wetland system in the Tai Lake region of China. Journal of Soils and Sediments 9(5), 433–442.

Liu, W., Lund, H., Mathiesen, B.V. and Zhang, X., 2011. Potential of renewable energy systems in China. Applied Energy 88(2), 518–525.

Lopez-Vazquez, C.M., Hooijmans, C.M., Brdjanovic, D., Gijzen, H.J. and van Loosdrecht, M.C.M., 2009. Temperature effects on glycogen accumulating organisms. Water Research 43(11), 2852–2864.

Ma, F.S., Zhou, B.H., Li, L. and Shi, C.H. 2012. Application of photovoltaic technology to rural sewage treatment. Water and wastewater (S1), 150–154.

Martín de la Vega, P.T., Martínez de Salazar, E., Jaramillo, M.A. and Cros, J., 2012. New contributions to the ORP & DO time profile characterization to improve biological nutrient removal. Bioresour Technol 114(0), 160–167.

Mino, T., van Loosdrecht, M.C.M. and Heijnen, J.J., 1998. Microbiology and biochemistry of the enhanced biological phosphate removal process. Water Research 32(11), 3193–3207.

Montgomery, M.A. and Elimelech, M., 2007. Water and Sanitation in Developing Countries: Including Health in the Equation. Environ Sci Technol 41(1), 17–24.

Oehmen, A., Lemos, P.C., Carvalho, G., Yuan, Z., Keller, J., Blackall, L.L. and Reis, M.A.M., 2007. Advances in enhanced biological phosphorus removal: From micro to macro scale. Water Research 41(11), 2271–2300.

Panswad, T., Doungchai, A. and Anotai, J., 2003. Temperature effect on microbial community of enhanced biological phosphorus removal system. Water Research 37(2), 409–415.

Ra, C., Lo, K. and Mavinic, D., 1999. Control of a swine manure treatment process using a specific feature of oxidation reduction potential. Bioresource Technology 70(2), 117–127.

Ruano, M.V., Ribes, J., Seco, A. and Ferrer, J., 2012. An advanced control strategy for biological nutrient removal in continuous systems based on pH and ORP sensors. Chemical Engineering Journal 183(0), 212–221.

Starkl, M., Brunner, N. and Stenström, T.-A., 2013. Why Do Water and Sanitation Systems for the Poor Still Fail? Policy Analysis in Economically Advanced Developing Countries. Environ Sci Technol 47(12), 6102–6110.

Tang, X.X., Ma, B., Xu Z. B., and Peng, Y.Z., 2012. Preices separation of organic removal and nitrification process and real-time control method of a autotrophic nitrogen removal process. CIESC Journal 63(11), 3666–3672.

Tu, Y. and Schuler, A.J., 2013. Low Acetate Concentrations Favor Polyphosphate-Accumulating Organisms over Glycogen-Accumulating Organisms in Enhanced Biological Phosphorus Removal from Wastewater. Environ Sci Technol 47(8), 3816–3824.

Wang, D.-b., Li, X.-m., Yang, Q., Zheng, W., Liu, Z.-y., Liu, Y.-l., Cao, J.-b., Yue, X., Shen, T.-t. and Zeng, G.-m., 2009. The probable metabolic relation between phosphate uptake and energy storages formations under single-stage oxic condition. Bioresource Technology 100(17), 4005–4011.

Wang, D., Li, X., Yang, Q., Zheng, W., Wu, Y., Zeng, T. and Zeng, G., 2012a. Improved biological phosphorus removal performance driven by the aerobic/extended-idle regime with propionate as the sole carbon source. Water Research 46(12), 3868–3878.

Wang, D., Yang, G., Li, X., Zheng, W., Wu, Y., Yang, Q. and Zeng, G., 2012b. Inducing mechanism of biological phosphorus removal driven by the aerobic/extended-idle regime. Biotechnology and Bioengineering 109(11), 2798–2807.

Whang, L.-M. and Park, J.K., 2006. Competition between Polyphosphate- and Glycogen-Accumulating Organisms in Enhanced-Biological-Phosphorus-Removal Systems: Effect of Temperature and Sludge Age. Water Environment Research 78(1), 4–11.

Ye, F. and Li, Y., 2009. Enhancement of nitrogen removal in towery hybrid constructed wetland to treat domestic wastewater for small rural communities. Ecological Engineering 35(7), 1043–1050.

Zeng, W., Peng, Y.Z. and Wang, S.Y., 2004. A two-stage SBR process for removal of organic substrate and nitrogen via nitrite-type nitrification-denitrification. Journal of Environmental Science and Health Part a-Toxic/Hazardous Substances & Environmental Engineering 39(8), 2229–2239.

Study on a multi-objective optimisation model for multi-reservoir ecological operation

L.N. Liu & W.L. Liu
Jiangxi Provincial Key Laboratory of Hydrology-Water Resources and Water Environment, Nanchang Institute of Technology, Nanchang, Jiangxi, China

ABSTRACT: The construction of reservoirs result in great changes of a river's flow regime, consequently altering the natural river ecosystem. In order to alleviate the influence of reservoirs on river ecosystems, a multi-objective reservoir ecological operation was proposed, from the perspective of maintaining the river ecosystem's health. Research results were summed up, and the basic concept of reservoir ecological operation was put forward. Based on it, a multi-objective mathematical model of multi-reservoir ecological operation was established by maximizing social, economic and environment benefits. Finally, according to the inherent nature of multi-objective optimisation problems, a novel approach was proposed to solve the formulated model of multi-objective ecological operation of reservoirs, which combined a Multi-Objective Particle Swarm Optimisation (MOPSO) algorithm with the non-dominated solutions evaluation approach. The proposed approach can offer the quantifiable benefits or costs among different objectives for the water managers, and can provide a useful tool for decision makers to solve multi-objective hydrology and water resources problems.

1 INTRODUCTION

The reservoir operation has a significant impact on river ecosystem, unreasonable operation resulting in the river siltation, the decrease of carrying capacity, the deterioration of water quality, and the disruption of the river ecosystem. However, the current reservoir operation is mainly designed for water supply, flood control or power generation, ignoring ecological demand both in reservoir area and downstream, which led to degradation of river ecosystem. In order to alleviate the influence of reservoirs on river ecosystem, one of the most effective measures is to carry out the reservoir ecological operation, maintaining the natural variability of a river's flow regime.

In many countries abroad, related studies have been done on ecological operation which have been the main means of river ecological restoration. Among these countries, America and Australia have a high research level, and the most researches have entered into the practice stage (Koel & Sparks, 2002, Richter et al., 2003, Jager & Smith, 2008), such as Murray-darling River in Australia and Colorado River and Tennessee River in America etc. (Petts, 1996, Hughes & Ziervogel, 1998, Baron, et al., 2002). These practices have proved that the reservoir ecological operation is the main means of river ecological restoration, and it also promoted the study of reservoir ecological operation. However, in China, limited by the corresponding research conditions and environment, ecological operation is basically in a stage of exploration.

At present, scholars at home and abroad have carried out a series of exploratory research on how to change the reservoir optimal operation mode, balancing the economic and ecological benefits of reservoirs, and have achieved abundant research results. From different perspectives, a lot of reservoir ecological operation models were presented (Dong, 2007, Hu et al., 2008, Ai & Fan, 2008), but among them the multi-objective operation model considering ecological operation was not much. In the meantime, from the perspective of solving techniques, previous studies for multi-objective optimisation problems often use the ε-constraint method, or the weighting method, which often fail in attaining a good Pareto front and may face problems whilst dealing with non-convex and discontinuous functions. Therefore, according to the engineering requirement and scientific problems of reservoir ecological operation, it is urgent to carry out the researches on multi-objective model and optimisation techniques for multi-reservoir ecological operation. In this study, we sum up research results and lay out the basic concept of ecological operation. Based on it, we present multi-objective mathematical model of multi-reservoir ecological operation. Finally, we propose a novel approach for multiple objective decision making, which combined a multi-objective particle swarm optimisation (MOPSO) algorithm with the non-dominated solutions evaluation approach. We employ MOPSO to generate a Pareto optimal set for a multi-objective ecological operation problem. Then, a simple but effective non-dominated solutions evaluation approach is presented to provide an opportunity for choosing the desired alternative from a set of Pareto-optimal solutions.

2 MULTI-OBJECTIVE MATHEMATICAL MODEL OF RESERVOIR ECOLOGICAL OPERATION

2.1 Definition of ecological operation

The concept of ecological operation appeared in recent years, but defining it is not uniform. Ecological operation in abroad has mostly been as a main method for eco-environmental goal in reservoir regulation. However, in China, some definitions for ecological operation were proposed from different perspectives which could be laid out as follows (Dong, 2007, Ai & Fan, 2008, Zhang et al., 2011): ecological regulation must first satisfy person's basic needs for water, further minimise the impact on the water environments in the reservoir area and downstream ecosystem caused by reservoir operation, in the meantime, achieve the benefit, by means of adjusting the relationship between ecological goal with other reservoir function targets during the comprehensive exploitation of reservoirs to promote harmony between human and river and keep on the economy development with healthy river ecosystem.

According to the definition, the connotations of an ecological operation can be expanded as follows: (1) ecological operation is only one part of the reservoir's comprehensive control, and cannot independently guide a reservoir's long-term operation separated from control system; (2) ecological operation goal is to improve the hydrological, hydraulic and water-environment conditions of reservoir area and downstream, further maintaining healthy river ecosystem; (3) ecological protection is an important goal of a multipurpose reservoir, and must give up priority to ensure water supply, electricity generation, flood control and other basic needs in the actual process (Zhang et al., 2011). Once the above needs are satisfied, we can practice ecological operation to protect river ecological functions.

2.2 Objective functions

The objective of the ecological operation model is to maximise social, economic and environment benefits by controlling the discharge process under the premise of the river ecological system health. The competing objectives are expressed as follows:

$$W(X) = \min_{x \in X}(f_1(X), f_2(X), \cdots, f_m(X))^T \quad (1)$$

where $x \in R^n$, $f_i : R^n \to R$; W is the reservoir benefit of comprehensive utilisation; f_i is the benefit obtained from the i target; m is the number of objectives; X is the feasible search space; $X = \{x_1, x_2, \ldots, x_n\}^T$ is the set of n-dimensional decision variables; R is the set of real numbers; R^n is an n-dimensional hyper-plane or space.

2.3 Constraints

The optimisation is subject to the following constraints:

2.3.1 Storage limits constraints

$$V_{i,\min} \leq V_{i,t} \leq V_{i,\max} \quad (2)$$

where $V_{i,\min}$ and $V_{i,\max}$ is the dead volume and the maximum volume of the reservoir i respectively; $V_{i,t}$ is the volume of the reservoir i at the tth month.

2.3.2 Storage continuity constraints

$$V_{i,t+1} = V_{i,t} + I_{i,t}T + R_{i,t}T - Q_{i,t}T - O_{i,t}T \quad (3)$$

where $V_{i,t+1}$ is the volume of the reservoir i at the (t + 1)th month; $I_{i,t}$ is the inflow discharge to the reservoir i from the other reservoirs during the tth month; $R_{i,t}$ is the nature inflow discharge to the reservoir i during the tth month; $O_{i,t}$ is the outflow discharge from the reservoir i at the tth month; $Q_{i,t}$ is the total demand water to be supplied at the ith reservoir.

2.3.3 The lower reaches ecological water process demand constraints

$$Q_{i,\min} \leq O_{i,t} \leq Q_{i,\max} \quad (4)$$

where, $Q_{i,\min}$ is the minimum ecological flow in the lower reaches of reservoir i at the tth month.

2.3.4 Non-negativity constraints

$$x_{ik}^l, V_{i,t} \geq 0 \quad (5)$$

3 METHODOLOGY

To handle the above formulated model of multi-objective ecological operation of reservoirs, in this study we propose a novel approach combining the MOPSO with the non-dominated solutions evaluation approach. The MOPSO is employed to generate a Pareto optimal set for a multi-objective ecological operation problem. Then, a simple but effective non-dominated solutions evaluation approach, based on k-means clustering and the decision-making approach, is adopted to provide an opportunity for choosing the desired alternative from a set of Pareto-optimal solutions.

3.1 MOPSO algorithm

The main algorithm consists of initialisation of population, evaluation, and reiterating the search on swarm by combining PSO operators with Pareto-dominance criteria. In this process, the particles are first evaluated and checked for a dominance relation among the swarm. The non-dominated solutions found are stored in an external repository, and are used to guide the search particles. It uses an external repository, in order to improve the performance of the algorithm to save computational time during optimisation. If the size of external repository exceeds the restricted limit, then it is reduced by using the crowded comparison operator, which gives the density measure of the existing particles in the function space. In addition, an efficient elitist-mutation strategy is employed for maintaining diversity in the population and for exploring the search space. The combination of these operators helps the algorithm to effectively propagate the search towards true Pareto optimal fronts in further generations. The detailed description of these operators can be found in Liu (2007). In order to handle the constrained optimisation problems, a constrained non-dominance scheme is adopted from Deb (2000).

3.2 Method of non-dominated solutions evaluation

To provide an opportunity for choosing the desired alternative from a set of Pareto-optimal solutions with well spread and wide coverage, we adopt k-means clustering method to reduce the large set of solutions to a few representative solutions. Then, the information entropy approach and the pseudo-weight vector approach are employed to facilitate final decision making.

3.2.1 K-means clustering method

K-means, proposed by MacQueen, is one of the most popular and effective clustering methods. The idea is to classify a given set of data into a certain number of clusters (assume k clusters) fixed a priori. The main procedure consists of two separate phases. The first phase is to define k centroids, one for each cluster. Considering different centroids causing different result, a better choice is to place them as far away from each other, by as much as possible. Then, the next phase is to take each point from a given data set and associate it to the nearest centroid. When no point is pending, the first step is completed and an early grouping is done. Next, we need to recalculate k new centroids as barycentres of the clusters resulting from the previous step. Once we find these k new centroids, a new binding has to be done between the same data set points and the new nearest centroid, generating a loop. During this loop, we may notice that the k centroids change their location in a step by step manner until no more changes are done, which indicates the convergence criterion for clustering. Finally, this algorithm aims at minimising a Euclidean distance function, which can be written in the following form:

$$f = \sum_{i=1}^{k} \sum_{x_i \in C_i} \sqrt{\sum_{l=1}^{q}(x_{il} - m_{jl})^2} \quad (6)$$

where C_i is a data set in the i-th cluster and m_j is the mean for that points over cluster j. The detailed algorithm of k-means can be found in Chen et al. (2007).

3.2.2 Pseudo-weight vector approach

Denote the maximum minimum and minimum values of i-th objective function by f_i^{max} and f_i^{min} from the obtained set of solutions, respectively. Then, the weight w_i for i-th objective function is calculated as follows (Deb, 2000):

$$w_i = \frac{(f_i^{max} - f_i(x))/(f_i^{max} - f_i^{min})}{\sum_{j=1}^{m}(f_j^{max} - f_j(x))/(f_j^{max} - f_j^{min})} \quad (7)$$

From this equation, we can see that the relative distance of the solution from the minimum (maximum) value in each objective function will be calculated by the above formula. Thus, the weight w_i for the i-th objective for the best solution is to be a maximum, and the sum of all weight components for a solution is equal to one. Once the weight vectors for all Pareto-optimal solutions are calculated, we can choose the non-dominated solution closer to a user-preferred weight vector for final decision making.

3.2.3 Entropy weight method

The information theory was established by Shannon in 1948, in which the information content is linked to entropy (Zhou & Li, 2012). In the information theory, entropy is the measurement of a degree of disorder for systems; it can quantify the valid degree of the information supplied by the data obtained. Therefore, information entropy is used to evaluate the order degree and the effectiveness of the obtained system information. Information entropy, marked by H, can be defined as:

$$H = -\sum_{i=1}^{n} p_i \ln p_i \quad (8)$$

where p_i is the frequency of i possible state.

Given n events and m objectives H_j, the entropy of the j objective, is defined as:

$$H_j = -\frac{1}{\ln(n)} \sum_{i=1}^{n} p_{ji} \ln p_{ji} \quad (9)$$

where x_{ij} means the standardised value of the j order evaluation indices of the i order evaluation event.

Suppose w_j is the relative important extent of the j objective compared to other objectives, w_i satisfies $0 \le w_i \le 1$, which is called the weight of the j objective. Objective weight vector $\mathbf{w}$ is characterized by $\mathbf{w} = \{w_1, w_2, \ldots, w_m\}$. Then, w_j is given by

$$w_j = (1 - H_j) / \sum_{i=1}^{m}(1 - H_i) \quad (10)$$

Based on the evaluation matrix of multiple indices and index weights, the superiority degree u_j, the composite measurement of multiple indices of the i assessment object, can be worked out as follows:

$$u_j = \sum_{i=1}^{m} w_i \mu_{ij} \quad (i=1,2,\cdots,m; j=1,2,\cdots,n) \quad (11)$$

where μ_{ij} is the normalised value of alternative j with regard to objective i.

Thus, superiority degrees can be obtained. According to the magnitude of the superiority degrees, the priority order of the i representative clustered Pareto-optimal solution is determined.

4 CONCLUSION

Reservoir ecological operation is a new field, involving river dynamics, ecology, hydrology and environment science etc. The strong multi-discipline joint made it more complexity. In this study, we discussed current research results and laid out the basic concept of ecological operation. Based on it, we build a multi-objective model for multi-reservoir ecological operation by maximising social, economic and environment benefits. Finally, according to the inherent nature of multi-objective optimisation problems, we propose a novel approach to solve handling the formulated multi-objective model, which combined MOPSO algorithm with the non-dominated solutions evaluation approach. We employ MOPSO to generate a Pareto optimal set for a multi-objective ecological operation problem. Then, we employ k-means clustering method to reduce the large set of solutions to a few representative solutions. To facilitate final decision making, the information entropy approach and the pseudo-weight vector approach are adopted to provide an opportunity for choosing the desired alternative from a set of Pareto-optimal solutions. The proposed approach can offer the quantifiable benefits or costs among different objectives for the water managers, and can provide a useful tool for decision-makers in aspects such as reservoir operation and integrated water resource management.

ACKNOWLEDGEMENTS

This work was financially supported by the National Natural Science Foundation of China (51309130), the Young Scientist Project of Jiangxi Provincial Department of Education (GJJ11254) and the Young Scientist Project of the Nanchang Institute of Technology (2010KJ001).

REFERENCES

Ai, X.S., Fan, W.T. 2008 .Study on reservoir ecological operation model. *Resources and Environment in the Yangtze Basin* 17:451–455.

Baron, J.S., Poff, N.L., Angermeier, P.L. et al., 2002. Meeting ecological and societal needs for freshwater. *Ecological Applications* 12:1247–1260.

Chen, Z.P., Ye, Z .L., Zheng, H.C., 2007. Fast Fractal Coding Technique Based on K-mean Clustering. *Journal of Image and Graphics* 12:586–591.

Deb, K., 2000. An efficient constraint handling method for genetic algorithms. *Computational Methods for Applied Mechanical Engineering* 18:311–318.

Dong Z.R., Sun D.Y., Zhao J.Y., 2007. Multi-objective ecological operation of reservoirs. *Water Resources and Hydropower Engineering* 38:28–32.

Hu, H.P., Liu, D.F., Tian, F.Q. et al., 2008. A method of ecological reservoir operation based-on ecological flow regime. *Advances in Water Scienc* 19:325–32.

Hughes, D.A. & Ziervogel, G., 1998. The inclusion of operating rules in a daily reservoir simulation model to determine ecological reserve releases for river maintenance. *Water SA* 24: pp. 293–302.

Jager, H.I. and Smith, B.T. 2008. Sustainable reservoir operation: can we generate hydropower and preserve ecosystem value. *River Research and Applications* 24:340–352.

Koel, T.M. and Sparks, R.E., 2002. Historical patterns of river stage and fish communities as criteria for operations of darns on the Illinois river. *River Research and Application* 18:3–19.

Liu, W.L., Dong, Z.C., Wang, D.Z. 2007. Hybrid intelligent algorithm and its application in optimal operation of feeding reservoir group. *Journal of Hydraulic Engineering*, 38:1437–1443.

Margarita, R.S., Carlos, A.C.C., 2006. Multi-Objective Particle Swarm Optimizers: A Survey of the State-of-the-Art. *International Journal of Computational Intelligence Research* 287–308.

Petts, G.E., 1996. Water allocation to protect river ecosystems. *Regulated Rivers: Research & Management* (12): 353–365.

Reddy, M.J., Kumar, D.N., 2007. Multi-objective particle swarm optimization for generating optimal trade-offs in reservoir operation. *Hydrol. Process* 21:2897–2909.

Richter, B.D., Mathews, R., Harrison D.L., 2003. Ecologically Sustainable Water Management: Managing River Flows for Ecological Integrity. *Ecological Applications* 13:206–224.

Zhang, H.B., Huang, Q., Qian, X. 2011. Connotation of reservoir ecological operation and its model framework. *Engineering Journal of Wuhan University* 44:427–433.

Zhou, J., Li X.B. 2012. Integrating unascertained measurement and information entropy theory to assess blastability of rock mass. *J. Cent. South Univ.* 19:1953–1960.

Legal thinking and countermeasures of environmental pollution in the course of rural urbanisation

GuoLin Zhang
Jilin Agriculture Science and Technology College

ABSTRACT: Rural environmental problems in the development of urbanisation have become the focus of social concern. The water pollution, soil pollution, air pollution, solid waste pollution of rural environment have caused ecological crisis. Governance to rural environmental pollution is one of the key links of urbanisation. Therefore, it is necessary to think of the law in the process of urbanization of rural environmental pollution problems.

Keywords: Rural Urbanisation; Environmental pollution; Law

On 9 July 9 2013, Premier Li Keqiang spoke again in Guangxi about promoting human-centred new urbanisation. Urbanisation is the engine and power of the development of China's present and future. Urbanisation has two core issues: first, the urbanisation of land, second, the urbanisation of the population. The urbanisation of these two elements is having a profound effect on China' economics, politics and social. New urbanisation emphasises human centred.

According to statistics, China's urbanisation rate is just over 52%. If calculated using household registered population, it is about 35%, far below the average level of developed countries, which is nearly 80 percent. Gap is the potential. Looking to the future, urbanisation is the great engine of economic growth in China.

With the development of urbanisation, the environmental problems become increasingly prominent, and have become the focus of attention of the whole society. The results of the environmental damage in rural areas not only harm the interests of farmers, but also directly relate to the problem of feeding each one of us, On rural environmental pollution leads to be too numerous to enumerate the food safety accident. At present, China's rural environmental pollution, soil pollution, air pollution, and solid waste pollution has caused the ecological crisis. All of these can lead to disruption of agricultural production, reducing the quality of agricultural products, and other issues. Therefore, we must attach great importance to rural environmental protection, actively carry out rural environmental pollution control work, and provide a good environment for China's agricultural development; this is our survival. Environmental problems in rural areas it is necessary to think of law.

1 THE IMPORTANCE OF THE RURAL ENVIRONMENT

1.1 *Impact on social life in rural environment*

With the acceleration of urbanisation, a growing number of rural and urban areas are faced with the same types of pollution. Whilst suffering from pollution, people still work in order to provide the vegetables, eggs and milk and other food to the nearby city. This requires a lot of labour. City suburb village even became the main residence of the big city. Part of migrant workers work in the city construction and everyday life sectors. Many of them gather in the city suburbs; A number of city white-collar workers, in order to alleviate the various aspects of pressure brought by rapid population growth in cities, also live in the area near the city with convenient transportation. So the rural outskirts of the city not only have the characteristics of urban pollution, but also combine the characteristics of rural environmental pollution. If pollution incidents are not properly addressed, they can easily generate social conflicts, and even cause social problems.

1.2 *Influence of rural environment to food safety*

Food safety incidents have occurred frequently in recent years, and have attracted the attention of the whole society. Rural as a major producer of food shouldering the 13 million people drink food problem. The merits of the rural environment directly affect peoples' health, life safety, and are related to the healthy development of future generations. Therefore, the rural environment cannot be ignored. In recent years there were "poisonous rice" and water pollution incidents, one after another. Excessive pesticides found in fruit and vegetables led to our exports experiencing problems, so the phenomenon vividly, These facts are all there to remind us that paying attention to the rural environment, reducing environmental pollution in rural areas, is important in order to protect ourselves, and to protect our future generations.

2 THE MAIN ENVIRONMENTAL POLLUTION IN RURAL AREAS

In recent years, China's environmental pollution has been getting worse, showing a trend that shifts from industrial to agricultural, and urban to rural migration. With the process of urbanisation, the increasing importance of China's rural environment, and rural environmental governing commitment is increasing. By summing found a wide variety of rural environmental pollution, complex structure, a very wide coverage.

2.1 *Industrial wastewater and pesticide pollution*

Environmental protection department as early as 2006 on statistics on the amount of chemical fertilizer and pesticide application, also conducted monitoring of organic pollution indicators. Although the application is large, and the indicators of normal, but fertilizer and pesticide pollution has an impact on water quality, and water reservoirs is indeed worrying. Water pollution in rural areas not only affects the local residents, but also the residents of the nearby cities. The extensive use of chemical fertilizers and pesticides over the years seriously pollutes surface water, groundwater, and coastal waters. Faced with increasingly complex environmental situation, rural water pollution, especially more and more pollution of drinking water sources in rural areas, has very serious consequences.

2.2 *Pesticide residues and heavy metal contamination of soil*

Environmental protection department in 2011 to monitor the pilot country 364 villages showed that soil samples exceeded the rate of 21.5% in rural areas. Rubbish dumps surrounding farmland, vegetable and businesses surrounding soil pollution is heavy. All this indicates that the soil contamination has become one of the priorities of rural environmental pollution. The quality of the soil directly affects the development of agriculture, rural construction, farmers' health, and have a critical impact on the environment.

2.3 *Atmospheric dust pollution*

In recent years, cities have to move out of some heavily polluted industries and enterprises. These companies have generated a lot of dust, but multiple alterations in the suburbs or rural areas around cities, undoubtedly brought enormous pressure to rural air quality.

Rural areas did not achieve a comprehensive collective heating, especially the winter of north is a major source of air pollution in rural areas, smoggy. China's rural straw burning phenomenon is always more serious, in addition to affecting sight, blocking traffic, but also a serious impact on the atmospheric environment in rural areas. Automobile exhaust emissions in some affluent areas also add new sources of pollution to the rural areas.

2.4 *Solid waste pollution*

Currently solid waste pollution in rural areas comes mainly from solid waste generated by the mining industry, plastic sheeting, rubbish and human and animal faeces. Rubbish is a problem of rural residents for a long time, but so far have comprehensive rubbish recycling system has not been established Because the rural population is scattered, almost no special rubbish collection, transportation, disposal, treatment system, and even some rural no designated waste sites, also does not have perfect laws and regulations on the management and control, livestock and poultry manure is an important source, the accumulation of a large number of faeces will produce odour, cause air pollution at the same time, also can cause the pollution of solid wastes, influencing the surrounding environment. The relocation of industrial enterprises, which are waste producers, to the rural areas has become an important source of rural solid waste pollution.

3 THE STATUS QUO OF CHINA'S RURAL ENVIRONMENTAL LEGISLATION

3.1 *The legislative system of China's rural environmental law*

Although the legal construction of China's environment has made considerable progress, there is still a large gap in comparison to the developed countries. Due to the environmental pollution events in recent years which continue to occur, the state has increased the environmental legislation and enforcement efforts. Through understanding and continuous efforts, China has formed a relatively scientific and rational legal framework, which has made certain achievements in the field of environmental law.

China's environmental legislation adopted the "Environmental Protection Law" is the main mode of comprehensive legislation, the legislation of the rural environment into "Water Pollution Prevention Law," and "Atmospheric Pollution Prevention Act" that apply to towns, rural general provisions, and "Pesticide Management Regulations", "irrigation water quality standards," specialized rural environment laws, regulations, standards, and so on.

Therefore, China has established a wide coverage, rich content, better environmental legal system, that is in the guide of the constitutional and, environmental protection law as the main body, environmental protection law, administrative regulations, departmental rules and regulations, local regulations, local government rules, environmental standards, as well as other departments of environmental protection laws and treaties.

3.2 Reflection on China's rural environmental legislative status

As the guiding ideas are backward, our understanding of agriculture has become more conservative. However, as the process of urbanisation and the occurrence of a series of rural environmental pollution accident, the rural environment is more and more attention. Therefore, changing the previous misconceptions that industrial pollution hazards are large, urban pollution is serious, and small agricultural pollution hazards, pollution is not serious in rural areas, to guide environmental legislation is necessary.

Rural environmental legislation is lack of independence. In our country for the rural environmental pollution control regulations, dispersed in different local regulations and not a complete law. Whilst the "environmental protection law" also on this aspect relates to not much content Therefore, it has not formed the fundamental, integrity principle, system, measures and means of processing waste, and other basic content. They do not even have the comprehensive, systematic guidance and promotion of the rural environmental pollution control. How effective coordination and smooth implementation becomes a more realistic problem. Therefore, we should establish a unified environmental pollution control in rural areas of the law is the fix.

4 THE LEGAL COUNTERMEASURES OF CHINA'S RURAL ENVIRONMENTAL CONTROL

4.1 To improve the rural environmental legislation

In view of the practical need of rural environmental protection laws and regulations, the author proposes specific rural environmental protection regulations, as China's existing legislation on environmental protection in rural areas the implementation regulations. This Ordinance may be for those special rural, the existing environmental legislation did not provide content to make comprehensive, specific provisions. The current law has been related to the content, and should, according to the specific implementation issues in rural areas, make corresponding regulations. When the time is ripe, then this can be an opportunity for the corresponding legal drafting or revising.

4.2 Strengthening the rural environmental law enforcement, supervision and inspection

First, improve the rural environmental supervision mechanism. China's rural township (town), most of them without the establishment of the environmental protection agency, foundation of environmental supervision is weak, resulting in the supervision, mode, means and effectiveness of passive. Therefore, must carry on the bold exploration and practice of the mode of operation, mode, rural environmental protection into the planning of new socialist countryside construction, establish and improve the rural environmental protection office, equipped with a full-time staff and the establishment of environmental protection, environmental protection work contact, timely communication policy, feedback information. By strengthening the organisation and management, it is possible to mobilise the enthusiasm of the cadres and the masses.

Second, the environmental indicators into the official examination system, the establishment of the government at all levels of environmental protection "accountability", let the government departments at all levels to shoulder the responsibility for environmental protection.

Third, the comprehensive use of legal means, administrative means and economic means, to improve the effectiveness of environmental law enforcement. The current environmental law of single administrative enforcement and direct control as the main means of law enforcement is not suitable for the countryside, the more dispersed production and the way of life, resulting in the rural environment law enforcement costs being high. Therefore, through the use of legal means, and administrative means at the same time, we should vigorously develop and use economic instruments for environmental protection to strengthen the policy guidance. Such as give preferential policies to the production and processing of green food and so on, Make the farmers to adopt environmentally friendly mode of production and life, thus greatly saving environmental law enforcement cost, improve the effectiveness of environmental law enforcement.

Finally, strengthen environmental enforcement and supervision by the masses, NPC and the CPPCC supervision, social supervision by public opinion combine enhanced environmental enforcement self-restraint.

4.3 To strengthen the judicial protection of the rural environment

First, establish rural environmental protection circuit court. At present, China has four environmental court, on the basis of this, according to the characteristics of environmental problems in rural areas, rural environmental cases in special set up reception centre of county court, the establishment of rural environmental protection circuit court, the rural environmental protection in criminal, civil, administrative cases and the implementation of "four in one" trial execution mode. According to the claims of the parties, grassroots, local trial including rural environmental pollution and damage aspects of the case. The trial and execution of unified rural environmental cases can use its concentration effect, produce larger public influence, thus play a declaration and educational justice, concrete and vivid case can improve the environmental legal awareness of farmers; Second, the establishment of rural environmental litigation costs, reduce corrosion, free system, ensure that farmers could get the environmental lawsuits, in order to address the actual need of our judicial system, the operation practice and environmental protection suit.

Although China has made some progress in the rural environment legislation, it is still insufficient compared to developed countries. Actively and effectively promoting the legal construction of rural environmental pollution control is the objective requirement of building a new socialist countryside, and is a necessary measure in order to control environmental pollution in rural areas, and is an important guarantee to promoting the development of rural urbanisation.

REFERENCES

Chen Chen. Legal countermeasures and Thoughts on rural environmental pollution in China. 2012. [Master dissertation]. Northwest Agriculture and Forestry University.

Liu Yingjie. Study on the legal problems of China's rural solid waste pollution prevention. 2011. [Master dissertation]. Beijing Jiaotong University.

Huang Xisheng. Duan Xiaobing. On the innovation of new rural construction and environmental legal system of socialism. Journal of Chongqing University. 2010. 16.

Gao Xiaolu. On Legal Problems of rural environmental pollution prevention and control. Contemporary Law 2009. (3).

Zhou Yuhua. Zheng Lei. Improve enforcement of our rural environment envisioned [J]. Northern Economy and Trade. 2007. (5).

The central research group. The status quo of China's rural environmental pollution and countermeasures [J]. Economic sector. 2008. (2).

Duan Bihua. 2009. New rural environmental protection and management. Golden Shield Press: 124–127

Hou Zhenhua. 2010. The new rural environment pollution prevention manual. Shenyang Publishing House: 110–117.

Motivation, electrical engineering and automation

The modelling and simulation of a doubly-fed induction motor

ZeYang Pei, ZhiJie Wang & Hao Xiao
School of Electrical Engineering, Shanghai DianJi University, Shanghai, China

ABSTRACT: This paper designs a back-to-back dual PWM converter, and it points out that the network side converter control objective is to ensure that the DC bus voltage constant and the input unit power factor can be controlled; machine side converter ensures that active and reactive power are completely controllable. Besides, it points out that the PWM converter's principle acted as the all control devices in trigger generator as well as its advantages. Finally, the network side converter based on grid voltage vector control and the machine side converter based on stator side chain vector control are made through discussion and analysis. Under the MATLAB/SIMULINK, a model is built and simulated, and further verified the theory of the right, to deepen the theoretical study.

Keywords: converter, doubly-fed induction motor, control

1 INTRODUCTION

General sense, double-fed induction wind generator of dual PWM converter is made up of two types of voltage source PWM converters which are all control devices (generally for IGBT) through DC bus connection, and generally a large capacitance is on the DC bus somewhere. This near the grid is referred to the net side converter. Near the doubly-fed induction motor converter called machine side converter. Dual PMW converter control can be divided into network side converter control and machine side converter control. Network side converter control is used to maintain the stability of the DC bus voltage, and to ensure the grid's unit input power factor. Doubly-fed induction motor stator power output is completely decoupled by machine side converter control, which is the independent control of active power and reactive power output. In this paper, on the basis of theoretical research using the MATLAB/SIMULINK simulation software, the study for the doubly-fed induction motor dual PWM converter control not only has important theoretical significance, but also is of great practical significance.

2 DOUBLY-FED WIND TURBINE MATHEMATICAL MODEL OF DUAL PWM CONVERTOR

Network side converter's mathematical model of vector control.

For parallel operation of doubly-fed induction motor, because of the stator windings on the almost infinite power grid, it can be thought of as the amplitude and frequency of the stator voltage is constant.

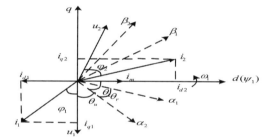

Figure 1. Machine side converter control vector diagram.

On the premise of ignoring stator winding resistance, the stator voltage vector and stator induction electromotive force vector superpose. According to the law of electromagnetic induction, the flux linkage vector is always ahead of induction electromotive force vector 90°. Therefore, the stator flux vector ψ_1 is ahead of the stator voltage vector u_1 90°. Based on these analysis, if choose the consistent of two-phase synchronous rotating dq coordinate system's d axis and the direction of the stator flux linkage vector, the rotation of the shaft system is consistent with the rotational speed of the stator magnetic field, so the q axis is with the stator voltage vector on a straight line. On this, the axis component of the stator flux linkage is 0, and at the same time the axis component of the stator voltage is 0, thus greatly simplifying the control model, and the back of the analysis and study have established a good foundation. This is the basic principle of stator flux, the vector diagram as shown in Figure 1.

The PWM converter which is the type of three-phase voltage is known in mathematical model of two-phase rotating coordinates at the speed of w. We can get the converter input current satisfied (1):

$$\begin{cases} L\dfrac{di_d}{dt} = -Ri_d + \omega L i_q + e_d - s_d u_{dc} \\ L\dfrac{di_q}{dt} = -Ri_q - \omega L i_d + e_q - s_q u_{dc} \end{cases} \quad (1)$$

We can make the following assumptions: Order $v_d = s_d u_{dc}$, $v_q = s_q u_{dc}$. Bring them into (1) we can get the following (2):

$$\begin{cases} L\dfrac{di_d}{dt} = -Ri_d + \omega L i_q + e_d - v_d \\ L\dfrac{di_q}{dt} = -Ri_q - \omega L i_d + e_q - v_q \end{cases} \quad (2)$$

Dq axis component of input converter current is controlled by many factors through (2), and it is the result of the interaction. We assume (3):

$$\begin{cases} v_d' = L\dfrac{di_d}{dt} + Ri_d \\ v_q' = L\dfrac{di_q}{dt} + Ri_q \end{cases} \quad (3)$$

$$\begin{cases} v_d'' = L\omega i_q \\ v_q'' = L\omega i_d \end{cases} \quad (4)$$

Bring (3) and (4) into (2) can get:

$$\begin{cases} v_d = -v_d' + v_d'' + e_d \\ v_q = -v_q' + v_q'' + e_q \end{cases} \quad (5)$$

where v_d', v_q' are the voltage decoupling items, v_d'', v_q'' are the voltage coupling compensation terms, e_d, e_q are voltage feedforward compensation terms. Through this method we can realise well the independent control of the converter input current, and the differential link existence in dynamic performance can also be improved.

In this article, the design of the network side converter uses power grid voltage oriented vector control strategy, and the two-phase rotating coordinate system d axis is oriented to the network voltage vector. It can get:

$$\begin{cases} e_d = U \\ e_q = 0 \end{cases} \quad (6)$$

In (6), U is the amplitude of voltage vector. Bring it into (5) can get:

$$\begin{cases} v_d = -v_d' + v_d'' + U \\ v_q = -v_q' + v_q'' + 0 \end{cases} \quad (7)$$

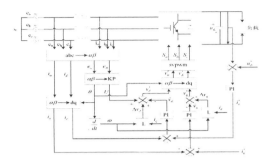

Figure 2. Net side converter control block diagram.

Above on, we can get the network side converter control block diagram as shown in Figure 2.

Machine side converter's mathematical model of vector control.

For parallel operation of doubly-fed induction motor stator, windings are on the almost infinite power grid, therefore we can think that the amplitude and frequency of the stator voltage are constant. Based on this point, a vector control model, which is based on stator flux, can be designed. Under the premise of the stator flux, based on the previous analysis and discussion, when doubly-fed induction motor is connected to the grid ideal, it has the following basic equations (8):

$$\begin{cases} r_1 = 0 \\ \psi_{d1} = \psi \\ \psi_{q1} = 0 \\ u_{d1} = 0 \\ u_{q1} = -U \end{cases} \quad (8)$$

r_1 is for the stator resistance, ψ_{d1}, ψ_{d2} are for the components of the stator flux linkage under the rotating coordinate system, u_{d1}, u_{q1} are for the stator voltages under the rotating coordinate system, U is for the grid voltage vector amplitude. Motor stator active power and reactive power are:

$$\begin{cases} P_1 = -Ui_{q1} \\ Q_1 = -Ui_{d1} \end{cases} \quad (9)$$

By above knowable, the stator active power and reactive power control relationship have become very simple; you just need to control the dq component of the stator current to control the stator output active power and reactive power.

According to what we have learned, that is knowledge of electrical machinery, under the condition of stator flux, we can get:

$$\begin{cases} 0 = -L_1 p i_{d1} + \omega_1 L_1 i_{q1} + L_m p i_{d2} - \omega_1 L_m i_{q2} \\ U = -L_1 p i_{q1} - \omega_1 L_1 i_{d1} + L_m p i_{q2} + \omega_1 L_m i_{d2} \\ u_{d2} = -L_m p i_{d1} + \omega_s L_m i_{q1} + (r_2 + L_2 p) i_{d2} - \omega_s L_2 i_{q2} \\ u_{q2} = -L_m p i_{q1} - \omega_s L_m i_{d1} + (r_2 + L_2 p) i_{q2} + \omega_s L_2 i_{d2} \end{cases} \quad (10)$$

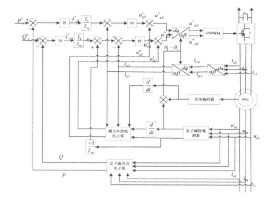

Figure 3. Machine side converter control block diagram.

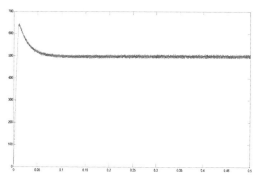

Figure 4. The voltage of DC bus capacitor.

And because of (11):

$$\begin{cases} \psi = L_1 i_{d1} - L_m i_{d2} \\ 0 = L_1 i_{q1} - L_m i_{q2} \end{cases} \quad (11)$$

we can get (12) through (11):

$$\begin{cases} i_{d2} = \dfrac{1}{L_m}(L_1 i_{d1} - \psi) \\ i_{q2} = \dfrac{L_1}{L_m} i_{q1} \end{cases} \quad (12)$$

$$\begin{cases} \psi_{d2} = a_1 \psi + a_2 i_{d2} \\ \psi_{q2} = a_2 i_{q2} \end{cases} \quad (13)$$

In them, $a_1 = -\dfrac{L_m}{L_1}, a_2 = L_2 - \dfrac{L_m^2}{L_1}$.

$$\begin{cases} u_{d2} = (r_2 + a_2 p)i_{d2} - a_2 \omega_s i_{q2} \\ u_{q2} = (r_2 + a_2 p)i_{q2} + a_1 \omega_s \psi + a_2 \omega_s i_{d2} \end{cases} \quad (14)$$

$$\begin{cases} u'_{d2} = (r_2 + a_2 p)i_{d2} \\ u'_{q2} = (r_2 + a_2 p)i_{q2} \end{cases} \quad (15)$$

$$\begin{cases} u''_{d2} = -a_2 \omega_s i_{q2} \\ u''_{q2} = a_1 \omega_s \psi + a_2 \omega_s i_{d2} \end{cases} \quad (16)$$

u'_{d2}, u'_{q2} are respectively for the rotor current decoupling of voltage in rotating coordinate system; u''_{d2}, u''_{q2} are respectively for the coupling compensation voltage of the rotor current under the rotating coordinate system.

By the above formula (14), (15), (16), the machine side converter vector control model can be designed in Figure 3.

3 DOUBLY-FED INDUCTION MOTOR SIMULATION ANALYSIS OF DUAL PWM CONVERTER CONTROL

From Figure 4, after a short climb, we can see net side current transformer DC voltage which stabilises near

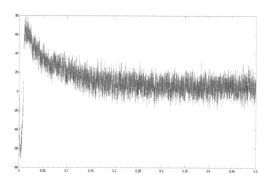

Figure 5. The net side current of the q axis component in rotating coordinate system.

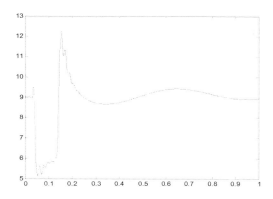

Figure 6. The output active power of stator.

the given capacitance voltage of 500 V. Network side converter control achieves the goal of a stable DC bus voltage. In Figure 5, by controlling the input current, we can see the q axis component, which can guarantee the grid unit input power factor.

From the Figure 6 and Figure 7, we can see that the stator output active power and reactive power can be decoupled, and it is easy to realise independent control.

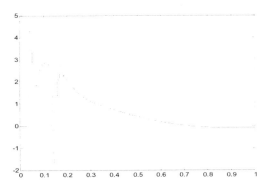

Figure 7. The output reactive power of stator.

4 CONCLUSION

In this paper, a doubly-fed induction motor simulation research was conducted on the dual PWM converter control.
(1) This paper briefly introduced the principle of vector control; the network side power grid voltage oriented vector control model of the converter was established. A control model was designed according to the network side converter control block diagram.
(2) This paper analysed the mathematical model of doubly-fed induction motor, established the machine side converter of the stator flux vector control model and designed the stator flux observer, and designed the machine side converter control block diagram.
(3) Setting the doubly-fed induction motor based on MATLAB/SIMULINK simulation model of dual PWM converter control, the simulation results show that the network side converter control and machine side converter control are set to achieve the established control target.

REFERENCES

Daqun Lu. Variable speed constant frequency doubly-fed wind power system control technology research: [master degree theses of master of]. Liaoning, Shenyang Institute, 2009.
Pena R, J.C. Clare. A doubly fed induction generator using back-to-back PWM converters and its application to variable-speed wind-energy generation. IEEE Electric Power Applications, 1996, 143(3): 231–241.
Yaxi Li, Xin Wu etc. The current situation and development trend of wind power, solar energy journal, 2004, 5(1): 6–7.
Yonggang Yao. Dual PWM converter and its control strategy research: [master degree theses of master of]. Henan, Henan University of science and Technology, 2009.
Yongshun Liu, Shuping Li, Rui Zhou. Ac variable frequency speed regulation system based on dual PWM technology. Relay, 2003, 31(10): 63–65.

About the author: Zeyang Pei (1990–), female, Xinzhou city of Shanxi province, student, master, major in the research of intelligent control and stored energy

Parameter inversion of horizontal multilayer

Jiang Tao Quan, Ling Ruan & Xin Tong
Key Laboratory of High-Voltage Field-test Technique of Electric Power Research Institute of Hubei Power Grid Corporation, Wuhan, China

Xi Shan Wen, Zhuo Hong Pan & Qi Yang
School of Electrical Engineering, Wuhan University, China

ABSTRACT: Soil parameters inversion is an important part of grounding grid design in which horizontal multilayer soils are commonly used. The root-mean-square error is defined as the objective function and exact closed-form analytical expressions for earth potentials due to current electrodes of four-point method in first layer of horizontal multilayer soils have been obtained. Different optimisation methods in standard optimisation software were introduced in order to implement parameters inversion of horizontal multilayer soils. The performance of these methods on inversed error and convergent rate was examined by comparative test cases. Results showed that nonlinearity of soil parameter inversion renders most methods stuck in the local optimum solutions, but satisfactory results have been obtained. The Levenberg-Marquardt method and trust region method are advantageous for better numerical stability and convergent rate.

1 INTRODUCTION

Generally speaking, grounding parameters like the grounding impedance and step/touch potential are determined by three factors: soil, grounding grid and excitation current. Especially as the potential gradient of earth surface near grounding electrodes is mainly a function of top layer soil resistivity. Therefore, the soil parameters inversion (SPI), aimed at acquisition and utilisation of the soil structure and composition by practical measured data, is an essential part of grounding grid design [1], [2].

In the real geological formation, electrical conductivity distribution of soil is inhomogeneous. This means that the models of soil have been more sophisticated than the homogeneous case. Other than the spherical model0, cylindrical form0 and finite volume structure0, horizontal multilayer soils (HMS) are much more widely used to simulate the real soil in grounding design 0, 0.

SPI is an optimisation problem of apparent resistivities measured by four-point methods 0 or other electrical prospecting methods. As shown in Fig. 1, parameters of HMS, such as the number of layers (n), resistivity (ρ_i), thickness (h_i) and depth (z_i) of the HMS layer, can be obtained by processing measurement data with optimum methods.

Different methods of SPI for HMS have been investigated in the past several decades. The curve matching method and steepest decent method were used as earth resistivity measurement interpretation techniques for

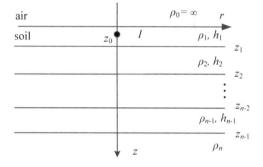

Figure 1. Horizontal n-layer soil. ρ_j and h_j is resistivity and thickness of layer j. z_j is z-axis coordinate of boundary of layer j and layer j + 1, j = 1, 2,..., n−1.

2-layer soil0. Moreover, the influence of overhead network and the earth-return path on measurement has been discussed. These methods and discussions were adopted by IEEE Standard 81-2012 and 80-20000, 0.

The theoretical model for a horizontally stratified n-layer earth structure with a surface point electrode which has been obtained by T. Takahashi0. Features of ρ-a curves in relation to earth parameters have been well documented and, moreover, ρ-a curves were used to approximate the measured data by Wenner method.

Based on the steepest descent method, the Marquardt method, the Newton method, the generalised inverse method and the quasi-Newton method,

J. Alamo 0 applied eight methods to 2-layer soils for a comprehensive comparison on accuracy and consumed time0. In Paper 0, 0, the classical image method was introduced for generating image groups for the calculations of apparent resistivity. The effect of errors in the measurements and in the apparent resistivity on the soil parameter estimates were evaluated by the Marquardt method.

The interpretation of the measured data obtained in the field is the most abstract and difficult part of SPI for HMS. The soil resistivity variation is large and sophisticated due to the heterogeneity of earth. In most cases, it is essential to establish a simplified equivalence of the earth structure. For applications in power engineering, the two or three layer models of HMS are accurate enough without too much mathematical theory involved. However, the most recent computer solutions for optimisation are available, and they can effectively estimate multilayer soil models for various measurement techniques. Those methods that have been devised for the solution to this problem can be classified into two broad categories based on the type of information supplied by the user:

1. Direct-search methods, which use only objective function values.
2. Gradient-based methods, which require estimates of the first-order derivative or second-order derivative of objective function.

This composite activity constitutes the process of formulating the engineering optimisation problem of SPI. Moreover, along with presenting the techniques, we attempt to elucidate their relative advantages and disadvantages wherever possible by presenting or citing the results of actual computational tests. Those methods should be characterised by robustness, number of function evaluations, and computer time to termination when applied to variations of test data.

For the tedium of numerical methods involved in optimisation algorithms, techniques of optimisation are intended primarily for computer implementation. Although the methodology has been developed with computers in mind, we do not delve into the details of program design and coding. Instead, our emphasis is on derivation of the objective function, on the numerical computation involved in selecting the appropriate techniques for parameters inversion of HMS. Those methods were characterised according to robustness, number of function evaluations, and computer time to termination when applied to a family of test problems.

In order to improve the robustness and precision of SPI for four-point method, a new method of objective function derivation and computation has been presented. Different direct-search and gradient-based optimisation methods used in standard optimisation software were introduced to solve the optimisation problem of parameters inversion for horizontal multilayer soils. Soil parameters inversion with constraints has been proposed to avoid unreasonable results and improve confidence of top and deep resistivity.

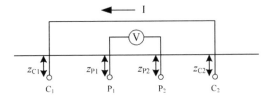

Figure 2. Configuration of four-point method.

2 OBJECTIVE FUNCTION OF SPI OF HMS

For the four-point configuration shown in Fig. 2, the potential difference of the potential probes is given by:

$$V_{P1}-V_{P2} = \phi_1(r_{C1}-r_{P1},z_{P1},z_{C1})-\phi_1(r_{C1}-r_{P2},z_{P2},z_{C1}) \\ -\phi_1(r_{C2}-r_{P1},z_{P1},z_{C2})+\phi_1(r_{C2}-r_{P2},z_{P2},z_{C2}) \quad (1)$$

Potential difference of P_1 and P_2 can be written as:

$$V_{P1}-V_{P2} = \frac{\rho_a I}{4\pi}(D_{C1-P2}+D_{C3-P4}-D_{C1-P3}-D_{C2-P4}) \quad (2)$$

where ρ_a is the apparent resistivity. D_{A-B} is defined as

$$D_{A-B} = \frac{1}{\sqrt{(r_A-r_B)^2+(z_A-z_B)^2}}+\frac{1}{\sqrt{(r_A-r_B)^2+(z_A+z_B)^2}} \quad (3)$$

By definition of apparent resistivity, we have:

$$\rho_a = \frac{4\pi(V_{P1}-V_{P2})}{I(D_{C1-P2}+D_{C3-P4}-D_{C1-P3}-D_{C2-P4})} \quad (4)$$

The objective function of SPI is expressed as the root-mean-square (RMS) error:

$$\min f_{RMS\_error}(\rho_1,\cdots,\rho_n,h_1,\cdots,h_{n-1}) = \sqrt{\frac{\sum_{i=1}^{m}\left(\frac{\rho_{ai}-\rho_{Mi}}{\rho_{Mi}}\right)^2}{m}} \quad (5)$$

where ρ_M is the m-groups measured data.

SPI belongs to small-scale optimisation problem with extreme-nonlinearity. The above derivation indicates that nonlinearity of SPI renders most optimised methods inefficient, and stuck in the local optimum solutions. Moreover, choosing proper initial solution is important for the final solution in these above methods. Therefore, we will consider examples from each class, since no one method or class of methods can be expected to uniformly solve all problems with equal efficiency. For instance, in some applications function evaluations are very time consuming; in other words, great accuracy in the final solution is desired. In some applications it is either impossible or else very time consuming to obtain analytical expressions for derivatives. In addition, analytical derivative development is prone to error. Consequently, different approximations must be employed if gradient-based techniques are to be used. This in turn may mean considerable experimentation to determine step sizes that strike the proper balance between round off and truncation errors. Clearly, it behooves the engineer to tailor the method used to the characteristics of the problem at hand.

3 OBJECTIVE FUNCTION OF SPI OF HMS

In next section, several comparative studies were surveyed and the selected results from a few of computational cases of SPI have been given, since painfully little of what is currently known of the performance of these methods on practical SPI problems has come from purely theoretical considerations.

Some methods of SPI were examined from primarily three perspectives: first, some methods are included because of their historical importance such as steepest decent method; second, numerous methods were thought to be of practical importance in SPI like Generic algorithm, BFGS quasi-Newton method and Levenberg-Marquardt method; third, some new methods are available for application in SPI, such as simplex search method and trust region method. As these methods were discussed, we include to the extent possible, remarks that delimit advantages and disadvantages of those methods. It is, of course, nearly impossible to be complete in this effort, and in addition we have to avoid extensive discussion of rate and region of convergence.

The methods introduced in this section are direct-search and gradient-based methods. Direct-search methods include:

1. Nelder-Mead simplex search method0 (NM in IMSL);
2. Generic algorithm (GA in MATLAB).

Gradient-based methods include:

1. Steepest decent method (SD in CDEGS 0, 0);
2. Levenberg-Marquardt method (CLM in CDEGS, LM in IMSL and MATLAB);
3. Conjugate gradient method (CG in IMSL and MATLAB);
4. BFGS quasi-Newton methods (BFGS in IMSL and MATLAB);
5. Trust regions method (TR in MKL and MATLAB).

4 COMPARATIVE TEST CASES

The ability of a particular algorithm to solve a wide variety of SPI problems are seldom specifically considered. This is truly regrettable, since most users consider robustness and ease of use to be at least equally as important as efficiency. It is important to recognise that the comparison of optimisation methods has qualitative as well as quantitative aspects: (1) CPU time, (2) terminal point accuracy, (3) number

Table 1. Apparent resistivities, Case I (GA 0, 0).

a (m)	1	3	6	8	15	20	40	60
ρ_a (Ω·m)	138	79	71	67	88	99	151	170

Table 2. Apparent resistivities, Case II (BFGS 0).

a (m)	1	2	3	4	6	10	12	14	20
ρ_a (Ω·m)	74.5	84.6	78.6	66.9	50.9	55.3	54.3	56.3	61.6

Table 3. Apparent resistivities, Case III.

a (m)	1	1.4	2	3.2	5	7	10	14	20
ρ_a (Ω·m)	208	194	164	140	112	94	78	66	56
a (m)	32	50	70	100	140	200	300	500	
ρ_a (Ω·m)	48	47	48	53	60	71	85	102	

Table 4. Initial parameters of SPI.

ρ_i (Ω·m)/h_i (m) RMS error	1	2	3	4
$n=3$, Table 1: 41.7%	150/1	15/5	500/∞	
$n=4$, Table 2: 18.4%	70/1	200/1	30/1	60/∞
$n=3$, Table 3: 20.3%	200/5	50/100	1500./∞	
$n=4$, Table 3: 35.5%	220/1	100/1	30/50	100/∞

of iterations, (4) robustness, (5) number of function evaluations. Robustness of the objective function has been discussed in section II and appendix, and the initial parameters have been given for comparison. So other than GA, the robustness of these methods was equalised. Most performance data are given in an effort to quantify algorithm speed or efficiency, and for simplification, only (1), (2) and (3) of these methods are compared in this section.

In this section, a is defined as the electrode spacing of Wenner configuration, and ρ_a is the apparent resistivity. For comparison of these methods, CDEGS, the commercial grounding computation software using SD and CLM0, and other research papers were introduced. The measured apparent resistivities was shown in Table 1-Table 3. The initial parameters can be found in Table 4. The estimation results were shown in Table 5–Table 10, and Fig. 3–Fig. 5. The comparison of RMS errors of inversions and numbers of iterations was listed in Table 11–Table 13.

Discussions for the comparison were shown as follows.

Table 5. Inversed parameters of Table 1 (n = 3).

method	ρ_1 ($\Omega\cdot$m)/ h_1 (m)	ρ_2 ($\Omega\cdot$m)/ h_2 (m)	ρ_3 ($\Omega\cdot$m)
TR, BFGS, LM, CLM: 2.7%	160.9/1.0	61.7/11.6	245.6
CG: 4.9%	355.4/0.5	68.4/14.5	263.1
NM: 2.6%	160.6/1.0	61.6/11.5	2445.6
SD: 2.7%	158.2/1.1	60.8/11.1	241.3
GA 0: 12.8%	164.5/1.2	71.6/10.6	203.7
GA 0: 44.0%	461.6/0.42	62.8/4.49	246.5

Table 6. Inversed parameters of Table 2 by 0.

n = 6	i	1	2	3	4	5	6
BFGS in	ρ_i ($\Omega\cdot$m)	68.0	627.9	7.3	387.3	7.0	125.4
0:3.1%	h_i (m)	1.1	0.3	1.2	2.6	3.2	∞

Table 7. Comparison of the result by different methods (Table 2, n = 4).

method	ρ_1 ($\Omega\cdot$m)/ h_1 (m)	ρ_2 ($\Omega\cdot$m)/ h_2 (m)	ρ_3 ($\Omega\cdot$m)/ h_3 (m)	ρ_4 ($\Omega\cdot$m)
TR: 2.7%	32.6/0.9	206.5/1.1	0.9/0.1	72.4
BFGS: 2.7%	31.8/0.4	200.9/1.1	10.2/1.0	72.4
LM: 2.7%	28.5/0.3	318.4/0.7	5.7/0.6	72.1
CG: 2.7%	37.2/0.4	202.7/1.1	5.4/0.6	72.7
NM: 3.2%	64.3/0.9	383.9/0.4	24.1/2.8	73.2
SD: 3.5%	70.0/1.0	188.7/0.9	15.3/1.7	74.0

Table 8. Inversed parameters of Table 3 (n = 3).

RMS error	Methods of this paper: 7.15%			SD by CDEGS: 7.16%		
i	1	2	3	1	2	3
ρ_i ($\Omega\cdot$m)	180.9	48.8	146.2	180.7	48.6	138.4
h_i (m)	3.7	135.6	∞	3.7	127.5	∞

Table 9. Comparison of the result by different methods (Table 3, n = 4).

method	ρ_1 ($\Omega\cdot$m)/ h_1 (m)	ρ_2 ($\Omega\cdot$m)/ h_2 (m)	ρ_3 ($\Omega\cdot$m)/ h_3 (m)	ρ_4 ($\Omega\cdot$m)
TR, CG, LM, CLM: 1.4%	219.5/1.4	103.5/6.0	44.1/101.9	132.0
BFGS: 1.9%	212.6/1.7	90.0/7.4	42.5/92.2	128.4
NM: 6.3%	184.5/2.8	97.2/2.1	48.2/127.2	140.7
SD: 1.5%	220.0/1.3	108.0/5.6	44.8/107.7	135.6

4.1 CASE I

Table 5 and Table 11 show that the inversed results of GA 00 have over 7% larger RMS errors than the presented methods and CDEGS. Also, Fig. 6 and Table 11 indicate that NM, SD, LM, CLM, BFGS, TR can obtain the similar result except CG. The comparison of TR and GA in Table 5 still involved that GA is

Table 10. Inversed parameters of Table 3 (n = 4).

n = 4 TR: 1.4%	i ρ_i ($\Omega\cdot$m) h_i (m)	1 219.6 1.4	2 103.5 6	3 44.1 101.9	4 132.0 ∞	
n = 4 SD of CDEGS: 4.7%	i ρ_i ($\Omega\cdot$m) h_i (m)	1 198.1 2.5	2 67.7 19.4	3 22.8 29.5	4 108.7 ∞	
n = 5 SD of CDEGS: 2.4%	i ρ_i ($\Omega\cdot$m) h_i (m)	1 212.1 1.6	2 105.1 4.2	3 53.7 24.3	4 27.6 37.8	5 118.7 ∞

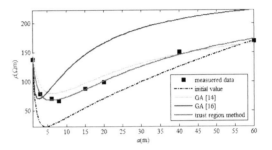

Figure 3. Apparent resistivity curves and measured data. (Tables 1, 5 and 6).

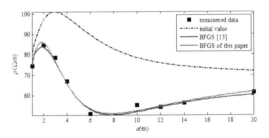

Figure 4. Apparent resistivity curves and measured data. (Tables 1, 5 and 6).

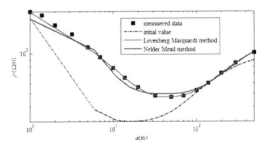

Figure 5. Apparent resistivity curves and measured data. (Tables 1, 5 and 6).

more likely prone to local optima under large nonlinear search space of the HMS parameter inversion. As the direct-search method used by MTALAB optimtool toolbox [22], NM obtain the global optima of Table 1.

Table 11. Comparison of different methods (Table 1).

method	NM	SD	CLM	LM	CG	BFGS	TR
RMS error (%)	2.5	2.7	2.7	2.7	4.9	2.7	2.7
number of iterations	548	168	36	14	109	82	15
CPU time (s)	4	2	1	1	2	2	1

Table 12. Comparison of different methods (Table 2).

method	NM	SD	CLM	LM	CG	BFGS	TR
RMS error (%)	3.2	3.5	2.7	2.7	2.7	2.7	2.7
number of iterations	980	119	507	106	414	64	39
CPU time (s)	8	2	2	1	4	3	1

Table 13. Comparison of different methods (Table 3, n = 4).

method	NM	SD	CLM	LM	CG	BFGS	TR
RMS error (%)	6.3	1.5	1.4	1.4	1.4	1.9	1.4
number of iterations	4470	489	39	16	556	232	85
CPU time (s)	12	2	1	1	6	3	2

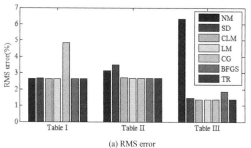

(a) RMS error

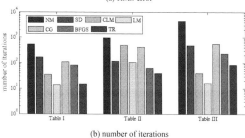

(b) number of iterations

Figure 6. Comparison of the performance of different methods.

Fig. 3 provides that the result of GA in 0 differs from practical experiences and leads to significant error to the measured data. So it clearly indicates that the computational method of apparent resistivity by Sunde's algorithm in 0was implemented correctly.

For the RMS error, we have GA > CG > SD, CLM, LM, BFGS, TR > NM, for the computational effort, LM < TR < CLM < BFGS < CG < SD < NM. The computational time consumed by all of these methods is less than 5 s.

4.2 CASE II

Table 6 shows that the result in 0 of 6-layer soil has 0.4% larger RMS error than 4-layer soil, obtained by using the same BFGS method. Moreover, the data in Table 2 have only 9 points which is inadequate for the inversion of 6-layer earth with 11 parameters. Even with 4-layer earth and the similar RMS error shown in Table 12, variance of results shown in Table 7 with dispersion has been obtained by using different approaches. In Table 7, parameters of the top three layers are quite different but very nearly the same resistivity of last layer was obtained. Because potential gradient of earth surface near grounding electrodes is mainly affected by top layer soil resistivity, a more stable inversed result is required to obtain a more likely explanation of top layer resistivity.

For the RMS error, we have LM, TR, CG, BFGS < NM < SD, and for the computational effort, we have TR, LM < SD, CLM < BFGS < CG < NM.

4.3 CASE III

As the experience of paper 0, the data in Table 3 is a typical 3-layer earth where estimated parameters were shown in Table 8 by using TR, BFGS, LM, SD, CG, NM. Considering increasing the layer number of the earth to four 0, more accurate results shown in Table 9 were obtained. Unlike Table 7, these methods shown in this paper can obtain much closed results than shown in Table 10. It indicates that adequate measured data is quite necessary for the numerical stability of the parameter inversion for HMS. As coincidence of different methods and inversed results has been assured, the results in Table 8 are more convincing.

For the RMS error, we have LM, CLM, TR, CG < SD < BFGS < NM, and, for the computational effort, LM < CLM < TR < BFGS < SD < CG < NM. When applied to the higher directional optimised problems, NM is more prone to local optima and consumes more CPU time.

4.4 Overall performance

Direct-search methods are not widely applicable as gradient-based methods for the performance indicators, such as the larger number of called function, more computing time and less accurate results. Since most users consider robustness and ease of use, some popular direct-search methods with heuristic inspiration, like GA, are more robust than the gradient methods because the initial value of SPI is not requires by these methods. But most of the time, these heuristic direct-search methods tend to be convergent to local optima.

If the initial value for the optimisation is adequately closed to the global minimum, most of the

127

gradient-based methods can obtain the global minimum solution. Paper 0 provided a selective method for initial value, but sometime the initial value could not lead to the global minimum solution for the bias of estimation in the inversion. On the knowledge of the authors, multi-methods and multi-initial values can be applied for a better solution.

For the gradient-based methods, LM and TR are superior to BFGS, CG, and SD for the numerical stability and convergent rate. So LM and TR are chosen for SPI with constraints.

5 CONCLUSION

A new algorithm for the objective function of four-point method considering buried depth in horizontal multilayer soils has been presented. The closed form expression of objective function and derivative of objective function to earth parameters for direct-search and gradient-based optimisation methods have been derived. Based on matrix pencil method, extra robustness and precision in apparent resistivity and derivatives computation have been achieved.

Different optimisation methods in standard optimisation software were introduced to implement parameters inversion of horizontal multilayer soils. The performance of simplex method, steepest decent method, Levenberg-Marquardt method, conjugate gradient method, BFGS quasi-Newton methods, trust region method on root-mean-square inversed error, number of iterations, and CPU time was examined by comparative test cases. Nonlinearity of soil parameter inversion renders most methods stuck in the local optimum solutions, but satisfactory results have been obtained. For all of these methods, Levenberg-Marquardt method and trust region method are superior to others for numerical stability and convergent rate.

REFERENCES

IEEE Guide for Measuring Earth Resistivity, Ground Impedance, and Earth Surface Potentials of a Ground System, IEEE Std. 81–2012, 2012.

IEEE Guide for Safety in AC Substation Grounding, IEEE Standard 80–2000, 2000.

J. Ma, F.P. Dawalibi, W.K. Daily, "Analysis of grounding systems in soils with hemispherical layering". *IEEE Trans. Power Del.*, vol. 8, no. 4, pp. 1773–1781, Oct 1993.

J. Ma, F.P. Dawalibi, "Analysis of grounding systems in soils with cylindrical soil volumes." *IEEE Trans. Power Del.*, vol. 15, no. 3, pp. 913–918, Jul 2000.

J. Ma, F.P. Dawalibi, "Analysis of grounding systems in soils with finite volumes of different resistivities". *IEEE Trans. Power Del.*, vol. 17, no. 2, pp. 596–602, Apr 2002.

Y.L. Chow, J.J. Yang, and K.D. Srivastava, "Complex images of aground electrode in layered soil," J. Appl. Phys., vol. 71, pp. 569–574, 1992.

Z. Li, W. Chen and J. Fan, "A novel mathematical modeling of grounding system buried in multilayer Earth," *IEEE Trans. Power Del.*, 21(3), 1267–1272, Jul. 2006.

F. A. Wenner, "Method of measuring earth resistivity," Bull. Nat. Bureau Std., Washington D.C., vol. 12, 1916.

F.P. Dawalibi, "Earth resistivity measurement interpretation techniques," IEEE Trans. Power Apparat. Syst., vol. 103, pp. 374–382, Feb. 1984.

T. Takahashi and T. Kawase, "Analysis of apparent resistivity in a multi-layer earth structure," *IEEE Trans. Power Del.*, vol. 5, pp. 604–612, Apr. 1990.

J. Alamo, "A comparison among eight different techniques to achieve an optimum estimation of electrical grounding parameters in two-layered earth," *IEEE Trans. Power Del.*, vol. 8, pp. 1890–1899, Oct. 1993.

P. J. Lagace, J. Fortin, and E. D. Crainic, "Interpretation of resistivity sounding measurement in N-layer soil using electrostatic images," *IEEE Trans. Power Del.*, vol. 11, no. 3, pp. 1349–1354, Jul. 1996.

P. J. Lagace, M. H. Vuong, M. Lefebvre, and J. Fortin, "Multilayer resistivity interpretation and error estimation using electrostatic images," *IEEE Trans. Power Del.*, vol. 21, no. 4, pp. 1954–1960, Oct. 2006.

B. Zhang, X. Cui, L. Li, and J. He, "Parameter estimation of horizontal multilayer earth by complex image method," *IEEE Trans. Power Del.*, vol. 20, no. 2, pt. 2, pp. 1394–1401, Apr. 2005.

I. F. Gonos and I. A. Stathopulos, "Estimation of multilayer soil parameters using genetic algorithms," *IEEE Trans. Power Del.*, vol. 20, no. 1, pp. 100–106, Jan. 2005.

E. D. Sunde, *Earth Conduction Effects in Transmission Systems*. New York: Dover, 1968.

W. P Calixto, L. M. Neto and M. Wu, "Parameters estimation of a horizontal multilayer soil using genetic algorithm," *IEEE Trans. Power Del.*, vol. 25, no. 3, pp. 1250–1257, Jul. 2010.

W. P. Calixto, A. P. Coimbra, B. Alvarenga, "3-D Soil Stratification Methodology for Geoelectrical Prospection," *IEEE Trans. Power Del.*, vol. 27, no. 3, pp. 1636–1643, Jul. 2012.

H. R. Seedher and J. K. Arora, "Estimation of two layer soil parameters using finite Wenner resistivity expressions," *IEEE Trans. Power Del.*, vol. 7, pp. 1213–1217, Jul. 1992.

Y. L. Chow, "Surface voltages and resistance of grounding systems of grid and rods in two-layer earth by the rapid Galerkin's moment method," *IEEE Trans. Power Del.*, vol. 12, pp. 179–185, Jan. 1997.

IMSL Fortran Library User's Guide MATH/LIBRARY Volume 2 of 2, Visual Numerics Inc, San Ramon, America, 2003.

Intel® Math Kernel Library Reference Manual, the Intel Inc, New York, America, 2013.

Optimization Toolbox™ User's Guide R2013a, the MathWorks Inc., Natick, America, 2013.

R. Mead and J.A. Nelder, "A simplex method for function minimization," *Computer. J*, vol. 7, no. 1, pp. 308–313, 1965.

Low frequency analysis of conductor networks-user's manual for computer program MALT, Safe Engineering Services & Technologies Ltd, Montreal, Canada, 1999.

Prediction of the performance and exhaust emission of a diesel engine based on Boost

Xue Guang Yang, Gu Yong Han, Hui Wang & Lei Zhang
Air force Logistics College, Xuzhou, China

Li Yong Huang
Great wall Automobile Company, Baoding, China

ABSTRACT: The VIBE two-zone combustion model and the zero-dimensional model of Woschni/Anisits are combined together with Boost software to achieve the synchronous prediction of the engine's performance and exhaust emission. In this paper, the impact of the engine's fuel supply advance angle on its performance and exhaust emission is analysed in detail. The related experimental validation is also conducted.

With the rapid development of the automobile industry, the global energy crisis is becoming increasingly severe. Relying on its excellent power performance and economic efficiency, the diesel engine has drawn more and more attention. There are mainly two methods regarding the R&D of the engine, including numerical simulation and trial, the former of which is divided into one-dimensional and multi-dimensional simulations. Compared to one-dimensional simulation, multi-dimensional simulation has the advantages of higher accuracy, more comprehensive computational data, etc., but it also has the disadvantages of longer modelling time. Moreover, the fact that each model of multi-dimensional simulation is only aimed at a single engine type is also a weak point that cannot be neglected. Therefore, greater importance has been attached to one-dimensional simulation by more and more R&D enterprises of engines. In this paper, a 4135 turbocharged direct-injection diesel engine was modelled on the basis of Boost, a kind of one-dimensional simulation software. In addition, the power performance, economical efficiency and exhaust emission of this engine, after its fuel supply advance angle is changed, are predicted and verified through experiments.

1 MODEL BUILDING AND DETERMINATION OF BOUNDARY CONDITIONS

1.1 Main parameters of engine

The main parameters of the engine are shown in Table 1, and the model of the engine is shown in Figure 1.

CO1 and CO2: Intercoolers; PL1: Pressure Stabilizing Chamber; PL2: Silencer; TC1: Turbine; Turbocharger CL1: Intercooler; C1-C4: Cylinders;

Table 1. Parameters of engine.

Parameter	Indicator
Model Number	4135
Structure	In-line 4-cylinder, 4-stroke, water-cooling, turbocharged inter-cooling
Cylinder diameter/mm	114
Stroke/mm	135
Exhaust Volume/L	8.3
Compression ratio	11
Combustion sequence	1-3-4-2

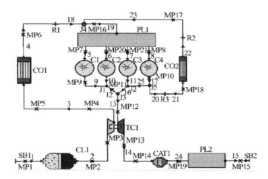

Figure 1. Model of engine.

MP1-MP21: Measuring Points; SB1-SB2: Boundary Conditions; CAT1: Catalytic Converter; R1-R3: Shutoff Valve; and J1-J4: Joints.

1.2 Selection of related computational model and determination of boundary conditions

A lot of engine-related parameters are required when Boost software is applied to building the engine model, such as each pipe's length, pipe diameter, friction coefficient, selection of cylinder heat transfer and combustion models, turbocharger coefficient settings, etc. The settings of the combustion models within the cylinder in Boost software have a significant impact on the accuracy of the calculations, so the settings of related combustion models should be described in detail.

1.3 Combustion models

Boost mainly consists of two combustion models, i.e., zero-dimensional and quasi-dimensional models, the former of which includes several combustion models, such as VIBE, Woschni/Anisits, the isobaric combustion model, etc., and the latter mainly includes the VIBE two-zone model, MCC, etc. The engine studied in this paper adopts the "pump-pipe-nozzle" fuel injection system. Furthermore, MCC of the quasi-dimensional model is specifically designed for the common rail engine type, so it can only utilise the VIBE two-zone model. The ignition point, combustion duration, the shape factor of m and the combustion quality factor of a for the VIBE two-zone model are required to be set and all of these parameters would change automatically, along with the changes of the fuel supply advance angle. However, in Woschni/Anisits of the zero-dimensional model, the parameters of ignition point, combustion duration, the shape factor of m, the combustion quality factor of a, etc., for the VIBE two-zone model can be calculated based on different fuel supply advance angles. Therefore, these two kinds of combustion models are combined together in this paper, so that the performance and exhaust emission of the engine can be predicted simultaneously by means of the cylinder pressure curves, measured from the trials that are taken as the boundary conditions.

1.3.1 Woschni/Anisits model

Model Calculation Principles: normally, a certain reference point is needed in order to calculate the data of other operating modes for diesel engines, whilst Woschni/Anisits exactly provides us with such a model. The ignition delay, combustion duration and other parameters of the reference point need to be set first, then according to which, the operational parameters of other operating modes will be calculated. Equations are shown as follows:

$$\Delta t = \Delta t_r \left(\frac{AF_r}{AF}\right)^{0.6} * \left(\frac{n}{n_r}\right)^{0.5} \qquad (1)$$

$$m = m_r * \left(\frac{P_{IV}}{P_{IV,r}}\right)\left(\frac{T_{IV,r}}{T_{IV}}\right) * \left(\frac{n}{n_r}\right) \qquad (2)$$

Wherein, the subscript "r" stands for the parameter values of the reference point; Δt represents the combustion duration; AF means the air-fuel ratio; n indicates the engine speed; m is the shape factor of VIBE; id stands for the ignition delay; and PIV and TIV are respectively the pressure and cylinder temperature at the moment when the intake valve is closed.

Model Parameter Settings: in this model, some parameters of the reference point, including ignition delay, combustion duration, shape factor, etc., need to be set. The combustion duration and shape factor values of the reference point can be deduced through entering the basic parameters of cylinder pressure, as well as related data of the reference point into the inbuilt "BURN" option of Boost.

1.3.2 The VIBE two-zone combustion model

Model Calculation Principles: the combustion percentage of the VIBE two-zone model is defined by VIBE functions, given that the temperature of the unburned area is the same as that of the combusted area. In addition, the unburned and combusted areas are respectively used in the calculations to take the place of the first law of thermodynamics. The specific formulas are shown as follows:

$$\frac{dM_b u_b}{d\alpha} = -p\frac{dV_b}{d\alpha} + \frac{dQ}{d\alpha} - \sum \frac{dQ_b}{d\alpha} + h_n \frac{dM_b}{d\alpha} \qquad (3)$$

$$\frac{dM_n u_n}{d\alpha} = -p\frac{dV_n}{d\alpha} - \sum \frac{dQ_n}{d\alpha} \qquad (4)$$

wherein, Q is the heat value of the fuel; M is the gas mass; V is the gas volume; P stands for the gas pressure; u represents the gas internal energy; α means the crank angle; h is the gas enthalpy; and the subscripts of b and n represent the combusted and the unburned areas respectively; the heat transfer between these two areas is negligible. Moreover, the changes of the total working volume must be equal to the volume of the engine cylinder V. So must the total volume of the combusted and unburned areas. Equations are shown as follows:

$$\frac{dV_b}{d\alpha} + \frac{dV_n}{d\alpha} = \frac{dV}{d\alpha} \qquad (5)$$

$$V_b + V_n = V \qquad (6)$$

wherein, V_b is the volume of the combusted area; V_n is the volume of the unburned area; and V is the total volume.

The total amount of the combusted mixture at each time step is determined by the user-set parameters of VIBE functions, such as the cylinder wall heat transfer loss, etc. Besides, the rest of the terms in this formula can be obtained through calculating those similar parameters in the VIBE single-zone model. Zeldovieh principle is applied to the NOx emission model, and the soot exhaust emission is calculated by the soot formation and oxidation theories of the internal-combustion engine cycles.

Model Parameter Settings: in the VIBE two-zone model, the four parameters of ignition point, combustion duration, the shape factor of m and the combustion quality factor of a need to be set, however, all of which will change when the fuel supply advance angle is changed. Therefore, different parameters of fuel supply advance angles obtained from Woschni/Anisits will be taken as its own parameter values.

Table 2. Equation of NOx Reaction.

Serial Number	Reaction Formula
1	$O_2 = 2O$
2	$O + N_2 = NO + N$
3	$N + O_2 = NO + O$

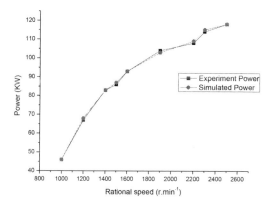

Figure 2. Comparison of power in external characteristic.

1.3.3 Emission model

The turbocharged diesel engine has a larger coefficient of excess air, so the emission values of soot and CO are both smaller. Thus, the emission of NOx becomes the main problem of the turbocharged diesel engine. The NOx emission model in Boost which will be described in the following section is established mainly based on the reaction theory proposed by Zevitch. The reaction of NOx is shown in Tab. 2.

2 COMPUTATIONAL RESULTS ANALYSIS AND RELATED EXPERIMENTAL VERIFICATION

2.1 Model experimental verification

The deviation is found to be lower than 5% by comparing the major parameters of power, torque, specific fuel consumption (SFC), air inflow, etc., so the simulation data can be considered accurate. The analysis of the simulated and the experimental results is shown as follows:

The fuel-injection quantity corresponding to each speed is gradually increased as the rotational speed is enlarged. The power is also increased. The maximum deviation between the experimental and the simulated data is less than 5% (see Fig. 2).

The torque firstly increases along with the increase in air inflow in each cycle, and when it reaches the maximum torque point of 1600 r/min, the exhaust valve is opened in order to keep the maximum supercharge ratio at the high rotation speed not excessively high. The pressure after the compressor and before the intercooler is basically kept at around 0.28 MPa.

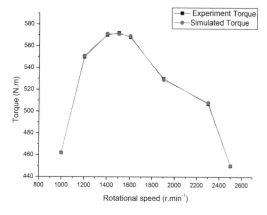

Figure 3. Comparison of torque in external characteristic.

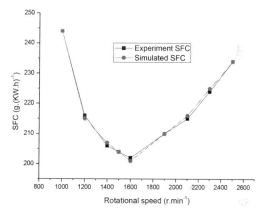

Figure 4. Comparison of SFC in external characteristic.

Afterwards, the torque is decreased gradually while the rotational speed is increased (see Fig. 3).

At the slow rotational speed ranging from 1000 to 1600 r/min, air-fuel ratio increases with the increase in the rotational speed. The combustion also becomes more complete, so its SFC declines with the increase in the rotational speed. However, when it reaches 1600 r/min, the absolute time of combustion is shortened along with the increase in the rotational speed, causing worsening of combustion, which will finally lead to the increase in SFC along with the increase in the rotational speed (see Fig. 4).

Comparison of Air Inflow in Each Cycle: The experimental and simulated mass inflow per cycle is compared as shown in Figure 5. When the exhaust valve is open at the engine's maximum torque point of 1600 r/min, the pressure after the press within the turbocharger is basically kept at approximately 0.28 MPa. However, as the rotational speed increases, the rate of air flow increases as well, resulting in the increase in friction loss. Thus, the mass inflow of each cycle declines comparatively.

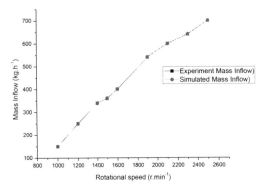

Figure 5. Comparison of mass inflow in external characteristic.

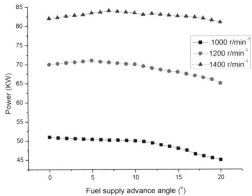

Figure 7. Relation curve of power with inject timing.

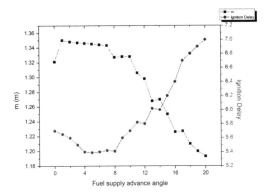

Figure 6. Relation Curve of m with ignition delay.

2.2 Analysis of computational results

2.2.1 Analysis of Woschni/Anisits' results

If the value of m increases, the amount of heat released at the early stage increases whilst the rate of pressure rise declines, resulting in mild combustion. On the contrary, if the value of m decreases, the heat released at the early stage is more and the rate of pressure rise is comparatively bigger, leading to fierce combustion. It can be seen from the formula of Woschni/Anisits' combustion model that the value of each fuel supply advance angle is mainly related to the temperature, pressure and ignition delay of the reference and operating points. As the temperature, pressure and other parameter values of the operating point vary not so much at the same rotational speed, the variation trend of m is opposite to that of ignition delay, i.e. ignition delay will increase along with the increase in the fuel supply advance angle. Moreover, the longer the ignition delay is, the more amount of fuel will be injected during ignition delay. Meanwhile, the rate of pressure rise at the early stage will become greater, while the value of m turns to be smaller. Figure 6 shows the variation trends of ignition delay with the value of m at 2500 r/min.

2.2.2 Related data analysis of optimized fuel supply advance angle

The fuel injection characteristic and the fuel injection (or supply) advance angle of the engines have a great impact on its performance, but it is difficult for the mechanical pump to control the fuel injection characteristic effectively. Therefore, the author in this paper only studies the optimization of the fuel injection (or supply) advance angle of engines. If fuel is injected prematurely, it will result in untimely ignition, so as to cause the increase in the compression negative work. Contrarily, if fuel is injected too late, then the fuel combusted at the expansion stage of the engine will be increased, too, which is also unfavourable to the improvement of the engine efficiency. Additionally, early fuel injection will not only lead to longer ignition delay, but also to excessive injected fuel during ignition delay, causing the increase in the maximum explosion pressure and flame temperature, the raising of the engine's mechanical load and thermal load, and the growth of NOx' emission ultimately. In terms of noise, with the extension of ignition delay, the maximum rate of pressure rise will also increase, thereby causing loud engine noise. Therefore, taking various aspects of the engine's performance into account, we believe that to optimize the engine's fuel injection advance angles under different working conditions is an indispensable part of the optimization of the engine parameters. In the following section, three working conditions of 1000, 1200 and 1400 r/min will be taken as an example to explain the optimization of the fuel injection advance angle. Figs. 7 and 8 respectively show the variation trends of the engine's power and torque with the changes of its fuel injection advance angles at these three rotational speeds. As can be seen from the figure, the variation trends of the engine's two performance indexes are both in curved shape. From the perspective of the engine's power, it can be found that the optimal fuel injection advance angles at these three speeds appear at 50, 80 and 100 respectively and increase as the rotational speed increases.

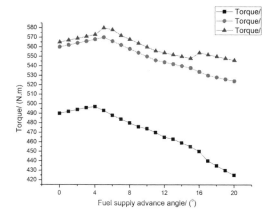

Figure 8. Relation curve of torque with inject timing.

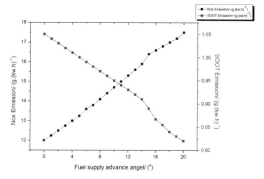

Figure 10. Relation of NOX and SOOT emissions.

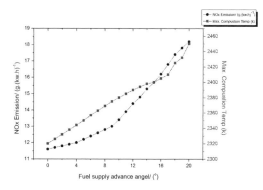

Figure 9. Relation of NOX emission and temperature.

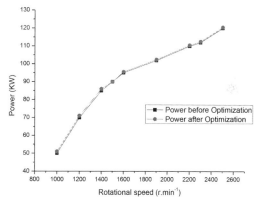

Figure 11. Power after optimization.

2.2.3 *Calculation and analysis of emission data*

The rotational speed of 1000 r/min is taken as an example in this section to analyse the emission data of the engine. Ignition delay is found to be prolonged as the fuel injection advance angle increases. In addition, the increase in the fuel injected during ignition delay will lead to the increase in the maximum combustion temperature. And the emission of NOX is positively correlated with the maximum temperature within the cylinder. Details about the changes can be seen from figure 9. Besides, the soot emission of the diesel engine is often opposite to that of NOX, as the diesel fuel is combusted more completely within the cylinder because of the better atomization and mixing of the diesel fuel thanks to the increase in the fuel injection advance angle. This is why the SOOT emission declines. Details about the changes can be seen from figure 10.

However, it is often difficult to take account of the emission of NOX and SOOT and the power and economic efficiency of the engine at the same time, solely by setting the advance angle. As can be seen from the figures, although the power of the diesel engine is improved to some extent after the fuel injection advance angle is optimized (see Figs. 11 and 12), however, the emission of NOX is significantly higher. Thus, to realize the flexible control of the fuel injection

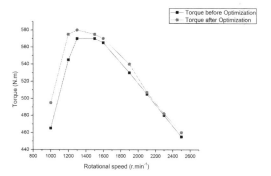

Figure 12. Torque moment after optimization.

characteristic and the free control of the fuel injection pressure of the diesel engine is an important means for us to consider both the economic characteristic of the diesel engine's power and the emission characteristics. The optimal fuel injection advance angle in terms of the engine's power and economic efficiency obtained in this study can provide a reference for the standardization of the electrically-controlled common rail engine type in the future, such as to decrease the emission of NOX by increasing the pilot injection at the fuel injection advance angle when the economic efficiency at each rotational speed is the optimal; or to reduce the emission of SOOT by increasing the fuel injection pressure.

3 CONCLUSIONS

The combination of the Woschni/Anisits and VIBE two-zone models in Boost can effectively predict the diesel engine's power and emission after the change of its fuel injection advance angle.

The optimal fuel injection advance angle of the mechanical pump's diesel engine is often determined by its eclectic results of power, economic efficiency and emission. The results of this study can serve as the guidance for the standardization of similar common-rail engine types to be developed in the future.

Changing the fuel injection advance angle of the diesel engine can only improve its performance within a certain range. And it is the only way to maximize its performance by adjusting the fuel injection characteristic and pressure of the diesel engine as per different working conditions.

REFERENCES

Dent, J. and P. S. Mehta (1981). "Phenomenological combustion model for a quiescent chamber diesel engine." Training 2014: 04–07.

Heywood, J. B. (1988). "Internal combustion engine fundamentals", McGraw-Hill New York.

Lefebvre, A. and J. Chin (1983). "Steady-state evaporation characteristics of hydrocarbon fuel drops." AIAA journal 21(10): 1437–1443.

Ramos, J. I. (1989). "Internal combustion engine modeling", Hemisphere Publishing Corporation New York.

# The study on the simulation model for power system

Jian Guo Zhu & Chang Lin Wang
College of Electrical Engineering and Information Science, Three Gorges University, China

ABSTRACT: This paper presents a developed simulation model for the whole dynamic process simulation of power system calculations. The model is composed of electric network model and generator or dynamic load model, contrary to traditional partitioned approach with implicit integration. This developed approach reduced the calculation amount of network, thus becoming more easily used in parallel computation of the whole system equations of power system.

1 INTRODUCTION

The traditional approach used widely in production-grade stability programs is partitioned approach with explicit integration. Its advantages are programming flexibility, simplicity, reliability, and robustness. Its principal disadvantage is susceptible to numerical instability. For a stiff system, a small time step is required throughout the solution period, dictated by the smallest time constant in the whole mathematical model of power system.

Simultaneous solution with implicit integration is another approach in which the state variables and the network variables are solved simultaneously. Its principal advantage are reliability, and robustness to numerical stability. Its disadvantages are programming inflexibility and complexity. For simultaneous solution, iterative solution using is the Newton-Raphson method, and the order of the whole mathematical model of power system is too great to be calculated at real time speed [1~4].

Here the developed method uses partitioned approach with implicit integration. Traditionally, this iterative solution considering saliency-effect solve the linear algebraic equations in complex domain. When there are no operations, the admittance matrix elements remain constant and unchanged according to the change of rotor angle. But the saliency effect and the rotor angle change in the calculation should be considered by modifying the generator injected current to the node of the network, which is unknown in this time step tn, and should be calculated by the node voltage of next time step time tn+1. In the developed approach, the rotor angle calculation realized in the generator model, and the linear algebraic equations of network be simply solved in complex domain. When there are no operations, the admittance matrix elements remain constant.

2 MATHEMATICAL MODEL AND SIMULATION ALGORITHM

The overall system equation, including the differential equation for all the devices and the combined algebraic equations for the devices and the network are expressed in the following general form comprising a set of first-order differential equations:

$$\dot{x} = f(x, V) \qquad (1)$$

and a set of algebraic equations:

$$I(x, V) = Y_N V \qquad (2)$$

with a set of known initial conditions (x_0, V_0)
where,
x = state vector of the system
V = bus voltage vector
I = current injection vector

In our work, the electric network and generating unit are modularized; the generating unit module consists of Generator rotor circuit equations, excitation system, acceleration or swing equation, prime mover governor, axes transformation, and the electric network module consist of Transmission network equations including stator equations and static loads, as figure 1 shows. The equations for each of the module can be expressed as the following:

The equations for synchronous machine module are:

$$\dot{x}_s = f_s(x_s, y_s) \qquad (3)$$

where,
x_s = stator vector of the generating units
y_s = currents and internal voltage of stator windings and rotor angle of the generator.

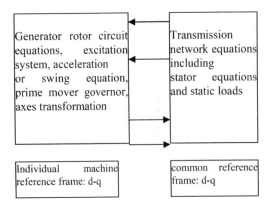

Figure 1. Structure of the complete power system model for transient stability analysis.

The equations for electric network can be expressed as:

$$\dot{I} = Y_N \dot{E} \quad (4)$$

where,
$\dot{I}$ = Current injection vector of the power system;
$\dot{E}$ = internal voltage of stator windings;
Y_N = the admittance matrix of network including synchronous machine reactance.

Here with an example, we introduce the method in detail.
For the 4 orders simplified practical model of synchronous machine, we have the equations of generator1 as follows:

$$u_d = E'_d + X'_q * i_q - r_a * i_d \quad (5)$$

$$u_q = E'_q - X'_d * i_d - r_a * i_q \quad (6)$$

$$T'_{d0} * pE'_q = E_f - E'_q - (X_d - X'_d) * i_d \quad (7)$$

$$T'_{q0} * pE'_d = -E'_d + (X_q - X'_q) * i_q \quad (8)$$

$$CV1 = \begin{bmatrix} \cos(\delta_1 - \delta_n) & -\sin(\delta_1 - \delta_n) \\ \sin(\delta_1 - \delta_n) & \cos(\delta_1 - \delta_n) \end{bmatrix} \quad (9)$$

$$T_m - T_e = TJ * \frac{d\omega}{dt} \quad (10)$$

where,
$E'd$ = direct-axis internal stator transient voltage;
$E'q$ = quadrature-axis internal stator transient voltage;
$X'd$ = direct-axis transient reactance;
$X'q$ = quadrature-axis transient reactance;
$T'd0$ = direct-axis open circuit transient time constant;
$T'q0$ = quadrature-axis open circuit transient time constant.
CV1: Transform the d and q axis variables from the coordinate axis of generator 1 into the coordinate

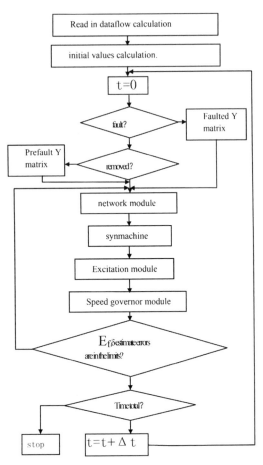

Figure 2. Flow chart of the module calculation process.

axis of generator n to solve the equations in the same coordinate axis.
Differentiate equations (7), (8), (9), and (10) with implicit integration method, and obtain the following differentiated equations in matrix form:

$$FA * E_{Fb} = FB * E_{Ff} + U_{Fm} \quad (11)$$

where,
FA, FB are coefficient matrix;
E_{Fb} state vectors at $t + \Delta t$;
E_{Ff} state vectors at t;
U_{Fm} is control vector.

Here, $U_{Fm} = [i_d; i_q; E_f; \delta_1; \delta_n]$, and equation (11) is the simulation model of generator units.
The network simulation module is (4).

3 SIMULATION RESULTS

Example is: when $t = 0.1$ s, three-phase fault occurs in a three generator units power system; when $t = 0.2$ s, the fault is removed. Simulation plots are shown as figure 3 to figure 4: in which δ (t) curves of G3-G1,

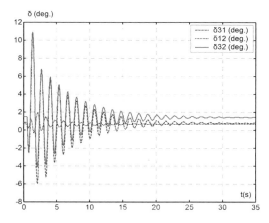

Figure 3. δ(t) curves betweem generators.

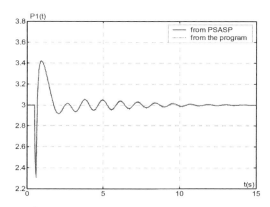

Figure 4. P(t) curve of generator 1.

G1-G2, G3-G2 simulated with the program are shown in Figure 3. P (t) curve of generator 1 simulated respectively with the program and with PSASP are shown in Figure 4.

4 CONCLUSION

This paper presents a developed simulation model for the whole dynamic process simulation of the power system calculations. The model is composed of the electric network model and the generator or dynamic load model, contrary to the traditional partitioned approach with implicit integration. This developed approach reduced the calculation amount of network, the traditional iterative solution considering saliency-effect solve linear algebraic equations in complex domain. When there are no operations, the admittance matrix elements remain constant and unchanged. But the saliency effect and the rotor angle change in the calculation should be considered by modifying the generator injected current to the node of the network, which is unknown in the step tn, and should be calculated by the node voltage of next step time tn+1. In the developed approach, the rotor angle calculation realized in the generator model, and the linear algebraic equations can be simply solved in complex domain. Thus, using this approach makes it more easily in parallel computation of the whole system equations of the power system.

REFERENCES

Yixin Ni, Shousun Chen. Dynamic Power System Theory and Analysis [M]. Qinghua University Press, 2001.

Xianshan Li, Chengming Wu, Xiangyong Hu. Full Scope Real-Time Simulation of Hydropower Plant for a Taining and Research Simulator[R], SM2-01, T&D-C0070, IEEE/PES T&D Aaia Pacific, August, 14–18, 2005 Dalian, China.

Xianshan Li, Xiangyong Hu, Changhong Deng. Modeling method studies of real-time electric simulation for large hydropower plant [J]. Proceedings of Power System Automation, 1999, 11(5): 76–80, 100.

Prabha Kundur. Power System Stability and Control [M].

Jiayu Huang, Liyi Cheng. Power System Digital Simulation [M]. 1995.

Junwei Chu, Daozhi Xia. Power System Analysis [M]. China Electric Power Press, 2002.

Design of a self-adaptive fuzzy PID control system for a supercritical once-through boiler

Dan Zhang & Dong Min Xi
Electric Power College, Inner Mongolia University of Technology, Hohhot, Inner Mongolia, China

ABSTRACT: A supercritical once-through boiler is the developmental direction of a large boiler, its control effect directly affects the economy, energy efficiency and environmental protection of thermal power plant. The fuel to water ratio is the major factor that has influence on the main steam temperature of the supercritical once-through boiler, so it is very important to master its operating characteristics and study its control methods. According to the operating characteristics of the supercritical once-through boiler, a self-adaptive fuzzy PID control system of fuel to water ratio, based on the fuzzy control and PID control, which was designed after the system decoupled. After simulation in MATLAB, it can be seen that the self-adaptive fuzzy PID control system adapts well to the dynamic characteristics of the supercritical once-through boiler and has a good robustness.

1 INTRODUCTION

In recent years, with the continuous development of China's wind power, nuclear power, and solar power generation, the thermal power is still the main part of the electric power industry. Compared to subcritical units, supercritical units have unparalleled advantages when taking into account the economy, energy saving, environmental protection and other aspects. Along with China's thermal power plant unit developing to supercritical pressure and large capacity, the supercritical once-through boiler, which is the only kind of boiler that can be applied to pressure above a supercritical level has become the developmental direction of large boiler of thermal power plant[1–3].

Owing to the dynamic characteristics of multi variety, large inertia, large delay, strong coupling, non-linear and time changeable, the main difficulties of the supercritical once-through boiler control process are: difficult to establish a precise mathematical model, not easy to solute the characteristics like nonlinear, large delay. As the parameters can't be changed with the object and the bad anti-interference ability, the conventional PID control method is no longer very applicable to the control of the supercritical once-through boiler. Fuzzy control, which depends on the model of language rules, can be well adapted to such systems.

Self-adaptive fuzzy PID control combines conventional PID control and fuzzy control. The inputs of the fuzzy controller are main steam temperature, pressure, eta, and the system adapts PID controller parameters through fuzzy controller's outputs to achieve PID parameters self-tuning. Self-adaptive fuzzy PID control enables the system adapt to the change of the dynamic characteristics due to the load and pressure change of once-through boiler, also has the advantages of fast response, strong anti-interference ability and small overshoot.

2 SELF-ADAPTIVE FUZZY PID CONTROL SYSTEM

The self-adaptive fuzzy PID controller is the core of the whole system; the structure block diagram of control system is shown in Figure 1. Two inputs of the controller are error (e) and error changes (ec). The processes of the fuzzy PID control system are as follows:

- First, the fuzzy control rules of e and ec to the three PID parameters should be made and saved according to the impact of three parameters of PID controller to the control effect.
- Second, the e and ec were continuously detected in the system running.
- Third, find the corresponding rules, and modify the three parameters Kp, Ki, Kd in timely.
- Thus, the system can adapt the changing of the control object[4–6].

The error (e) is given by Equation 1, where r is the given target variable, and y is the system output variable:

$$e(t) = r(t) - y(t) \qquad (1)$$

The error changes (ec) is:

$$ec(t) = \frac{de(t)}{dt} \qquad (2)$$

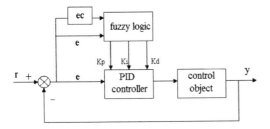

Figure 1. Block diagram of self-adaptive fuzzy PID control system.

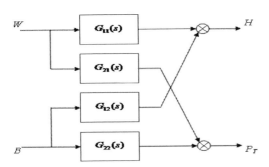

Figure 2. Mathematic model of supercritical once-through boiler.

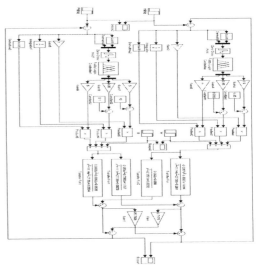

Figure 3. SIMULINK diagram of system on full load.

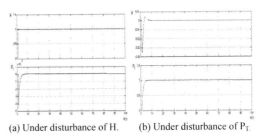

(a) Under disturbance of H. (b) Under disturbance of P_T.

Figure 4. Dynamic characteristics of self-adaptive fuzzy PID control on full load.

3 SIMULATION RESEARCH

3.1 Mathematic model

The supercritical once-through boiler can be shorted into a coupled dual-input dual-output mathematical model when considering the ratio of fuel to water[7], which is shown in Figure 2.

In which B = Fuel; W = Water; P_T = Main steam pressure; H = Micro superheated steam enthalpy.

3.2 Self-adaptive fuzzy PID control on full load

According to the actual operation parameters of production site, the transfer function of the dynamic mathematical model of the supercritical once-through boiler on 100% load after short[7] were given in Equation 3.

$$\begin{cases} G_{11}(s) = \dfrac{H(s)}{W(s)} = \dfrac{-0.0367s^2 - 0.3207s - 1.014}{s^3 + 11.14s^2 + 27.56s + 0.291} \\ G_{12}(s) = \dfrac{H(s)}{B(s)} = \dfrac{0.053s^2 + 0.7583s + 1.797}{s^3 + 11.1s^2 + 27.08s + 0.0929} \\ G_{21}(s) = \dfrac{P_T(s)}{W(s)} = \dfrac{0.82s + 0.058}{s^2 + 0.3501s + 0.0232} \\ G_{22}(s) = \dfrac{P_T(s)}{B(s)} = \dfrac{0.0005s^2 + 0.0082s + 0.0526}{s^3 + 11.14s^2 + 27.56s + 0.2604} \end{cases} \quad (3)$$

Get the mathematical model of supercritical once-through boiler on 100% load decoupled by the method of former compensation[8], and simulat the system in SIMULINK of MATLAB[9–11]; the SIMULINK block diagram of the self-adaptive fuzzy PID control system is shown in Figure 3.

The dynamic characteristics under the unit step disturbance of given value of H and PT are shown in Figure 4.

The Figure 4 shows that: under the regulation of the self-adaptive fuzzy PID controller, the output can quickly follow the given, and the coupling influence between micro superheated steam enthalpy and main steam pressure is very weak.

3.3 Self-adaptive fuzzy PID control on changing load

The supercritical once-through boiler is a multi-variable, strong coupling, time-varying control object; as the system load changes, the model parameters will also change.

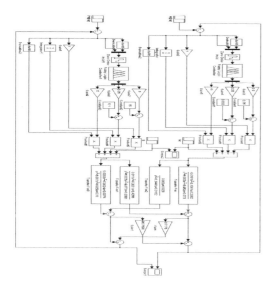

Figure 5. SIMULINK diagram of system on changing load.

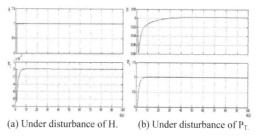

(a) Under disturbance of H.   (b) Under disturbance of P_T.

Figure 7. Dynamic characteristics of PID control on full load.

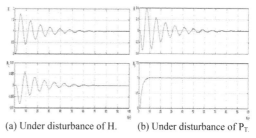

(a) Under disturbance of H.   (b) Under disturbance of P_T.

Figure 8. Dynamic characteristics of PID control on changing load.

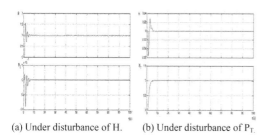

(a) Under disturbance of H.   (b) Under disturbance of P_T.

Figure 6. Dynamic characteristics of self-adaptive fuzzy PID control on changing load.

Equation 4 shows the transfer function on 70% load.

$$\begin{cases} G_{11}(s) = \dfrac{H(s)}{W(s)} = \dfrac{-0.0101s^2 - 0.1001s - 0.2002}{s^3 + 6.003s^2 + 8.962s + 0.073} \\ G_{12}(s) = \dfrac{H(s)}{B(s)} = \dfrac{0.011s^2 + 0.2031s + 0.4266}{s^3 + 6.025s^2 + 9.073s + 0.0331} \\ G_{21}(s) = \dfrac{P_T(s)}{W(s)} = \dfrac{0.003s + 0.028}{s^2 + 0.1441s + 0.0102} \\ G_{22}(s) = \dfrac{P_T(s)}{B(s)} = \dfrac{0.0003s^2 + 0.024s + 0.0074}{s^3 + 6.001s^2 + 8.929s + 0.203} \end{cases} \quad (4)$$

Get the system of 70% load controlled under the coupling parameters and fuzzy PID controller of 100% load; the SIMULINK block diagram is shown in Figure 5.

The dynamic characteristics under the unit step disturbance of given value of H and P_T on changing load are shown in Figure 6.

When the system load mutated, the model parameters changed as follows, as can be seen from Figure 6; the action of the control system is also very fast, the stability of the system is maintained, the coupling degree of the system is also very small, and the control requirements are achieved.

3.4 PID control

In order to compare the control effect of the self-adaptive fuzzy PID control system, the simulation of the conventional PID control system of full load and changing load are made, which are shown in Figure 7 and Figure 8, respectively.

It can be seen from Figure 7 and Figure 8 that, when the load changes, the PID control effect turns bad, the adjusting time is too long and the overshoot is too big.

4 CONCLUSION

Comparing the simulation results of the self-adaptive PID control system and the conventional PID control system, it is obvious that the control effect of the self-adaptive PID control is much better than the conventional one the self-adaptive PID controller is more suitable for the system that parameters are easy changing and overcomes the weakness of the conventional PID controller. Therefore, in the design of the control system of fuel to water ratio for the supercritical once-through boiler, the self-adaptive PID control can achieve the control demand more easily and has a good robustness.

But there are some difficulties in the construction of the fuzzy PID controller, such as the fuzzy logic rules which are hard to make, and the ratio of actual variable to fuzzy variable is difficult to determine.

ACKNOWLEDGMENT

This paper is supported by the priority science research project of Inner Mongolia University of Technology, grant ZD201235; Natural Science Foundation of Inner Mongolia, grant 2013MS0919; Inner Mongolia Higher Scientific Research Project, grant NJZY13103.

AUTHORS

Zhang Dan (1990–), female, postgraduate student at Electric Power College, Inner Mongolia University of Technology; main research interest on thermal power plant control and high voltage in power grid. E-mail: zhangdan1024@163.com.

Xi Dongmin (1974–), corresponding author, male, postgraduate tutor, the associate professor at Electric Power College, Inner Mongolia University of Technology, engaged in teaching and research on power system and thermal automatic control. E-mail: xdmin1501@sina.com.

REFERENCES

Liu Shijun. 2013. Control Strategy of Supercritical Unit[J]. *Energy Conservation & Environmental Protection* (11): 58–59.
Li Wengang. 2013. Design and simulation for main steam temperature control of 600 MW supercricital unit[D]. Master Thesis of Shanxi University.
Li Rong-mei. 2007. The research on the temperature control strategy of 600MW Supercritical Once-through Boiler[D]. Master Thesis of North China Electric Power University.
Liu Xiaohe & Guan Ping & Liu Lihua. 2011. *Adaptive Control Theory and Applications*[M]. Beijing: Science Press.
Qiao Zhijie & Wang Weiqing. 2008. Design of Self-adaptiveFuzzy PID Control System and its Computer Simulation[J]. *Automation and Instrumentation* 23(1): 26–29.
Meng Yu & Peng Xiaohua & Zhang Hao. 2006. Self-adaptive Adjusting of Fuzzy-PID Controller and Its Simulation Study[J]. *Mechanical Engineering and Automation* (6): 92–96.
Wang Ya-shun. 2008. The New Research on control of ratio of fuel to water for supercritical once-through boiler[D], Master Thesis of North China Electric Power University.
Bian Lixiu & Zhou Junxia & Zhao Jinsong & Yang Jianmeng. 2001. *Thermal Control System*[M]. Beijing: China Electric Power Press.
Lan Yan-ting & Chen Xiao-dong. 2012. Design of Fuzzy PID Auto-tuning Controller[J], *Mechanical Engineering and Automation* (3): 125–126.
Han Cheng-hao & Zhao Ding-xuan. 2012. Eletro-hydraulic servo-system design based on ruzz-adaptive PID control algorithm[J]. *Manufacturing Automation* 34(4): 11–13.
Zhang Fengli & Li Tai. 2011. Self-adaptive Fuzzy PID Control System and its Simulation[J]. *Journal of Luohe Vocational Technology College* 10(5): 34–35.

The development and utilization of resources

A study on the comparison of the several typical processes for dealing with vanadium titanium magnetite resources

Jie Qin, Gong Guo Liu, Zhan Jun Li & Jian Ling Qi
Pangang Group Research Institute Co. Ltd., State Key Laboratory of Vanadium and Titanium Resources Comprehensive Utilization, Panzhihua, Sichuan, China

ABSTRACT: Processes of dealing with vanadium titanium magnetite have been reviewed by three different enterprises, i.e. The New Zealand Steel, Highveld Steel & Vanadium Corp Ltd in South Africa and Panzhihua Iron and Steel Corporation in China. A process of rotary kiln direct reduction + EAF smelting DRI + recovering vanadium from vanadium bearing hot metal is adopted by the New Zealand Steel and Highveld Corporation. However, this process only recovers the Fe and V elements in the iron ore, whilst the Ti element in the EAF slag is difficult to extract because of the low grade TiO_2. A process of rotary hearth furnace (RHF) direct reduction + EAF deep reduction + vanadium recovery is employed by Panzhihua Iron and Steel Corporation, and there are no other materials except for DRI and reductants that are added into EAF, resulting in above 45% TiO_2 content in the slag, which is higher than that of the other corporations. Moreover, the Ti element in the slag could be extracted by the traditional sulfuric acid method. By comparison to the former process, the latter explores a new way to recover the Fe, V and Ti elements in the vanadium titanium magnetite. Furthermore, the direct reduction processes described in this article are constructive for comprehensive recovery of the vanadium titanium magnetite, as well as other similar composite ore resources.

1 INTRODUCTION

Vanadium titanium magnetite is a kind of symbiotic complex ore which contains Fe, V, Ti and other valuable elements. At present, the BF-BOF process is the main way to smelt the magnetite, and although the Fe and V elements could be recovered, most of Ti element in presence of the slag is hard to extract under present technical conditions.

In order to make full use of the resources, some different smelting processes have been applied to New Zealand Steel and Highveld Steel & Vanadium Corp. Ltd. That have achieved better recovery of vanadium and titanium elements, but poorer recovery of titanium elements.

In the 1970s to 1980s, the researchers conducted tests to seek new processes of smelting vanadium titanium magnetite in China, and failed to realize industrial production due to various reasons[1]. Based on the former research achievements and absorbing the relevant mature processes, Pangang proposed a new process by means of rotary hearth furnace (RHF)—electric arc furnace (EAF) to deal with the magnetite, and a pilot operation has been put in place in order to verify the feasibility of this technology. Up to date, the pilot operation has produced titanium slag with about 45% TiO_2 and vanadium slag with around 12% V_2O_5.

The processes used by New Zealand Steel, Highveld Steel & Vanadium Corp Ltd., as well as the pilot operation of Pangang are summarized in this article, comparing the characteristics of three processes.

2 NEW ZEALAND STEEL

New Zealand Steel was founded in 1965; it is a unique company that uses iron sands to produce steel products. Main components of the vanadium titanium magnetite produced by New Zealand Steel are shown in Table 1.

Table 1. Compositions of the vanadium titanium magnetite/mass%.

TFe	SiO_2	CaO	MgO	Al_2O_3	V_2O_5	TiO_2
57.43	3.5	1.0	3.0	4.2	0.51	8.0

2.1 Iron making process

Iron making process by New Zealand Steel is shown in Figure 1.

Along with coal and limestone, the primary concentrate (iron sand) is heated and dried in one of four multi-hearth furnaces at around 650°C, then

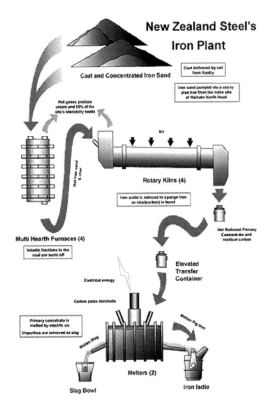

Figure 1. Iron making process of New Zealand Steel.

Table 2. Compositions of RPCC[2].

Compositions	TFe	MFe	FeO	TiO$_2$	V$_2$O$_5$
Mass/%	64.5	51.6	16.58	8.9	0.57

Compositions	SiO$_2$	CaO	MgO	Al$_2$O$_3$	η
Mass/%	5.3	2.6	3.4	5.35	80

fed into the four reduction kilns, where it is converted to 80% metallic iron. Gas temperature in the kilns is about 1100°C and discharge temperature is 900~1000°C. Rotary kilns' products are called RPCC; the components of RPCC are shown in Table 2.

Upon departure from the kilns, RPCC is introduced to two rectangle electric melters in order to achieve the final reduction of iron oxides and to produce the liquid iron. The melter has two tapping holes and two slag notches. Every four hours, the hot metal is tapped at the temperature of around 1500°C.

In the smelting process, the oxidized scale or iron concentrate should be added into the molten steel to restrain the over reduction of SiO$_2$ and TiO$_2$ in the slag, by managing to control the Si and Ti content in hot metal. Usually the FeO content in the slag is controlled 3% or so. The compositions of slag and hot metal are shown in Tables 3 and 4, respectively.

Table 3. Compositions of slag[2]/mass%.

TiO$_2$	SiO$_2$	Al$_2$O$_3$	CaO	V$_2$O$_5$	MgO	FeO
33	19.48	16.5	9.8	0.2	14	3

Table 4. Compositions of hot metal[3]/mass%.

C	Si	Mn	S	P	V	Ti	Fe
3.40	0.11	0.33	0.06	0.06	0.42	0.16	95.50

Table 5. Compositions of vanadium slag/mass%.

TFe	V$_2$O$_5$	TiO$_2$
34	15.5	13.3

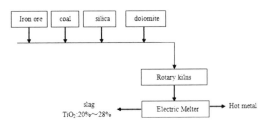

Figure 2. Iron making process of Highveld Steel.

2.2 Vanadium recovery from hot metal

The vanadium bearing hot metal in ladle is transported to vanadium recovery station where the temperature is around 1380~1420°C. A stainless steel straight pipe lined with refractory is used to inject oxygen into the hot metal, Oxygen blowing time for 8~12 min, and at the same time an immersed lance is used to spray N2 onto the hot metal. At the blowing end, the vanadium slag formed on the surface of liquid metal will be removed by skimmer. The whole cycle lasts for 26~45 mins and the vanadium oxidation ratio reaches 60~70%; the recovery is around 40~50%. The compositions of vanadium slag are shown in Table 5.

3 HIGHVELD STEEL & VANADIUM CORP. LTD

3.1 Iron making process

Highveld Steel & Vanadium Corp. Ltd. is the largest vanadium producer and exporter in South Africa. The compositions of iron ore used by the corporation are shown in Table 6, and the iron making process can been seen in Figure 2.

Table 6. Compositions of iron ore/mass%.

TFe	V$_2$O$_5$	TiO$_2$	SiO$_2$	Al$_2$O$_3$	Cr$_2$O$_3$
53~57	1.4~1.9	12~15	1.0~1.8	2.5~3.5	0.15~0.6

Table 7. Compositions of vanadium titanium magnetite used by the pilot/mass%.

TFe	SiO$_2$	CaO	MgO	Al$_2$O$_3$	V$_2$O$_5$	TiO$_2$
53.5	3.22	0.93	2.77	3.87	0.60	12.68

Table 8. Compositions of vanadium slag/mass%.

TFe	V$_2$O$_5$	MgO	SiO$_2$
24~35	~12	12~20	10~20

The basic raw materials for iron making, i.e. iron ore, coal and fluxes, are sized to meet process requirements before being delivered to the iron plants, then blended in the rational proportions and fed to each of the pre-reduction kilns that are fired with pulverized coal, and heated to about 1100°C[4].

The hot pre-reduced charge with MFe content around 60% leaving the kilns and then is hoisted in refractory – lined containers to the charging floor, which is above the closed-top electric smelting furnaces. The materials are melted using 7 furnaces to separate the slag and hot metal, whose smelting cycle is about 3.5~4.0h; the resulting vanadium content in the molten iron is 1.22%, and the resulting TiO$_2$ content in the slag is 32%.

3.2 Recovery of vanadium by using the shaking ladle process

Shaking ladle process is a method that works through injecting oxygen into vanadium bearing hot metal, in order to oxidize vanadium element and form the vanadium slag. Shaking ladle provides good conditions for mixing and facilitating oxygen injecting into the metal under low pressure, leading to achieving better vanadium oxidation rate.

The hot metal is fed into shaking ladle, and some pig iron and scrap are added into ladle as coolants. Oxygen is injected into hot metal accompanied with shaking of the ladle. Oxygen supplying pressure is 0.15~0.25 MPa and injecting time is 52 mins. After cutting off oxygen, the ladle should keep shaking for 7~8 mins, whilst some amount of coal is added into the metal to reduce the content of FeO in the slag[5]. At the end of ladle shaking, the vanadium slag is recovered and semi-steel is transported to a converter. The V$_2$O$_5$ content in vanadium slag reaches more than 25%.

4 UTILIZATION OF VANADIUM TITANIUM MAGNETITE IN THE PILOT PLANT CONSTRUCTED BY PANGANG

4.1 Iron making process

According to characters of resources in Pan-xi area, Pangang devotes a lot of time to study the new technologies on direct reduction of vanadium titanium magnetite. On the bases of carrying out a series of technology research and industrial experiments, Pangang invested 290 million yuan in constructing a pilot plant in 2009, in order to deal with vanadium titanium magnetite. The products of this pilot plant include 52000 tons of pig iron with low carbon content, 5900 tons of vanadium slag and 24000 tons of titanium slag. The process of the pilot plant is showed in Figure 3 and the compositions of iron ore concentrate are seen in Table 7[6].

Vanadium titanium magnetite concentrate, coal powder and binder are mixed with together according to certain proportion, and then the mixture is fed into the briquetter to produce pellets. The pellets are charged into RHF after drying and stay in the furnace for 15~40 min, giving rise to metallized pellet and the metallization rate beyond 85%. Then the pellets are conducted deep reduction by means of electric arc furnace where the pellets smelt and the titanium slag are separated from vanadium bearing hot metal. After desulfurization and vanadium recovery, the vanadium bearing hot metal finally is produced the ingot.

In order to improve the grade of titanium slag, non-flux is added into the electric arc furnace during the smelting process. Now that vanadium reduction rate reaches more than 80%, MFe content in the slag is usually below 2%, and the TiO$_2$ content in the slag is about 45%.

4.2 Ladle vanadium recovery

After desulfurization, the vanadium bearing hot metal is carried to vanadium recovery station. The method is simple: injecting oxygen from surface of hot metal and supplying nitrogen at the bottom of the ladle. At present, the V$_2$O$_5$ content in vanadium slag could reaches more than 12% and vanadium oxidation ratio is approximately 80%. The composition of vanadium slag is shown in Table 8.

4.3 Vanadium recovery and titanium recovery from the vanadium slag and titanium slag

For the vanadium slag, the qualified flake V2O5 products which meet GB3283-87 has been produced by selecting appropriate roasting schedule, as well as vanadium precipitation technology.

Using the titanium slag as the material, a sulfuric acid process is applied in order to produce titanium dioxide products, and the titanium recovery reaches more than 80%.

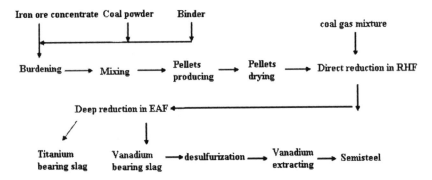

Figure 3. Process of the pilot in Pangang.

5 COMPARISON BETWEEN PANGANG, NEW ZEALAND STEEL AND HIGHVELD STEEL & VANADIUM CORP LTD

5.1 *The main direct reduction facilities are different*

New Zealand Steel and Highveld are all applying the rotary kiln as the direct reduction facilities, whilst Pangang selects RHF in order to reduce the magnetite.

Researches show that the magnetite has special characters compared to normal iron ore, such as complex mineralogical structure, which during the direct reduction process results in higher smelting temperature, longer reaction time, as well as in serious expansion and pulverization of the pellets.

Since RHF has high reduction temperature features, short reaction time (10–30 min) and the pellets staying at the bottom of the hearth still, it means that RHF is more suitable than rotary kiln for dealing with the vanadium titanium magnetite.

5.2 *The products are different*

The TiO_2 content in the titanium slag produced by New Zealand Steel and Highveld is around 30%~35%, which is lower than that of Pangang. Because there is nothing added into EAF during the melting process except for reductants, the TiO_2 content in slag reaches above 45% and it can be used as the material for the TiO_2 products through sulfate process.

6 CONCLUSIONS

A process of rotary kilns + EAF + vanadium recovery is chosen by New Zealand Steel and Highveld, but the TiO_2 content in the slag is only 30%~35% which is difficult to utilize at present. For the vanadium recovery process, the oxidation ratio is 60%~70% and the recovery ratio is 40%~50%.

A process of RHF + EAF + Desulphurization + vanadium recovery is selected by Pangang; the TiO_2 content in the slag is beyond 45%, and the resulting titanium slag can be used as the material to produce TiO_2 white meeting PTA121 using sulfate process. For the vanadium recovery process, the oxidation ratio reaches above 70%, and the resulting V_2O_5 content is more than 12%.

REFERENCES

Liu Gongguo. Study on utilization technology of vanadium titanium magnetite based on the rotary hearth furnace direct reduction process [J], Research on Iron and Steel, 2012, 40(2): 4–7.

Bian Derang. Smelting vanadium titanium magnetite by EAF in New Zealand [J].Vanadium Titanium, 1993, (5): 19–21.

New Zealand Steel. The Science of Steel [EB/OL]. http://www.nzsteel.co.nz/about-new-zealand-steel/student-information-/the-science-of-steel#, 2012-08-16.

Highveld Steel. Iron Making [EB/OL]. http://www.evrazhighveld.co.za/ironmaking.asp,2012-08-18.

Jinzhou iron alloy factory. Ferrovanadium and vanadium oxide [J], Ferro-alloys,1983, (3): 50–55.

Qinjie. Study on Recovery of High Temperature Waste Gas from Rotary Hearth Furnace [J], Mining and Metallurgy, 2011, 20(4): 86–88.

Energy and Environmental Engineering – Wu (Ed.)
© 2015 Taylor & Francis Group, London, ISBN 978-1-138-02665-0

Internal architecture analysis and modeling of point-bar—A case study of PI2 formation, X area

Jia Yi Wu, Shang Ming Shi, Xin Feng, Xiao Xiong Wu & Xian Li Du
Geoearth Institute, Northeast Petroleum University, Daqing, China

Yi Bao
Wireline Logging Company of Daqing Drilling and Exploration Engineering Corporation, Daqing, China

ABSTRACT: With the enhanced exploration and development of the oilfield, the internal architecture of the reservoir becomes one of the main factors controlling the distribution of the remaining oil. As the rich sand region of the composite channel sand body in the meandering river, the point-bar still controls a large amount of remaining oil by the internal lateral accretion shale beddings. According to the internal architecture analysis, this paper models the internal lateral accretion shale bedding within the point-bar and its physical property parameters, so as to provide a reliable geology model for the study of numerical reservoir simulation and remaining oil distribution.

1 INTRODUCTION

The internal architecture of the reservoir is one of the main factors controlling the distribution of the remaining oil in later-period oilfield development. In recent years, with the enhanced exploration and development of oilfields, the study of the internal architecture in composite channel sand body in the meandering river has made a great progress. As the rich sand region of the composite channel sand body in the meandering river, point-bar still controls a large amount of remaining oil by the internal lateral accretion shale beddings. Recognising the extent of the internal architecture of point-bar determines the overall exploitation effect of the meandering river deposits, which plays an important role in recognising and exploiting the remaining oil controlled by lateral accretion shale beddings. To keep the high and stable oilfield production, we need to enhance the internal architecture of the point-bar.

The main content of analysing the internal architecture of point-bar includes: the lateral acceleration scale, the shale intercalation scale and the dip of lateral accretion bedding, and so on. The internal architecture analysis is, in essence, based on the lateral accretion and shale intercalation recognition of single well, to predict the scale of lateral accretion and construct the internal architecture model of point-bar by the method of the internal architecture pattern of point-bar.

2 SINGLE WELL INTERPRETATION OF POINT-BAR

The well log response model of lateral accretion is developed through core calibration and well logging. Figure 1 is the core profile of well B*-3**-JP25, which shows 3 lateral accretion beddings and 2 shale intercalation barriers in the single channel deposit of PI2 formation. Compared with the meandering river deposit, the lateral accretion bedding is thinner ranging from 0.1 m to 0.3 m, which shows obvious return but low separations of micro-electrode log curves and slight return of natural gamma log curve and spontaneous potential log curve (some intercalation barriers show poor return of natural gamma log curve).

3 INTERNAL ARCHITECTURE ANALYSIS OF POINT-BAR

Lateral accretion bedding is the basic sedimentary construction unit of point-bar sediment. The lateral accretion beddings in a point-bar have different scales, which depend on the strength of hydro dynamism that the accretion sedimented. Shale intercalation barriers deposited between lateral accretion beddings, the contact surface between lateral accretion and shale intercalation is called lateral accretion surface. In order to make reasonable analysis, this paper makes study on the dip and dip angle of shale intercalation and

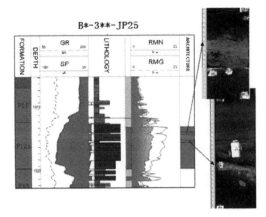

Figure 1. Lateral accretion beddings and shale intercalation barriers of point-bar (well B*-3**-JP25).

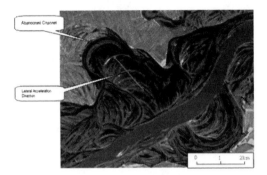

Figure 2. Composite pattern of point-bar and abandoned channel in the meandering river (satellite photograph of Songhuajiang).

the interspacing and width of lateral accretion, so as to construct the quantitative distribution mode of the point-bar's internal architecture.

3.1 Dip of shale intercalation

The shale intercalation is always inclined to the direction of abandoned channel. Based on modern depositional model, the direction of lateral acceleration in point-bar points to the concave bank of abandoned channel (Figure 2, Figure 3).

3.2 Dip angle of shale intercalation

Due to the inducement of periodic flood, fine-grained suspended matters deposited in point-bar as shale intercalation to a certain point of distribution between two floodplain sedimentary events. According to the sectional deposit distribution of shale intercalation dip angle, this paper divides shale intercalation dip angle into two modes: gentle-steep mode and gentle-steep-gentle mode (Figure 4).

Predecessors make the conclusion that there is a good exponential relationship between width to depth

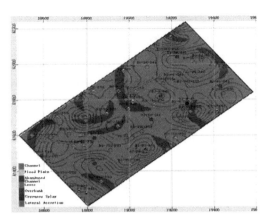

Figure 3. Distribution profile of shale intercalation dip in X area.

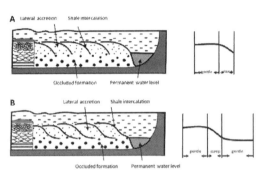

Figure 4. Two modes of shale intercalation dip angle (A. gentle-steep mode, B. gentle-steep-gentle mode).

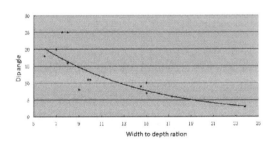

Figure 5. Width to depth ration and dip angle of shale intercalation relationship diagram.

ration of river and dip angle of shale intercalation (Figure 5), based on the materials of the meandering river outcrops and modern sedimentary characteristics. As shown in formula 1, the correlation coefficient can reach up to 0.9.

$$y = 32.966\exp(-0.0966x) \quad (1)$$

In the formula, y – dip angle of shale intercalation, x – width to depth of river.

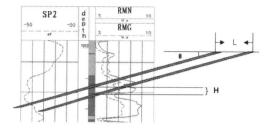

Figure 6. Interspacing identification cartogram of shale intercalation.

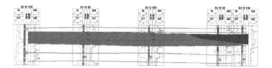

Figure 7. Interwell distribution profile of shale intercalation barriers in PI2 formation.

3.3 Interspacing of shale intercalation

The horizontal interspacing of shale intercalation can be calculated using the thickness of lateral accretion in single well. Two shale intercalation barriers extend from their dip angles and intersect the upper surface of point-bar. The distance between two intersection points on plane ΔL is the horizontal interspacing of shale intercalation (Figure 6). By using this method, this paper concludes that the interspacing of shale intercalation in the study zone ranges from 20 m to 30 m, and in some places may reach up to 40 m.

3.4 Horizontal width of single lateral accretion

The horizontal width of single lateral accretion is about two thirds of river bank full width. River bank full width can be calculated using the Leeder formula (1973),

$$\text{Log}(w) = 1.54\text{Log}(h) + 0.83 \qquad (2)$$

In the formula, w – river bank full width, m, h – river bank full depth.

According to the calculation, the average river bank full width of PI2 formation is about 130 m, so it can be estimated that the horizontal width of single lateral accretion is about 87 m.

Based on the interpretation of shale intercalation barriers in single well and guided by the quantitative distribution mode of point-bar internal architecture, this paper make a prediction of the interwell distribution of shale intercalation barriers (Figure 7), so as to provide scientific basis of an integral exploitation adjustment scheme and remaining oil development in study zone.

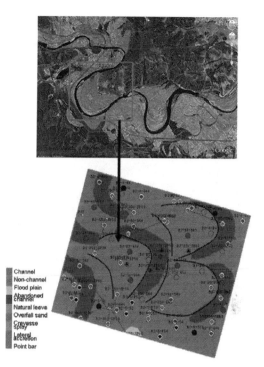

Figure 8. Plane mode of lateral point-bar in the meandering river.

4 LATERAL ACCRETION MODELLING

In recent years, the techniques of lateral accretion modelling have been becoming a hot topic. The internal shale intercalation barriers of point-bar are modelled by using the Direct software, as follows:

(1) Based on the two dimensional plane microfacies distribution map and the point-bar lateral accretion development mode, the start and end boundary lines of lateral accretion surface projected in 2D space are mapped, which represent the length, width and gradient development mode of shale intercalation barriers (Figure 8).

(2) According to the plane regional scale of point-bar, three dimensional coordinates of lateral accretion surface points are taken from well points as the original data of lateral accretion in three dimensional space modelling, by regarding single point-bar as the study unit (Figure 9).

(3) The three dimensional model of lateral accretion surface is established by using shale intercalation development information including the interspacing, dip and dip angle as simulation parameters and combining them with the well point materials and 3D stratum model (grid coordinates) (Figure 10).

(4) According to the multiwall architecture profiles and the 3D comprehensive analysis, the lateral accretion surface model can be verified and perfect.

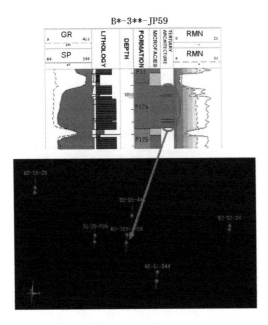

Figure 9. Three dimensional view of lateral accretion superface points.

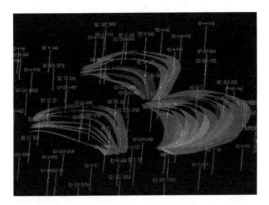

Figure 10. Three dimensional view of lateral accretion surface.

(5) On the basis of architecture identification and modelling, the intercalation thickness distribution map can be established by gridding the intercalation thickness of single well points. By adding the intercalation thickness distribution map and 3D architecture interface together, the spatial distribution of shale intercalation can be set up, so as to build the architecture mode by embedding the shale intercalation mode into microfacies mode (Figure 11).

By using the same method, this paper establishes the architecture mode of other areas and reproduces the spatial distribution of lateral accretion and shale intercalation of point-bar (Figure 12 – Figure 13). In conclusion, lateral acceleration points to the concave

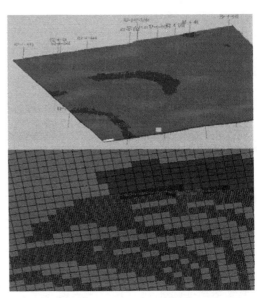

Figure 11. Architecture mode of inspection well area.

Figure 12. 3D fence diagram of point-bar internal architecture in PI2 formation.

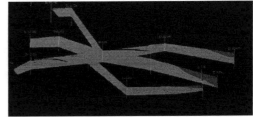

Figure 13. 3D fence diagram of point-bar internal architecture in PI2 formation.

bank of abandoned channel and extends to the two thirds of river thickness.

5 3D RESERVOIR PARAMETER MODELLING

Within the constraint of architecture model, this paper establishes the porosity, permeability and oil saturation models (Figure 14 – Figure 16) with the method of sequential Gaussian simulation. The result indicates that the parameter models are controlled, obviously, by

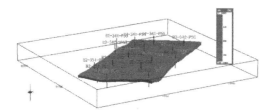

Figure 14. Three dimensional display of porosity model.

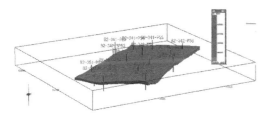

Figure 15. Three dimensional display of permeability model.

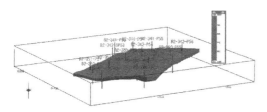

Figure 16. Three dimensional display of oil saturation model.

the architecture model. The physical properties and oil-bearing strata of lateral accretion beddings are better than those of subsea sand bodies.

6 CONCLUSIONS

According to the internal architecture analysis, this paper models the internal lateral accretion shale bedding within point-bar and its physical property parameters, so as to provide reliable geology model for the study of numerical reservoir simulation and remaining oil distribution.

REFERENCES

Yue Dali, Wu Shenghe, Wen Lifeng, et al. (2013). Physical simulation of remaining oil distribution controlled by reservoir architecture within point-bar of meandering river. China Sciencepaper Online 8(5):473–475.

Lu Ying, Zhao Yingying. (2010). The tracking and prediction method of point-bar lateral body. Oil-gasfield surface engineering 29(5):103–104.

Zhai Zhiwei. (2011). Study on the construction structure of point bar sandbody and its control action on the residual oil distribution: a case study from the Northeast layer-recombination pilot area No.2 of Daqing oilfield. Northeast Petroleum University 25–33.

Miall A. D. (1985). Architectural elements analysis: a new method of facies analysis applied to fluvial deposits. Earth Sci Rev 22(2):261–308.

Miall A. D. (1996). The geology of fluvial deposits: sedimentary facies, basin analysis and petroleum geology. New York: Springer-Verlag 20–25.

Yue Dali. (2006). The study on architecture analysis and remaining oil distribution patterns of meandering river reservoir: a case study of Guantao formation, Gudao oilfield. China University of Petroleum 40–55.

Wu Shenghe. (2010). Reservoir characterization and modeling. Bejing: Petroleum Industry Press 50–65.

Bai Zhenqiang, Wang Qinghua, Du Qinglong, et al. (2008). Numerical reservoir simulation and remaining oil distribution patterns based on 3D reservoir architecture model. China Univ Petrol: Nat Sci 32(2):68–74.

Influence of mined-out area on blast vibration effect in mine's underground stope

Guo Sheng Zhong
Department of Architecture and Civil Engineering, Huizhou University, Huizhou, China

Li Ping Ao
Huizhou University Library, Huizhou, China

ABSTRACT: Middle-deep hole blast vibration effect was monitored under a mined-out area by the speed sensor in a metal mine. The effect of the maximal section dose and mined-out area on Peak Particle Velocity (*PPV*) was analysed. Research results show that: it is favourable for the stability of the mine stope and underground structures by controlling the maximal section dose; the existence of mined-out area, the blast seismic wave was diffracted and the energy was dissipated, and it was a certain attenuation for *PPV* far away from the blast site; in the case of the same, the *PPV* value behind the main blast free surface was significantly larger than the corresponding laterally the blast free surface.

Keywords: Mined-out area, blast vibration effect, peak particle velocity, middle-deep hole blast

1 INTRODUCTION

Due to the large-scale disorder group mining activities, the main ore body in a metal mine left hundreds of mined-out areas, leaving a deadly danger to safety production of the mine. Under the blast hole dynamite explosion loading, the blast hole medium crushed failure, blast vibration effect for distance far medium. These factors, such as charge, charging structure, stemming and initiation, geological, mechanical conditions of ore and rock, and the terrain conditions of blast site, have effect on blast vibration effect. The parameters of blast vibration effect, such as the peak particle velocity (*PPV*) or the equivalent acceleration, the maximal section dose and the distance of blast source to measuring point, has been paid great attention to by researchers at home and abroad [1]. In recent years, some researchers did a meaningful exploration for the topography of blast site influence on blast vibration effect [2–5]. In view of the mine blast conditions, few people have studied the existing complex goaf influence on blast vibration effect. Combined with the mining engineering practice, the existing complex goaf influence on blast vibration effect based on a large data of middle-deep hole blast vibration monitoring has been explored in order to prevent the surrounding rock collapse caused by large area and stope structural instability mining disasters caused by the blast vibration effect.

2 THE CAVING OF MIDDLE-DEEP HOLE BLAST IN THE MINE'S STOPE

2.1 *The geological structure and the mechanical parameters of mine*

The mine's stope in 1202 m of 1$^{\#}$ ore body is composed of 34 lines residual column and roof pillar. The main minerals are sphalerite and galena, the ore is body thick, the inclination angle is 40 to 70 degree, good stability.

In the stope area, the average thickness is 35 m, the 60 to 65 degree, ore bulk density 2.9 t/m^3, $f = 6\sim 8$, with an average grade of Pb + Zn 10.37%. The upper and under plate is crystalline limestone, $f = 5\sim 8$, mineral rock is medium firm [6].

The geological structure of the mine's stope is shown in Figure 1, with the six deep-hole drilling roadway, perpendicular to ore body. The average width between columns is 10 m, the sides of the column for mined-out area, and the upper plate ore rock near the ore body collapsed to be mined-out area.

In the columns, four drilling roadway are arranged in 1213 m, 1224 m, 1237 m and 1250 m. The average width of the pillar is about 30 m, under which the mined-out area, and two drilling roadways are arranged in 1258 m.

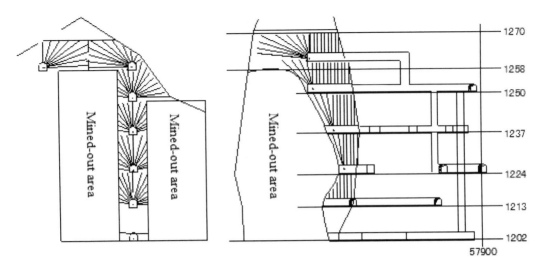

Figure 1. Layout of stope drilling roadway section and fan-shaped blast hole (unit: m).

Table 1. Detonator section, using region and initiating charge in the same section.

Detonator section	Drilling roadway	Initiating charge/Kg	Detonator section	Drilling roadway	Initiating charge/Kg
section 2	1237 1250	1876.4	Section 8	1237 1250	1331.6
section 3	1213 1224	1042.3	Section 10	1237 1250	1488.3
Section 4	1237 1250	1167.0	Section 11	1258; 33[#] 1258; 34[#]	2282.7
Section 5	1213 1224	1027.0	Section 12	1258; 33[#] 1258; 34[#]	1813.0
Section 6	1237 1250	1292.7	Section 13	1258; 33[#] 1258; 34[#]	1133.2
Section 7	1213 1224	750.9	Section 14	1258; 33[#] 1258; 34[#]	481.1

2.2 The drilling and blast parameters of the mine

The fan-shaped middle-deep holes were drilled using the drilling machine in the drilling roadway, with the blast hole diameter of 60 mm. For the arrangement of fan-shaped blast holes in the vertical plane, the spacing in the adjacent blast hole is 1.2 m. The other fan-shaped blast holes were arranged with inclining plane in the drilling roadway ends, and the vertical distance from the bottom of the row plane blast holes to another adjacent row plane was generally controlled 1.8~2.5 m.

The 2[#] rock explosive was filled blast holes by the charge device, where the linear charge density of the blast hole was 2.83 kg/m. Two non-electric millisecond detonators, which were used to initiate the explosive, were arranged at the bottom of the each blast hole. For the blast holes in the vertical row plane, the orifice plug length was 1.5m. To remove the blockage, the remaining length of the blast holes was filled with explosives. For the blast holes in the inclining plane, the orifice plug length ranged from 1 to 1.5 m.

For the blast holes of the inclining row plane in the same drilling roadway, there was a charge unit for each 2 to 3 rows of the blast holes. In the unit, the first row blast hole was near the main blast free surfaces, and the remaining length of the blast holes was filled with explosives. Of all the blast holes of other row planes in the unit, only a half of them near the bottom was filled with explosives in the remaining length only.

There are several complex mined-out areas around the ore rock in the mine's stope. Using the existing mined-out area as the compensation space of the blast initial free surface and caving ore, reasonable arrangement of millisecond detonator sections, with millisecond and extruding blast technology in the same section, under an ignition blast all ore rock in the stope were forced caving. The detonator section, the using region and the initiating charge are listed in Table 1.

Blast caving ore rock total were 34357t, and total consumptions were: ①15686.2 kg explosives; ②1452 detonators; ③540 m of detonating cord; ④6759.7 m of blast hole. Explosive consumption was 0.48 kg/t.

3 MONITORING OF BLAST VIBRATION IN THE MINE'S STOPE

3.1 Monitoring equipment

Many blast vibration engineering practices show that: the value of peak particle velocity (*PPV*) caused by blast, is closely related to the degree of the structure damage and among vertical component of t peak particle velocity (*PPV*) in three directions is larger. Thus, the speed sensors were used in this test, with blast vibration self-recording instrument to monitor the peak particle velocity (*PPV*) of the vertical direction in the roadway floor, and the peak particle velocity (*PPV*) may be used to estimate the changes of blast vibration effect.

3.2 Monitoring site layout and monitoring result

On the basis of the assessment results of blast vibration effect range, ten monitoring sites were provided to monitor vibration effect of forced caving in the mine's stope. Monitoring sites 1 to 7 were arranged in the 1202 m level of in the roadway, and monitoring sites 1 and 2 were located on the right behind the main free surface of the fan-shaped blast hole. Monitoring sites 3 to 7 were in a side of the main free surface of the fan-shaped blast hole, and mined-out area located in these five monitoring sites and caving scope. The position relationship between these seven monitoring sites in the 1202 m level and the caving ore scope is shown in Figure 2. Monitoring sites 8 to 10 were arranged in 1250 m level the under plate of the roadway, and these three monitoring sites were right behind the main free surface of the fan-shaped blast hole.

In this field test, seven monitoring sites recorded the useful signal. Due to the reasons for the recorder itself or monitoring site being away from the explosion centre, there were three monitoring sites with no results, and these three recorders were not triggered. All monitoring results of peak particle velocity (*PPV*) are shown in Table 2.

4 INFLUENCE OF MINED-OUT AREA ON BLAST VIBRATION EFFECT IN MINE'S UNDERGROUND STOPE

Sodev Formula reveals the empirical relationship between the peak particle velocity (*PPV*) and the maximal section dose (Q), the distance of blast source to monitoring site (R), the site coefficient (k), the blast vibration attenuation coefficient (a). By many blast vibration monitoring in the mine's stope, it obtained that $k = 100$ and $a = 1.5$ under the mining conditions.

Let $k = 100$ and $a = 1.5$, and respectively put the maximal section dose ($Q_1 = 2282.7$ kg, $Q_2 = 1876.4$ kg and $Q_3 = 481.1$ kg) in this blast caving test into Sodev Formula. The change trend of the *PPV* with the R under the different Q values can be obtained; the *PPV* of 7 monitoring sites are shown in Figure 3.

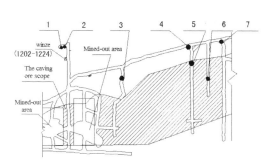

Figure 2. Position of 1202 m level monitoring sites and the caving ore scope.

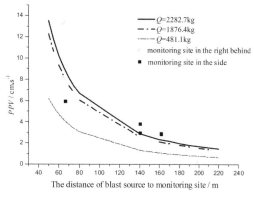

Figure 3. The calculated *PPV* value and the monitored *PPV* value.

Table 2. Monitoring results of blast vibration peak particle velocity (*PPV*).

Monitoring site	Distance to explosion centre /m	Voltage /v	Frequency /Hz	Sensor sensitivity / v·cm^{-1}·s	*PPV*/cm·s^{-1}
1	64.19	4.883	54.39	0.367	13.31
2	64.53	4.414	55.50	0.271	8.144
3	66.34	3.203	60.90	0.269	5.954
4	140.11	4.5699	31.23	0.600	3.808
5	140.11	3.5549	49.21	0.600	2.962
6	161.75	3.3199	27.23	0.577	2.877
7	197.75	0.2350	19.69	0.568	0.207

Considering the influencing factors on the *PPV*, such as the distance of blast source to monitoring site, the position relationship between the monitoring sites and the main free surface of fan-shaped blast hole caving ore, the occurred condition of the mined-out area between the monitoring site and the explosion centre, the following conclusions can be obtained as:

First, under the same section dose and the same R value, when the region is less than 165 m away from the explosion centre (that is $R < 165$), the *PPV* of the mine's underground stope is greater than the corresponding value of the open environment.

Second, under the same Q value (the maximal section dose) and the same R value (the distance of blast source to monitoring site), the *PPV* of the right behind main blast free surface in the fan-shaped blast hole caving ore was obviously greater than the corresponding value of one side of the main blast free surface.

Third, the occurred mined-out area between the monitoring site and the explosion centre, resulted in the diffraction of seismic waves and the partial dissipation of its energy which led to the different degree attenuation of the *PPV* in the monitoring site. For example, monitoring site 1 and 2 were both located behind the main blast free surface and roughly the same distance away from the explosion centre. Because of one chute in front of monitoring site 2, the *PPV* of monitoring site 2 was obviously less than the corresponding value of monitoring site 1, as shown in Figure 3.

5 CONCLUSIONS

(1) When blast caving in the underground mined-out areas, the dose of millisecond detonator initiation in the same section, the occurrence conditions of mined-out area in the ore rock and the space relationship of monitoring site and the main blast free surface, have influence on the peak particle velocity (*PPV*).

(2) In the region adjacent to the explosion centre, with the same dose blast and the same distance to the blast source, the *PPV* value of underground site is significantly greater than the corresponding value of the open pit mining. Properly controlling the maximal section dose in the underground blast caving would be advantageous to the stability of the mine's stope and underground structures.

(3) In the existing mined-out area between the monitoring site and the blast source, the seismic wave would produce the diffraction and partly waste the energy, and the *PPV* value of the monitoring site would be reduced by varying degrees.

(4) In the same conditions, due to the blast of the fan-shaped blast hole caving, the *PPV* value of the right behind the main blast free surface would be obviously greater than the corresponding value of the side of the free surface.

ACKNOWLEDGEMENTS

Financial support for this work was provided by the National Natural Science Foundation of China (No.51064009), Higher School Talent Introduction Project of Guangdong Province (No.$A_4$13.0210) and Professor & Doctoral Scientific Research Foundation Project of Huizhou University (No.$C_5$13.0202), is gratefully acknowledged.

REFERENCES

Huateng Chen, Qiang Niu, Shengyu Tan. A Handbook for Blast Calculation [M]. Shenyang: Liaoning Science and Technology Publishing House, 2011.

Meng Hu. Earthquake Effect of Large Diameter & Long Hole Blast in Anqing Copper Mine [J]. China Copper Engineering, 2009, 25(3): 10–12.

Huoran Sun, Haijun Xiao, Yunsen Wang. Study on the Stability of Footrill under Dynamic Blast Load [J]. Metal Mine, 2011, 32(12): 15–18.

Yucheng Shi, Lanmin Wang, Xuewen Lin. Effect of Blast Induced Ground vibration in Loess Sites [J]. Chinese Journal of Rock Mechanics and Engineering, 2013, 32(11): 1933–1938.

Singh P K. Blast Vibration Damage to Underground Coal Mines from Adjacent to Open-pit Blast [J]. Rock Mechanics & Mining Sciences, 2008, 45(3): 959–973.

Baiyin Nonferrous Metals Company & Central South University. Mining Technology of Low Dilution and Loss for Group Mining in the Complex Mined-out area of Changba Lead-zinc Mine [R]. Changsha: Central South University Publishing House, 2010.

Energy and Environmental Engineering – Wu (Ed.)
© 2015 Taylor & Francis Group, London, ISBN 978-1-138-02665-0

Overburden movement in a shallow coal seam with great panel width

Yang Li, Shan He Hao, Yuan Jin Guo, Wei Jun Hu, Bin Jiang & Xin Yu E
School of Resource and Safety Engineering, China University of Mining and Technology (Beijing), Beijing, China

ABSTRACT: Statistics were gained and analysed in order to obtain the law of overburden movement by using nine borehole extensometers and twenty-seven monitoring cats. The computer numerical model was established to simulate movement of overburden in shallow coal seam with great panel width, through which the movement features of overlying strata were observed in different face advanced distance. It is proved that fracture zone and curve subsidence zone can be precisely distinguished and the separating deformation between two zones can clearly be seen. The overburden movement continued for three weeks and the mining influencing range of overburden varied from 330 metres to 420 metres. The overburden movement between strata with different hardness appeared in the form of separation.

1 INTRODUCTION

A wide range of coal seam longwall mining can disturb the movement of overburden strata, or even destroy its aquifers, which can lead to serious permeability and aggravation of the ecological environment and its destruction. Thus it is absolutely essential to master the laws of overburden strata movement for the early warning of water inrush disasters and protection of groundwater resources (Peng 2006).

In China, economic development still heavily relies on coal, which is the main energy source. The coal-dominated energy structure would not be changed for the future 50 years. Last several years, the focus of China's energy development had shifted from east to west. Most of the western coal field is thick coal seam with simple structures, shallow buried depth (less than 200 m), thin bedrocks, and thick unconsolidated layers. The fissure caused by underground mining often lasts from the roofs to the surface, which would change the runoff conditions of surface waters and groundwater, and make them connect with the underground space. Pouring water and sand accidents had occurred repeatedly in some of the western coalfields, causing enormous economic losses and casualties. (Lir et al. 2012).

The same problems exist in America. The United States is rich in coal resources, with a large amount of production and exporting. America's proved reserves of coal resources were 442.9 billion tons, ranking in the second place in the world after China. At present, although the U.S. mining industry had passed its peak, it still remains the local pillar industry in some places that have coalfields. Also, longwall mining kept the most effective, productive mining methods, whose production accounted for 53% of the current underground production (Li 2012). Up to January 2011, 46 longwall panels existed in America. The mining depth ranged from 101 m to 640 m, mostly from 100 m to 400 m, with an average depth of 310 m. The average panel length increased from 140 m in 1976 to 313 m in 2011. Panel length gradually increased, and the longest panel length reached 457 m in recent years. The average length of mining area increased from 1128 m in 1976 to 3089 m in 2011, most of which were longer than 915 m, and the longest was 5791 m. In America, 17% of the longwall panels are buried within 200 m, and 13% of the panels are longer than 400 m. Thus it is necessary to study the overburden movement in shallow coal seam with great panel width.

2 ENGINEERING SITUATION

The area studied in this paper is located in the Appalachia Coalfield, in the east of the United States, where most of the coal seams are buried shallowly. However, the concept of shallow coal seams does not exist in America's underground longwall mining. The panel is located at the junction of Pennsylvania and West Virginia. With a shallow buried depth of 575 ft (175 m), the overburden strata are mostly bedrock, and the panel width can reach up to 1430 ft (435 m), which is common in the United States, but rarely seen in China. Thus this study will provide practical experience and theoretical basis for western China's super long panel in shallow coal seam mining.

The panels studied in this paper are: Panels B5 and B6, whose thickness of overburden varies from 600 ft (180 m) to 900 ft (270 m). The length of mining area in Panel B5 and B6 is 12000 ft (3600 m) and 5700 ft (1710 m), respectively. Panels B5 and B6, have the same panel width of 1433 ft (435 m), as well as the discharging sluice and lower sluice with the same width of

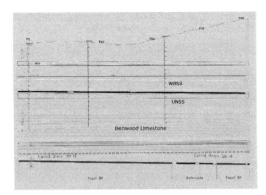

Figure 1. The cross-section of borehole extensometer installation.

16 ft (5 m). The width of the chain pillar between Panels B5 and B6 is 16 ft (5 m), the average withdrawing speed is 30–50 ft/day (9–15 m/day).

3 OVERBURDEN MOVEMENT OBSERVED BY BOREHOLE EXTENSOMETER

3.1 Borehole extensometer installation

In order to observe the overburden displacement perpendicular to the mining direction in different positions, nine extensometers were installed in three boreholes, each with a diameter of 6 (150 mm). Drill #1 (B1) was located in the centre of Panel B6 mining area. B2 was installed on 1/4 position of the panel width. The edge of Panel B6 mining area saw B3. Drilling locations and their structures are shown in Figures 2–8.

The nine extensometers were installed in three boreholes to detect the overburden movement associated with mining activity. Each hole contained three extensometers, which were equipped with six observation anchors, respectively.

3.2 Movement Analysis of Drilling extensometer in Panel B6

(1) Drilling Extensometer B1

As is shown in Figure 2, the anchors (1.1.1) to (1.1.6) were six lower observation anchors, which started to move when they were 600 ft (180 m) away from the panel. The borehole suffered a sudden drop, as the panel has just passed through it. In particular, anchors (1.1.6) and (1.1.4) dropped by 0.2 in (5 mm), 1.4 in (35 mm), respectively. As the panel was 100 ft (30 m) or less away from the borehole, the movement accelerated at the speed of 1.58×10^{-3} ft per 1ft panel advance. As the panel was 700 ft (210 m) or more away from the borehole, the displacement of observation anchors was stable, varying from 6.5 in (165 mm) to 12.5 in (318 mm). The largest displacement of the observation anchor was 8–13 in (203–330 mm), which occurred after the panel has passed through 100 ft (30 m). The distance of sudden drop was 4–6 in (100–150 mm).

For the middle extensometers, when the borehole was 700 ft (210 m) or less away from panel, six observation anchors moved at the speed of 1.58×10^{-3} ft per 1 ft panel advance. As the panel was just below the borehole, the speed increased to 1.14×10^{-3} ft/ft. Anchor (1.2.6) suffered a sudden 1 in drop, whilst the others suffered 1.4 in drop. As the panel was 200 ft (60 m) or less away from the borehole, observation anchor (1.2.6) moved at the speed of 1.21×10^{-3} ft/ft., whereas the others moved at the speed of 1.54×10^{-3} ft/ft. As the panel was 550 ft (168 m) or more passing through the borehole, the displacement of observation anchors was stable, varying from 6.5 in to 8.5 in. The largest displacement of observation anchor was 7.5–8.5 in, which occurred after the panel has passed through 175 ft (54 m).

The four anchors (1.3.1–1.3.4) started to move when they were 700 ft (210 m) away from the panel. Meanwhile, anchors (1.3.5) and (1.3.6) began to move when the panel was 150 ft (45 m) and 10 ft (3 m) away from the borehole, respectively. The initial speed was very insignificant, which was 8.3×10^{-6} ft/ft. However, it increased to 8.0×10^{-5} ft/ft. when the panel was 300 ft (90 m) away from the borehole. Meanwhile, all anchors, except anchor (1.3.6) dropped 0–0.98 in suddenly when the panel was right beneath them. As the panel was at the distance of 200 ft (60 m) when passing through them, the speed increased to 9.72×10^{-4} ft/ft. As the panel was 400 ft (120 m) away when passing through the borehole, the displacement of observation anchors was varying from 0.1 in (2.5 mm) to 5.6 in (142 mm). The largest displacement of an observation anchor was 0.2–6 in (5–152 mm), which occurred after the panel has passed through 200 ft (60 m).

The displacement of all of the eighteen anchors had rebounded when the panel was passing through 100–150 ft (30–45 m). Deeper extensometers rebound larger, and shallow extensometers rebound smaller.

It seemed that the three lower anchors installed in sandstone group moved as a whole. They were separated from the anchors (1.1.4–1.1.6) which have been installed in shale and limestone. Anchors (1.1.4–1.1.6) acted as a whole too. As the depth increased, the three upper observation anchors (1.3.4–1.3.6) have separated. Thus, there were three large separations. The first one was located between anchor (1.1.3) and anchor (1.1.4), the second was seen between anchor (1.2.6) and anchor (1.3.1), and the third between anchor (1.3.3) and anchor (1.3.4). The strata 70 ft (21 m) below the surface were disturbed at the same time.

(2) Drilling Extensometer B2

As shown in Figure 3, and mentioned earlier that the statistics obtained by anchors (2.1.1–2.1.6) installed in the lower part are not reliable. The terminal mining of Panel B6 indicates that the displacement or anchors reached, or even exceeded 20 in, the maximum allowable deformation for observing anchor.

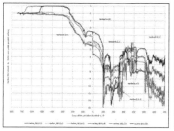 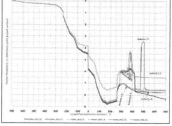 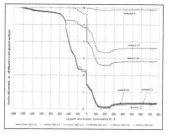

Figure 2. Measured displacement for borehole 1, 3 extensometer.

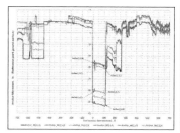 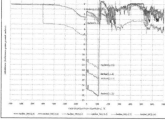 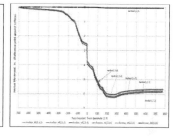

Figure 3. Measured displacement for borehole 2, 3 extensometer.

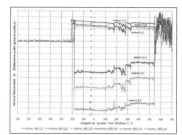 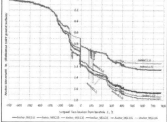 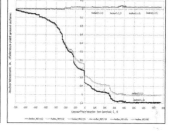

Figure 4. Measured displacement for borehole 3, 3 extensometer.

The data obtained by anchors (2.2.1–2.2.6) in the middle part are unreliable, but it is possibly exceptional in terms of anchors (2.2.1) and (2.2.3). When they were 700 ft away from the panel, anchors (2.2.1) and (2.2.3) began to move. When the distance between the anchors and panel has varied from 0–300 ft (90 m), the displacement speed of the observation anchor increased to 6.39×10^{-4} ft/ft. Furthermore, the speed increased to 1.14×10^{-3} ft/ft. when the panel was passing through the anchors. Particularly, the two anchors dropped by 13–14.2 in suddenly when the panel was just below them. As the panel passed through them by a distance of 110 ft (33 m), the maximum displacement for anchors (2.2.3) and (2.2.1) was 14.5 in (368 mm) and 18 in (457 mm), respectively.

Among the upper anchors (2.3.1–2.3.6), anchor (2.3.6), which was the closest to the surface, kept still before the panel arrived. Its movement wasn't observed until the panel passed through it. The other five anchors acted as a whole, they started to move at the speed of 6.25×10^{-5} ft/ft. as the panel was 700 ft (210 m) away from them. When the panel was 300 ft (90 m) away from the borehole, the speed increased to 6.67×10^{-4} ft/ft. Furthermore, as the panel passed through the boreholes, the speed increased to 9.58×10^{-4} ft/ft. When the panel was right beneath them, the anchors suffered a sudden 1.5 in (38 mm) drop. The displacement of the anchors began to stabilise when the panel was 600 ft (180 m) away when passing through the anchors. The largest displacement was 6.3 in (160 mm).

There was a separation between the observation anchor (2.2.6) and anchor (2.3.1), probably located between white sandstone and dark limestone. The separation was also spotted by anchor (2.3.5) and anchor (2.3.6), which may have been located between gray shale and moderate gray shale which. The strata 24 ft below the topsoil were not disturbed.

(3) Drilling Extensometer B3

As shown in Figure 4, the data obtained by the six anchors (3.1.1–3.1.6) are not reliable, because they were not affected by mining activity as the panel was passing through.

The six middle anchors started to move at the speed of 4.2×10^{-5} ft/ft. when the panel was 700 ft (210 m) away from the anchors. When the panel was 300 ft

(90 m) away from the borehole, the speed has increased to 1.27×10^{-4}–2.25×10^{-4} ft/ft. When the panel was right beneath them, the displacement of anchors varied from 0.82–1.18 in, and the sudden drop varied from 0.7–1.2 in (18–30 m). The largest displacement of the anchors varied from 1.17–1.75 in (30–45 mm) when the panel was 700 ft (210 m) away passing through the anchors. There was a separation of 0.5 in (12.5 mm) between the upper two anchors (3.2.5 and 3.2.6) and the lower two anchors (3.2.1 and 3.2.4).

Anchor 3.3.6, the closest to the surface among the upper six anchors (3.3.1–3.3.6), kept still until the panel was 200 ft (60 m) away. When the panel was 700 ft (210 m) away, the other five anchors moved as a whole at the speed of 4.2×10^{-5} ft/ft., which increased to 1.39×10^{-4} ft/ft. when the panel was 300 ft (90 m) away. The speed decreased to 0.4×10^{-4} ft/ft. when the panel passed through. When the panel was right beneath them, the anchors suffered a sudden 1.4 in (36 mm) drop. The displacement of the anchors began to stabilise when the panel was 400 ft (120 m) away when passing through the anchors. The largest displacement was 1.1 in (28 mm).

It seemed that the separation between white sandstone and brown sandstone was observed. The strata starting from the bottom of anchor (3.2.1) (white sandstone) to the central anchor (3.3.4) (gray shale) seemed to be a whole rock, and there was no separation between them. The strata 50 ft (15 m) below the topsoil were not disturbed.

4 CONCLUSIONS

(1) As the panel was beneath the observation anchor, the movement speed of the lower part of the anchor has accelerated. The fracture zone and the curve subsidence zone can be distinguished by the fourteenth layer (UN sandstone group). The height of the caving zone and the fracture zone is 214 ft (65 m), and the height of curve subsidence zone and topsoil is 300 ft (91 m) and 35 ft (10 mm,) respectively.
(2) As the panel was 700 ft (210 m) away from the observation anchor, the strata began to move. The movement speed accelerated when the distance between the panel and the observation anchor was 300 ft (90 m) or less. As the panel was beneath the observation anchor, the anchor suffered a sudden drop, which indicates that the strata would move rapidly while mining. The overburden movement stopped when the panel was 400–700 ft (120–210 m) away.
(3) There were eighteen observation anchors installed above the seam at the range of 206–228 ft (63–69 m). Beyond the range, the overburden movement between strata with different hardness appears in the form of separation, such as the separation between sandstone and shale, and limestone and shale.
(4) The overburden movement continued for three weeks.

ACKNOWLEDGEMENT

Supported by the Fundamental Research Funds for the Central Universities (2013QZ03); National Natural Science Foundation of China (NO.U1361209); National Undergraduate Innovation Program.

REFERENCES

Peng S.S. (2006). Longwall Mining. 2nd ed. Morgantown, WV: Peng S.S. publisher, pp. 46–55.
Peng, S.S. (2008). Coal Mine Ground Control. 3rd ed. Morgantown, WV: Peng S.S. publisher, pp. 319–330.
Li Y. (2011). "Overburden movement in solid waste rock cemented backfill mining methods." MeitanXuebao/Journal of the China Coal Society. 36(S1): 370–374.
Li Y., Qiu B. (2012). "Investigation into key strata movement impact to overburden movement in cemented backfill mining method Source." Proceedings of International Conference on Advances in Computational Modeling and Simulation. V. 31. Amsterdam, Netherlands: Elsevier Ltd, pp. 727–733.
Wang J.C. (2009). The Theory and Technique on the Thick Coal Seam Mining. Beijing, China: Metallurgical Industry Press, pp. 147–155.

Energy and Environmental Engineering – Wu (Ed.)
© 2015 Taylor & Francis Group, London, ISBN 978-1-138-02665-0

Optimisation of flocculation conditions for aqueous extracts of Apocynum Venetum leaves using chitosan by response surface methodology

Hua Sheng Yang, Wen Jun Gao, Yong Ming Luo & Lu Wu
Jiangxi University of Traditional Chinese Medicine, Nanchang, Jiangxi, China

ABSTRACT: A screening experiment with Fractional Factorial Design (FFD) and Response Surface Methodology (RSM) with Box-Behnken Design (BBD) was used for the flocculation technology of aqueous extracts of *Apocynum venetum* leaves by chitosan. The optimum operating conditions were finally obtained using a desirability function. A 2^{7-3} fractional factorial design was initially employed and it was found that chitosan dosage, flocculation temperature and the time of slow mixing were the most important variables that affected the removal of extracta sicca. RSM was used to further investigate the optimal conditions for these factors on flocculation. Analysis of variance and other relevant tests confirmed the validity of the suggested model. The optimum chitosan dosage, flocculation temperature, and the time of slow mixing level was 10.91 mg/g, 42.96°C and 12 min, respectively. Under these conditions, the removal of extracta sicca was 49.28%, which agreed with the predicted value of 49.11%.

Keywords: *Apocynum venetum*, flocculation, chitosan, response surface methodology, traditional Chinese medicine

1 INTRODUCTION

Apocynum venetum, called Luobuma in Chinese, is a wild shrub widely distributed throughout mid and northwestern China, whose leaves have been used as a traditional Chinese medicine for a long time. *Apocynum venetum* leaves have been proved to exhibit a lot of activities, which were used for the treatment of cardiac disease, hypertension, nephritis, and other diseases [1, 2].

The *Apocynum venetum* is rich in flavonoids [3]. The dominant compounds in flavonoids of *Apocynum venetum*, hyperoside have been proven by pharmacological experiments to lower blood pressure and exert an anti-inflammatory effect. The quality control of hyperoside and total flavonoids in *Apocynum venetum* is the principal factor.

Generally, *Apocynum venetum* leaves are fluxed in water to obtain their extract, which is a suspension with brownish colloidal particles, so, this extract must be clarified prior to commercialisation. Water extraction and alcohol precipitation method is often adopted in the purification of traditional Chinese medicine. However, this approach has the drawback such as ethanol consumption, long production cycle, high production cost, fewer of active components. One of the other possibilities for clarification is by means of flocculation. Inorganic flocculants such as alum and iron chloride are efficient, but are required in high doses and result in contamination with aluminium or iron. Compared with inorganic flocculant, organic polymer flocculants are harmless, more efficient, and environmentally friendly. In recent years, the biopolymers like chitin, guar gum, alginic acid, or starch have been widely used in flocculation. Of these, chitosan has been shown to be an effective flocculant for suspended solids [4, 5].

This work aims to study flocculation conditions for aqueous extracts of *Apocynum venetum* leaves using chitosan. Several important factors on the flocculation behaviour were screened by FFD, and the optimum extraction conditions were obtained by RSM with desirability function.

2 MATERIALS AND METHODS

2.1 Materials and reagents

Apocynum venetum leaves were obtained from a local chemist (Nanchang, China). Chitosan (viscosimetric molar mass = 105 kDa; degree of deacetylation = 85.61 ± 4.15%) was purchased from Sinopharm Chemical Reagent Co., Ltd (Shanghai, China). Reference substance of rutin (10080-20707) and hyperoside (111521-200303) were obtained from the National Institute for the Control of Pharmaceuticals and Biological Products (Beijing, China). Acetonitrile and phosphoric acid (chromatographic grade) were obtained from local chemical suppliers; all other chemicals were of analytical grade. Deionised water was purified by a Milli-Q academic water purification system (Millipore, Bedford, MA, USA).

2.2 Apparatus

The HPLC-UV analysis was performed using Agilent 1200 series HPLC (Agilent, USA). UV-1700 spectrophotometer (Shimadzu Corporation, Japan) was used for total flavonoids analysis of samples.

2.3 Determination of flavonoids

The content of total flavonoids was determined by the colorimetric method with some modifications [6]. 1mL diluted solution containing flavonoids, 5 mL of 80% (v/v) ethanol and 1 mL of 5% (w/w) NaNO$_2$ were mixed for 6 min, and then 1 mL of 10% AlCl$_3$ (w/w) was added and mixed; 6 min later, 10 mL of 1 mol/L NaOH was added. With 15 min standing, the absorbance of the solution was measured at 510 nm with UV-1700 spectrophotometer against the same mixture, without the sample as a blank. A calibration curve was prepared with a standard solution of rutin (0.2, 0.4, 0.6, 0.8, 1.0, 1.2 mg/ml). The calibration curve ($Y = 11.08X + 0.0022$, where y is absorbance value of sample, and x is sample concentration) ranged 0.2–1.2 mg/ml ($R = 0.9999$). The content of total flavonoids was expressed as milligram rutin equivalent per dry weight material (mg/g).

2.4 Determination of hyperoside

Hyperoside was analysed by high performance liquid chromatography (HPLC) on an equipment Agilent 1200 series HPLC (Agilent, USA). Kromasil C18 column (250 mm × 4.6 mm, 5μm) from AKZO NOBEL Company was used. A mixture of acetonitril and 0.2%-phosphoric acid (15:85) was used as the mobile phase at the detection wavelength of 256 nm. The flow rate of mobile phase was set at 1 mL/min, the injection volume was 10μL, and the column temperature was maintained at 35°C. The concentration of the hyperoside was determined from standard curves made with known concentrations [7]. The calibration curve (y = 27.25x −4.12, where y is value of peak area, x is sample concentration) ranged 0.036–0.364 mg/mL (r = 0.9998). The average recovery was 98.12%, 101.00%, and 99.19%, respectively and R.S.D. was 0.54%, 1.13%, and 1.35%, respectively. The chromatographic condition is suitable for determination of hyperoside in aqueous extracts of *Apocynum venetum* leaves.

2.5 Preparation of aqueous extracts

Extraction is the first step in the flocculation process. *Apocynum venetum* leaves were ground in a grinder to obtain a fine powder (particle diameter: 0.2–0.9 mm). Samples of 1.0 kg were refluxed twice with 14 L distilled water for 1.5 h. The extraction was separated from insoluble residue through a paper filter. The filtrate was concentrated to obtain a decoction of *Apocynum venetum* leaves with a concentration of 0.167 g medicinal materials/mL suspension and prepared for analysis and weight measurement of the extracta sicca.

The contents of flavonoids and hyperoside were standardised on an amount 3.67%, 0.38%, respectively. The yield of extracta sicca was 28.95%.

2.6 Flocculation experiments

The chitosan solution, with a concentration of 10 mg/ml (w/v), was obtained by dissolving chitosan powder in 1.0 vol. % acetic acid solution and magnetic stirring for at least 24 hours.

Flocculation was carried on according to the experimental design. The aqueous extracts of *Apocynum venetum* leaves (60 mL) was placed in a 100 mL beaker and heated to a designed temperature in a water bath, different volume of the chitosan solution was added, the sample was immediately stirred at a fast speed for designed time, followed by a slow stirring for another time, and then allowed to stand for some time to settle. After flocculation, the suspension was separated from insoluble residue through a paper filter and the filtrate was analysed. The removal of extracta sicca (RMOES), retention rate of total flavonoids (RTROTF) and retention rate of Hyperoside (RTROHP), were evaluated employing Eq. 1 and 2.

$$Removal(\%) = \frac{M_f}{M_i} \times 100 \quad (1)$$

$$Removal(\%) = \frac{M_i - M_f}{M_i} \times 100 \quad (2)$$

where M_i and M_f are the initial and final parameter value of the aqueous extracts of *Apocynum venetum* leaves, respectively.

2.7 Analysis of effective variables

These experiments were designed and analysed by Design-Expert 7.1.3. Two level, seven-factor blocks were constructed considering chitosan dosage (A), flocculation temperature (B), fast mixing time (C), fast mixing speed (D), slow mixing time (E), slow mixing speed (F), settling time (G) as independent factors. To find the significant factors among the independent factors involved in the flocculation, a fractional factorial design (2^{7-3}) with 16 runs was utilised (Table 1). The range was determined based on preliminary studies and literature review. Experiments were performed in randomised order according to the run number that was arranged by the software.

2.8 Optimisation of flocculation conditions

Analysis of the fractional factorial experiments showed that the chitosan dosage (A), flocculation temperature (B), and the time of slow mixing (E) were significant independent factors on RMOES. However, all the facts had little effect on the RTROTF and RTROHP. Therefore, in the next experiment, we chose to study the effects of the chitosan dosage, the flocculation temperature, the time of slowly mixing on

Table 1. Experimental conditions used for 2^{7-3} fractional factorial design, and the corresponding response measured.

Standard no.	Run	Factors							Response variables		
		A	B	C	D	E	F	G	RMOES	RTROTF	RTROHP
1	16	2	30	0.5	200	8	30	20	35.12	85.56	83.45
2	14	12	30	0.5	200	20	30	120	51.56	78.12	75.12
3	12	2	70	0.5	200	20	80	20	28.49	88.12	85.56
4	11	12	70	0.5	200	8	80	120	45.32	85.56	84.56
5	10	2	30	2	200	20	80	120	28.67	87.56	85.12
6	15	12	30	2	200	8	80	20	54.32	87.56	85.65
7	13	2	70	2	200	8	30	120	30.58	83.23	81.48
8	4	12	70	2	200	20	30	20	45.58	86.12	84.35
9	2	2	30	0.5	500	8	80	120	30.23	79.56	78.89
10	1	12	30	0.5	500	20	80	20	44.45	86.89	86.22
11	6	2	70	0.5	500	20	30	120	23.12	86.65	84.32
12	9	12	70	0.5	500	8	30	20	46.56	81.36	83.26
13	7	2	30	2	500	20	30	20	28.25	90.12	88.68
14	8	12	30	2	500	8	30	120	55.12	84.68	75.36
15	3	2	70	2	500	8	80	20	28.79	88.23	89.32
16	5	12	70	2	500	20	80	120	45.56	83.89	79.53

Table 2. Regression coefficients and their significance level.

Source	RMOES		RTROTF		RTROHP	
	Coefficient	p-Value	Coefficient	p-Value	Coefficient	p-Value
Constant	38.86		85.20		83.18	
A	9.70	0.0019	−0.93	0.4042	−1.42	0.1573
B	−2.11	0.0382	0.19	0.8465	0.87	0.3094
C	0.75	0.2185	1.22	0.3009	0.51	0.5130
D	−1.10	0.1225	−0.03	0.9775	0.02	0.9801
E	−1.90	0.0465	0.73	0.4944	0.43	0.5699
F	−0.63	0.2764	0.72	0.5009	1.18	0.2086
G	−0.09	0.8556	−1.54	0.2230	−2.63	0.0548
AB	−0.70	0.2423	−0.23	0.8159	0.30	0.6860
AC	0.84	0.1878	0.07	0.9466	−1.04	0.2469
AD	0.46	0.3903	−0.04	0.9685	−0.68	0.4000
AE	–	–	−1.25	0.2930	−0.88	0.3027
AF	−0.52	0.3467	0.98	0.3826	1.06	0.2419
AG	0.92	0.1626	0.33	0.7418	−0.48	0.5317
BD	0.36	0.4906	–	–	–	–
R^2	0.9966		0.8451		0.9488	
Adj-R^2	0.9749		−0.1619		0.6158	
Adeq precision	20.169		3.423		6.293	

RMOES. In this experimental design, a BBD, the most widely used approach of RSM, was used to optimise the three independent variables (factors). The BBD was designed and analysed by Design-Expert 7.1.3. On the basis of the experimental data, a second-order polynomial model corresponding to the BBD was fitted to correlate the relationship between the independent variables and the response.

3 RESULTS AND DISCUSSION

3.1 Identifying effective variables

The designed values for the independent variables and experimental responses were shown in Table 1. Test for significance of regression coefficients was shown in Table 2. Table 2 showed that terms such as C, D, F, G had little effect on the retention rate and removal. Considering the time and laborious factor, we fixed the fast mixing time, fast mixing speed, slow mixing speed, settling time at 0.5 min, 200 rpm, 30 rpm, 20 min, respectively. As shown in Table 2, the terms such as A, B, and E were more significant on removal of extracta sicca. Therefore, A (chitosan dosage), B (flocculation temperature,) and C (the time of slow mixing) turned out to influence the flocculation of aqueous extracts of *Apocynum venetum* leaves. On the other hand, the retention rate of active ingredients maintained constant in the flocculation process; all the independent variables (flocculation parameters) had no effect on the retention rate of total flavonoids and hyperoside. A negative "Pred R^2", which was −8.9149, and −2.2788, respectively, implied that the overall mean was a better predictor of the response than the current model.

Factors with a negligible effect on response at a significance level of 95% were screened out. Therefore, chitosan dosage, flocculation temperature and the time of slow mixing turned out to influence RMOES, and were selected for optimisation in the next experimental design.

Table 3. Experimental conditions for the BBD and the corresponding responses measured.

Standard No.	Run	Factors A^a	B^b	C^c	Response variables
1	7	2	30	14	27.27
2	2	12	30	14	43.13
3	5	2	70	14	22.75
4	10	12	70	14	37.69
5	13	2	50	8	29.77
6	3	12	50	8	43.91
7	16	2	50	20	24.42
8	12	12	50	20	42.63
9	14	7	30	8	50.04
10	17	7	70	8	43.67
11	4	7	30	20	45.33
12	1	7	70	20	42.86
13	8	7	50	14	50.01
14	15	7	50	14	49.24
15	11	7	50	14	51.79
16	6	7	50	14	53.23
17	9	7	50	14	51.44

A^a: chitosan dosage; B^b: flocculation temperature; C^c: slow mixing time.

3.2 Using RSM for optimisation

The flocculation of aqueous extracts of *Apocynum venetum* leaves was optimised through RSM approach. All 17 of the designed experiments were conducted for optimising the three individual parameters in the current Box-Behnken design. Table 4 showed the experimental conditions and the results of RMOES. By applying multiple regression analysis on the experimental data, the response variable and the test variables were related by the following second-order polynomial equation:

$$RMOES = -4.566 + 9.261A + 0.802B + 0.345C \\ -0.002AB + 0.034AC + 0.008BC - 0.574A^2 - \\ 0.010B^2 - 0.044C^2 \quad (3)$$

To determine whether or not the quadratic model was significant, the statistical significance of regression equation was checked by F-test, and ANOVA for response surface quadratic polynomial model were summarised in Table 4. The P-value of the model was smaller than 0.0001, which indicated that the model was suitable for use in this experiment. In the present study, the value of 30.459 indicated an adequate signal. This model could be used to navigate the design space.

Three-dimensional response surface plots, as presented in Figs. 1–3, were very useful to see interaction effects of the factors on the responses [8]. The regression coefficients and the corresponding P-values were also presented in Table 4. From the P-values of each model term, it could be concluded that all the independent variables studied (A, B, C) and four quadratic terms (A^2, B^2, C^2) significantly affected RMOES. However, the analysis showed the interactions between two arbitrary parameters were insignificant. The results of the study also represented

Table 4. Analysis of variance for the response surface quadratic model.

Source	Sum of square	d.f.	Mean square	F	p-Value
Model	1565.225	9	173.913	117.298	<0.0001
A	498.490	1	498.490	336.213	<0.0001
B	44.180	1	44.180	29.797	0.0009
C	18.45281	1	18.452	12.445	0.0096
AB	0.211	1	0.211	0.142	0.7168
AC	4.141	1	4.141	2.793	0.1386
BC	3.802	1	3.802	2.564	0.1533
A^2	868.523	1	868.523	585.787	<0.0001
B^2	69.738	1	69.738	47.036	0.0002
C^2	10.741	1	10.741	7.245	0.0310
Residual	10.378	7	1.482		
Lack of fit	0.6111	3	0.203	0.083	0.9655
Pure error	9.767	4	2.441		
Cor total	1575.603	16			
R^2			0.9934		
Adj-R^2			0.9849		
Adeq precision			30.459		

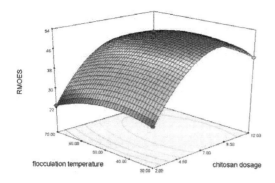

Figure 1. Effect of chitosan dosage and flocculation temperature on RMOES.

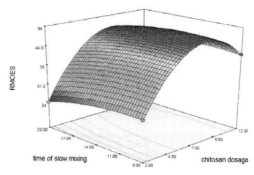

Figure 2. Effect of chitosan dosage and the time of slow mixing on RMOE.

that the chitosan dosage and flocculation temperature were the most significant parameters which influenced RMOES followed by slow mixing time.

The effects of chitosan dosage and flocculation temperature on RMOES can be seen from Figures 1 and 2. Figures 1 and 2 representing the maximum RMOES were achieved with the chitosan dosage between 10 mg/g and 11 mg/g, but it decreased when the chitosan dosage was kept at levels higher than 11 mg/g. To understand this phenomenon, the flocculation mechanism of chitosan has to be considered. It is well-known that particles in aqueous dispersion become charged. The charge, measurable by zeta potential, is an important parameter for the stability of dispersion. If the particles have a large potential, they will repel each other, and the dispersion is stable. If the particles have low potential, then there is no force to prevent the particles from coming together, the dispersion is unstable and particles aggregate. It was clear that overdosing of chitosan resulted in dispersion restabilisation. This phenomenon is commonly observed with polyelectrolyte flocculants.

The effect of temperature on RMOES using chitosan was investigated and represented in Figures 1 and 3. The nature of the curve shows that the removal was continuously increasing with increasing temperature. Also, the descending curve was observed after 43°C. Therefore, optimum temperature was attained at 43°C for chitosan. The highest removal was found to be 49.11%. In general, the viscosity of the flow decreases with increasing temperature, which is conducive to the diffusion of flocculant molecular. On the other hand, the probability of contact between the colloid particles increases to accelerate the agglomeration by ccharge neutralization and adsorption bridging action between chitosan and impurity particles. However, high temperature would make the chitosan chain broken and degraded; what is more, the average shear rate of the flow increase at high temperature, causing floc destruction.

For the flocculation of water-extract solution of *Apocynum venetum*, two types of mixing (rapid and slow) may be applied. In flocculation process, rapid mixing is related to the diffusion of the flocculants, and

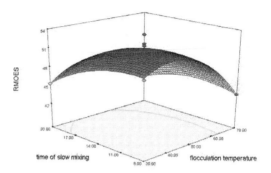

Figure 3. Effect of flocculation temperature and the time of slow mixing on RMOES.

slow mixing is related to the flocculant size. Adequate slow time must be provided to allow production of particles of sufficiently large size to permit their efficient removal in sedimentation process. If the slow mixing time is too short, then, the probability of contact between the colloid particles would decrease, which is not conducive to the formation of flocculant.

3.3 Optimisation of flocculation parameters and validation of the model

In this study, our purpose was only to clarify the extract, so it was important to minimise the flocculation of active ingredients. In other words, to find the conditions which give the maximum retention rate of active ingredients and maximum RMOES. The optimal values of the selected variables were obtained by solving the regression equation (Eq.3) using Design-Expert 7.1.3 software. The software predicted the optimum chitosan dosage, flocculation temperature, and the time of slow mixing level was 10.91 mg/g, 42.96°C, and 12 min, respectively. The software predicted RMOES was 49.11%.

However, considering the operability in actual production, the optimal conditions could be modified as follows: chitosan dosage of 11 mg/g, flocculation temperature of 43°C, and the time of slow mixing of 12 min, respectively. Under the modified conditions, the RMOES was 49.28 ± 0.43 mg/g (N = 3), which was close to the predicted value. On the other hand, the retention rate of total flavonoids, hyperoside was (85.28 ± 8.45)%, (83.28±7.56)%, respectively, indicating that the model was adequate for the flocculation process and flocculation had little effect on retention rate of total flavonoids and hyperoside.

4 CONCLUSIONS

The statistical analysis of FFD, RSM and desirability function enabled us to screen several significant progress parameters and find out the optimum flocculation conditions. Our results show that chitosan is a potentially useful flocculant for water-extract solution of *Apocynum venetum*. Further research is needed in order to evaluate the influence of other factors (such as initial pH and extract concentration) for chitosan flocculation and, therefore, to optimise the flocculation method.

ACKNOWLEDGMENT

The work was supported by the project of National Natural Science Foundation of China (Grant No. 81360647), the technical platform for the new drug creation and development and new drug-delivery system of TCM (2009ZX09310-005), the Education Research project of Science and Technology Foundation, Jiangxi Province, RP China (Grant No. GJJ13602)

REFERENCES

Ministry of Public Health of the People's Republic of China, Pharmacopoeia of the People's Republic of China, vol. 1, 2010, p. 196

Grundmann, J. Nakajima, K. Kamata, S. Seo and V. Butterweck, Kaempferol from the leaves of *Apocynum venetum* possesses anxiolytic activities in the elevated plus maze test in mice. Phytomedicine. 4(2009) 295–302

T. Yokozawa and T. Nakagawa, Inhibitory effects of Luobuma tea and its components against glucose-mediated protein damage, Food Chem. Toxicol. 6(2004) 975–981

P.S. Strand, K.M. Vårum and K. Ostgaard, Interactions between chitosan and bacterial suspensions: adsorption and flocculation, Colloid. Surf. B 27 (1) (2003)71–81

M. Mihai and E.S. Dragan, Chitosan based nonstoichiometric polyelectrolyte complexes as specialized flocculants, Colloids and Surfaces A: Physicochem. Eng. Aspects 346 (2009) 39–46

Z.L. Sheng, P.F. Wan, C.L. Dong and Y.H. Li, Optimization of total flavonoids content extracted from Flos Populi using response surface methodology. Ind. Crops Prod. 43 (2013) 778–786

Ministry of Public Health of the People's Republic of China, Pharmacopoeia of the People's Republic of China, vol. 1, 2010, p. 196

Y. Sun, J. Liu and J.F. Kennedy, Extraction optimization of antioxidant polysaccharides from the fruiting bodies of Chroogomphis rutilus (Schaeff. Fr.) O.K. Miller by Box–Behnken statistical design. Carbohyd. Polym. 82 (2010) 209–214

Energy and Environmental Engineering – Wu (Ed.)
© 2015 Taylor & Francis Group, London, ISBN 978-1-138-02665-0

Mechanisms of multi-thermal fluid huff and puff recovery for offshore heavy oil reservoirs and its influential law

Qing Jun Du, Jian Hou, Ji Chao Wang, Li Na Shi, Guang Fu Zhang, Xiao Ning Li & Shan Peng Li
China University of Petroleum, Qingdao, China

ABSTRACT: Multi-thermal fluid huff and puff is an emerging technology for offshore heavy oil reservoirs, and multi-thermal fluids which are composed of several components including hot water, CO_2 and N_2. In this paper, a novel mathematical model of multi-thermal fluid huff and puff is established. Taking a typical reservoir numerical simulation model for example, yield increase mechanisms and influence of sensitive parameters on the development effect of multi-thermal fluid huff and puff technology are investigated thoroughly using the proposed model. The results show that the main mechanisms of multi-thermal fluid huff and puff recovery are viscosity reduction by heating and dissolving of CO_2, as well as pressure maintenance and cleanup with N_2. Furthermore, the main influential factors include injection intensity, time for next cycle, injection temperature and increase rate of cyclic injection volume.

Heavy oil resources of Bohai Sea are widely distributed with high developing risk, which are mainly developed by the horizontal well technique (Wehunt et al., 2003). Due to the limitation of platform space, economy, environmental protection and security (Xu et al., 2013, Liu et al., 2012), it is very difficult to implement the conventional thermal recovery methods, such as steam stimulation and steam flooding (Gates 2010, Zhao et al., 2014). Multi-thermal fluid huff and puff is an efficient thermal recovery technique, and the multi-thermal fluids are composed of several components including hot water, CO_2, and N_2 (Zhong et al., 2013). Except for the yield increase mechanisms of conventional thermal recovery methods and gas flooding, the synergistic effect between different components has a great impact on the improvement of production performance (Chu 1985, Doscher, 1983, Closmann, 1984). Field tests also indicate that multi-thermal fluid huff and puff technology can improve single well production to a large extent, which is 1.5 to 3 times that of conventional thermal methods. It is well known that CO_2 is the main component of greenhouse gases, so generalisation of the technology mentioned above is helpful to reduce the greenhouse effect. A pilot test of multi-thermal fluid huff and puff technology has been launched in the Bohai heavy oil reservoirs, and the yield-increasing effect is very striking. However, due to the complexity of yield increase mechanisms, there are great differences in multi-cycle scheme design and methods to predict development effect compared to the conventional thermal methods. Furthermore, it is easier for gas channelling to happen in the procedure of oilfield development. As a result of the above-mentioned problems, this paper presents a novel mathematical model for multi-thermal fluid huff and puff simulation, and on this basis, yield increase mechanisms and influence of sensitive parameters on the development effect of multi-thermal fluid huff and puff technology are investigated thoroughly.

1 MATHEMATICAL MODEL OF MULTI-THERMAL FLUID HUFF AND PUFF

In order to establish the mathematical model of multi-thermal fluid huff and puff, the following assumptions are made: a system of oil-gas-water three-phase fluids exists simultaneously in the reservoir, which is made up of four components including oil, water, CO_2 and N_2; distribution of different components between the phases follows the equilibrium principle; the flow of multiphase fluid obeys Darcy's law and no chemical reactions occur in the process of seepage; kinetic energy and viscous force except for heat energy are neglected; a non-isothermal process is considered; oil viscosity and the oil-gas-water three-phase relative permeability curve are influenced by reservoir temperature.

Mass conservation equation

$$\nabla \cdot \left[\sum_{j=o,w,g} \rho_j x_{ij} \frac{kk_{rj}}{\mu_j} \nabla(p_j - \rho_j gD) \right] + \sum_{j=o,w,g} q_j x_{ij}
= \sum_{j=o,w,g} \frac{\partial}{\partial t}(\varphi \rho_j S_j x_{ij}) \quad (i = \text{oil, water, } CO_2, N_2) \quad (1)$$

where ρ_j = reservoir density of phase j (oil, water, gas), kg/m³; x_{ij} = mole fraction of i component in phase j; k = absolute permeability, m²; k_{rj} = relative permeability of phase j; μ_j = viscosity of phase j, Pa·s; p_j = pressure of phase j, Pa; g = gravitational acceleration, m/s²; D = reservoir depth whose positive direction is downward, m; q_j = mass of phase j injected or produced at unit time and unit volume under reservoir conditions, kg/(m³·s); ϕ = porosity, fraction; S_j = saturation of phase j.

1.1 Energy conservation equation

$$\nabla \bullet (\lambda_r \nabla T) + \nabla \bullet \left\{ \sum_{j=o,w,g} \left[\sum_{\text{oil, water, CO}_2, N_2} \rho_j H_{ij} x_{ij} \frac{k k_{rj}}{\mu_j} \nabla (p_j - \rho_j g D) \right] \right\}$$
$$-\bar{q}_l + \bar{q}_h = (1-\varphi)\frac{\partial}{\partial t}(\rho_r C_r T) + \varphi \frac{\partial}{\partial t}\left[\sum_{j=o,w,g} \left(\sum_{\text{oil, water, CO}_2, N_2} x_{ij} \rho_j S_j U_{ij} \right) \right]$$
(2)

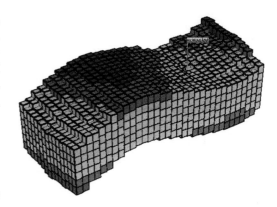

Figure 1. The typical single-well numerical simulation model.

where λ_r = reservoir thermal conductivity, kJ/(s·m·°C); H_{ij} = enthalpy of per unit mass of i component in phase j, kJ/kg; $\bar{q}_l$ = energy spoiled associating with the top and bottom layers at unit time and unit volume, kJ/(m³·s); $\bar{q}_h$ = I/O energy at unit time and unit volume (I is positive, O is negative), kJ/(m³·s); ρ_r = rock density, kg/m³; C_r = calorific receptivity of reservoir rock, kJ/(kg/°C); U_{ij} = internal energy of per unit mass of i component in phase j, kJ/kg.

Table 1. Basic parameters of the geological model.

Parameter	Value	Parameter	Value
Effective thickness, m	6	permeability, $10^{-3} \mu m^2$	3600
Porosity	0.372	temperature, °C	56
Horizontal well length, m	270	viscosity, mPa·s	449

1.2 Viscosity equation

The relationship between oil viscosity and temperature satisfies the Walther viscosity-temperature law, which is expressed as follows:

$$\lg \lg(v_{od} + 0.8) = -n \lg(\bar{T}/\bar{T}_1) + \lg \lg(v_{od1} + 0.8) \quad (5)$$

$$v_{od} = \mu_{od} / \rho_{od} \quad (6)$$

where $\bar{T}$ and $\bar{T}_1$ are temperature, K; v_{od} and v_{od1} are kinematic viscosity of dead oil at temperature $\bar{T}$ and $\bar{T}_1$, respectively, 10^{-6} m²/s; μ_{od} = dynamic viscosity of dead oil, mPa·s; ρ_{od} = density of dead oil, g/cm³; n = a constant quantity.

Table 2. Production effect of three-cycle multi-thermal fluid huff and puff.

Production cycle	Cumulative oil production, $10^4 m^3$	Periodic oil-water ratio
1	2.84	7.11
2	4.50	4.14
3	5.58	2.71

2 MECHANISMS OF MULTI-THERMAL FLUID HUFF AND PUFF RECOVERY

2.1 Numerical simulation model

Using the actual petro-physical properties of the Bohai Oil Region, a typical geological model for horizontal well is constructed. The research domain has evolved from fluvial deposits with high porosity, high permeability and strong heterogeneity. Oil density and viscosity under formation conditions are large, and sensitive to variation of temperature. A pilot test of multi-thermal fluid huff and puff technology has been performed by stage from 2008, which has improved the oil recovery rate to a great extent. A typical single-well numerical simulation model is established based on the reservoir characteristics of test area, as shown in Figure 1. The basic parameters of the geological model are listed in Table 1.

The basic scheme of the single-well numerical simulation model in oilfield is established on the basis of history matching the production performance. In this study, injection rate of hot water is 165 m³/d; injection rate of air is 57600 m³/d; injection temperature is 240°C; liquid production rate is 120 m³/d; injection strength is 14 m³/m and soak time is limited to 3 days. Moreover, the huff and puff will be terminated when daily oil production decreases to 35 m³/d. Based on the basic scheme, flow simulation of threecycle multi-thermal fluid huff and puff has been performed.

Table 2 presents the production effect of three-cycle multi-thermal fluid huff and puff. As can be seen, the performance of huff and puff is good when periodic oil-gas ratio is larger than 2.5, and the yield increase performance has been valid for a longer time. However, with the increase of huff-puff cycle, the

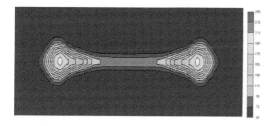

Figure 2. The planar temperature distribution around horizontal wellbore after shut-in for soaking.

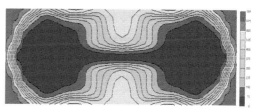

Figure 4. The Planar viscosity distribution around horizontal wellbore after shut-in for soaking.

Table 3. Viscosity-reduction effect of crude oil by dissolving CO_2.

Initial viscosity, mPa·s	Viscosity after dissolving CO_2, mPa·s
1~9	0.5~0.9
10~100	1~3
100~600	3~5
1000~9000	15~160

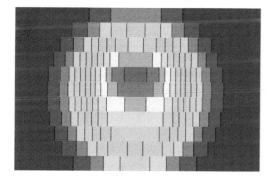

Figure 3. Temperature distribution at toe end of horizontal wellbore after production for 30 days.

residual oil saturation, reservoir pressure and productivity decrease gradually in the domain swept by multi-thermal fluids.

2.2 Mechanisms of multi-thermal fluid huff and puff recovery

2.2.1 Viscosity reduction by heating

Because of high sensitivity to the variation of temperature, crude oil viscosity can be reduced significantly as temperature rises. The higher the oil viscosity is, the larger the reduction amplitude is (Sun et al., 2011). Both the liquidity of crude oil and the oil-water mobility ratio can be improved to a great extent by reduction of oil viscosity, which results in a higher sweep efficiency and a better development effect of heavy oil reservoirs. Meanwhile, this is the main mechanism of enhanced oil recovery by thermal methods.

An air chamber will be formed once the gases in multi-thermal fluids system are injected into the formation. An upward protective layer caused by gas overlap phenomena can reduce the heat loss and raise the utilisation ratio of heat energy due to a lower thermal conductivity. Figure 2 shows the planar temperature distribution along the horizontal wellbore when the process of soaking well is ended. Figure 3 presents the vertical temperature distribution at the toe end of horizontal wellbore when production is performed for 30 days. The results show that temperature extension scope of the multi-thermal fluid is higher at the toe and root end of horizontal wellbore, where a circular air clamber with elevated temperature will be constructed, which centres on the horizontal well along the vertical direction. In addition, strong heat carrying and dispersal abilities of the multi-thermal fluid system make a high heating scope both in the planar and vertical directions. Furthermore, variation of temperature is relatively stable and temperature difference between the inner and outer boundaries is small, indicating that the heat energy has been utilised effectively. The planar oil viscosity distribution of horizontal well when shut-in for soaking is terminated is shown in Figure 4. As can be seen, the reduction scope and degree of oil viscosity caused by multi-thermal fluid huff and puff is very good.

2.2.2 Viscosity reduction by dissolving CO_2

It is the synergistic effect between heating and dissolving gases that leads to the reduction of oil viscosity when performing the multi-thermal fluid huff and puff. Compared with other components of the multi-thermal fluid, it is much easier for CO_2 to be dissolved in oil, and the larger the quantity of the dissolved CO_2 is, the more remarkable viscosity-reduction effect is, which plays a major role in viscosity reduction by dissolving gases (Zhong et al., 2013, Liu et al., 2010, Sun et al., 2013).

The light hydrocarbon components of crude oil will be augmented after dissolving CO_2, which causes the reduction of oil viscosity significantly. Table 3 shows the viscosity-reduction effect of crude oil by dissolving CO_2. It can be seen that the larger initial viscosity of the crude oil, the greater the reduction degree is.

Figure 5 presents the planar and vertical molar concentration distribution of CO_2 in crude oil around horizontal wellbore after shut-in for soaking. Combined with the viscosity distribution of CO_2 in crude oil shown as Figure 6, distribution law of CO_2 in crude

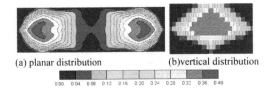

(a) planar distribution (b) vertical distribution

Figure 5. The planar and vertical molar concentration distribution of CO_2 in crude oil.

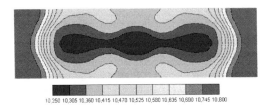

Figure 6. The planar pressure distribution around horizontal wellbore after being produced for 30 days.

oil is the same as that of oil viscosity distribution. The higher the molar concentration of CO_2, the lower the viscosity of crude oil is.

2.3 *Maintaining pressure and cleanup by N_2*

Compressibility of N_2 is three times that of CO_2, which is helpful to the displacement of crude oil. The formation volume factor of N_2 drops evenly with the increase in pressure, thus contributing to maintaining the reservoir pressure and greatly improving oil recovery. In addition, in view that the conductivity of N_2 is lower than that of steam as well as other reservoir fluids, it is usually taken as a heat shield to slow down the heat loss. The mechanism is of vital importance to production by heating, which can increase the heat efficiency effectively and improve the development effect.

Figure 6 shows the planar pressure distribution around horizontal wellbore after being produced for 30 days. It indicates that a better effect of pressure maintenance can be achieved with a gentle transition zone. The main reason is that parts of elastic energy can be stored owing to the existence of N_2 in multi-thermal fluid system, which is beneficial to the pressure maintenance and cleanup to a certain extent.

3 STUDY ON INFLUENTIAL LAW OF PRODUCTION PERFORMANCE

When performing the multi-thermal fluid huff and puff technology, lots of injection-production parameters will influence the development effect. So it is of great importance to study the influential law of sensitive parameters on production performance of multi-thermal fluid huff and puff. The injection-production parameters discussed in this paper mostly

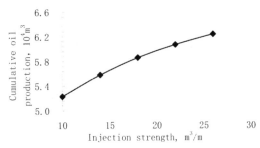

Figure 7. Cumulative oil production vs. the increase of injection strength.

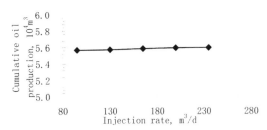

Figure 8. Cumulative oil production vs. the increase of injection rate.

include injection strength, injection rate, injection temperature, soak time, liquid production rate, time to next cycle, and increase rate of cyclic injection volume.

3.1 *Injection strength*

For the multi-thermal fluid system in horizontal well, injection strength is defined as the cycle injection volume per unit length of lateral segments. The injection strengths are: 10, 14, 18, 22 and 26 m^3/m, respectively, and the remaining parameters are identical to those of the basic scheme. Influence of different injection strengths on cumulative oil production after three cycles is simulated. The result is shown as Figure 7.

From Figure 7, it can be seen that, with the increase of injection strength, higher cycle injection volume and a higher maintenance degree of heat energy and reservoir pressure will be achieved, which leads to the improvement of oil-phase mobility, thus increasing the cumulative oil production The increase rate of cumulative oil production drops gradually when cycle injection strength higher than 18 m^3/m.

3.2 *Injection rate*

The injection rates are: 130, 165, 200 and 235 m^3/d, respectively, and the remaining parameters are identical to those of the basic scheme. Influence of different injection rates on cumulative oil production after three cycles is simulated. The result is shown as Figure 8. It demonstrates that injection rate has little effect on cumulative oil production. As injection rate increases,

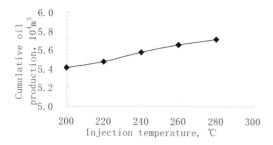

Figure 9. Cumulative oil production vs. the increase of injection temperature.

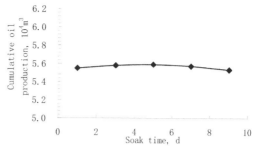

Figure 10. Cumulative oil production vs. the increase of Soak time.

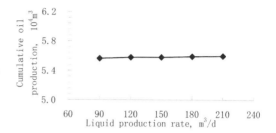

Figure 11. Cumulative oil production vs. the increase of liquid production rate.

cumulative oil recovery raises slightly. The main reason is explained as follows: when injection strength is identical, heat loss is relatively higher and the heat energy around the horizontal wellbore will be reduced with the decrease of injection rate, which influence the development effect of the multi-thermal fluid huff and puff technology. Therefore, considering the injection capacity constraints of gas-injection device, the injection rate should be as high as possible.

3.3 Injection temperature

The injection temperatures are: 200, 220, 240, 260 and 280°C, respectively, and the remaining parameters are identical to those of the basic scheme. The influence of different injection temperatures on cumulative oil production after three cycles is simulated, and the result is shown as Figure 9. As can be seen, with the increase of temperature, quantities of gas-injection will be raised significantly when keeping the cycle water-injection, which results in a higher reservoir pressure and larger cumulative oil production.

3.4 Soak time

The soak times are: 1, 3, 5, 7 and 9d, respectively, and the remaining parameters are identical to those of the basic scheme. Influence of different soak times on cumulative oil production after three cycles is simulated, and the result is shown as Figure 10. It shows that development effect is the best when soak time equals to 3–5 days. At the beginning of shut-in for soaking, the heated area increases gradually as heat of the multi-thermal fluid system transfers along the radial direction. However, when soak time reaches a certain value, the heat loss is accelerated and heat around the horizontal wellbore is reduced, which leads to a smaller cycle cumulative oil production.

3.5 Liquid production rate

Liquid production rate affects the drop speed of bottomhole pressure and has certain effect on the validity of thermal recovery methods. The liquid production rates are: 90, 120, 150, 180 and 210 m³/d, respectively. The remaining parameters are identical to those of the basic scheme. Influence of different liquid production rates on cumulative oil production after three cycles is simulated, and the result is shown as Figure 11. It can be seen that when liquid production rate is higher, reservoir energy can be fully utilised with a lower heat loss, which causes a slight increase in cumulative oil production. In short, liquid production rate has little effect on the improvement of cumulative oil production.

3.6 Time for next cycle

In this study, daily oil production is taken as the termination condition and the way to determine the time for next cycle. The daily oil productions are: 30, 35, 38, 42 and 46 m³/d, respectively. The remaining parameters are identical to those of the basic scheme. Influence of different daily oil production on cumulative oil production after three cycles is simulated, and the result is shown as Figure 12. As seen, cumulative oil production decreases with the increase of daily oil production used as the termination condition. Compared with cold production technique, the effect of thermal recovery is better when daily oil production is slightly lower, or close to that of cold production.

3.7 Increase rate of cyclic injection volume

Increase rate of cyclic injection volume is defined as the increment percentage of cyclic injection volume compared with that of the previous cycle. At the initial stage of multi-thermal fluid huff and puff, the scope swept by heat is so narrow that there is a greater difficulty for gas-injection, resulting in a small quantity of gas-injection. As cycles of steam stimulation increase, the heated scope becomes higher and more heat needs to be injected into the formation, thus the cycle injection volume should be raised gradually.

Cumulative injection volume for three cycles is limited to a constant value of 12,000 m³; the increase rate of cyclic injection volumes are: 0%, 5%, 10%, 15% and 20%, respectively. The remaining parameters are identical to those of the basic scheme. Influence of different increase rate of cyclic injection volume on cumulative oil production after three cycles is simulated. The result is shown as Figure 13. It demonstrates that the cumulative oil production becomes larger as the increase rate of cyclic injection volume increases. However, when the increase rate is higher than 10%, the increment degree of cumulative oil production drops gradually. In short, the best develop effect will be obtained when the increase rate of cyclic injection volume is between 10 and 15%.

On the basis of the above analysis, the influence of different sensitive parameters on production performance of the multi-thermal fluid huff and puff technique is various; in order to determine the main influential factors of the development effect, eq. 11 and 12 are used to calculate the variation coefficient when sensitive parameters differ from each other.

$$C_V = \frac{S}{|\bar{y}|} \quad (11)$$

$$S = \sqrt{\frac{1}{n}\sum_{i=1}^{n}(y_i - \bar{y})^2} \quad (12)$$

where C_v = the variation coefficient of samples; S = standard deviation of samples; $|\bar{y}|$ = the mean value of samples; n = the total number of samples; and y_i = the value of the ith sample.

The result is shown as Figure 14. It can be concluded that the main influential factors determining the development effect of the multi-thermal fluid huff and puff technique include injection strength, time for next cycle, injection temperature and increase rate of cyclic injection volume.

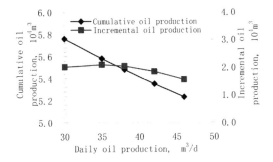

Figure 12. Cumulative oil production vs. the increase of daily oil production.

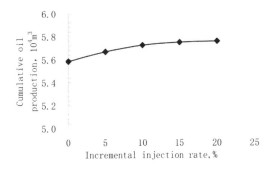

Figure 13. Cumulative oil production vs. the increase of cyclic injection volume.

4 CONCLUSION

Considering the physicochemical properties of multi-thermal fluids, a novel mathematical model of multi-thermal fluid huff and puff is established in this paper.

Taking a typical reservoir numerical simulation model for example, yield increase mechanisms and influence of sensitive parameters on the development effect of multi-thermal fluid huff and puff technology are investigated using the proposed model. Results show that the main mechanisms of multi-thermal fluid huff and puff recovery are viscosity reduction by heating and dissolving CO_2, as well as pressure maintenance and cleanup using N_2. Furthermore, the

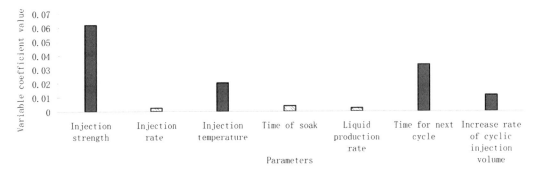

Figure 14. Variable coefficient values of different parameters.

main sensitive parameters which influence the development effect include injection intensity, time for next cycle, injection temperature and increase rate of cyclic injection volume.

ACKNOWLEDGEMENTS

The writers greatly appreciate the financial support of the National Natural Science Foundation of China (Grant No.11102236), and the Fundamental Research Funds for the Central Universities (Grant No.14CX02035A, 13CX05016A).

REFERENCES

Chu, C. 1985. State of the art review of steam flood field projects. Journal of petroleum technology 37: 1887–1902.

Closmann, P.J. 1984. Steam zone growth in cylindrical channels. Old SPE Journal 24: 481–483.

Doscher, T., Ghassemi, F. 1983. The influence of oil viscosity and thickness on the steam drive. Journal of Petroleum Technology 35: 291–298.

Gates, I.D. 2010. Solvent-aided steam-assisted gravity drainage in thin oil sand reservoirs. Journal of Petroleum Science and Engineering 74: 138–146.

Liu, D., Zhao, C.M., Su, Y.C. 2012. New Research and Application of High Efficient Development Technology for Offshore Heavy Oil in China. OTC23015.

Liu, Y. G., Yang, H. L., Zhao, L.C. 2010. Improve offshore heavy oil recovery by compound stimulation technology involved thermal, gas and chemical methods. Paper presented at the Offshore Technology Conference, Houston, Texas, 3–6 May 2010. OTC 20907.

Sun, Y.T., Zhao, L.C., Lin, T. 2013. Case study: thermal enhance Bohai offshore heavy oil recovery by co-stimulation of steam and gases. Paper presented at the SPE Heavy Oil Conference Canada, Calgary, Alberta, Canada, 11–13 June 2013. SPE 165410.

Sun, Y.T., Zhao, L.C., Lin, T. 2011. Enhance offshore heavy oil recovery by cyclic steam-gas-chemical co-stimulation. Paper presented at the SPE Heavy Oil Conference and Exhibition, Kuwait City, Kuwait, 12–14 December 2011. SPE 149831.

Wehunt, C.D., Burke, N.E., Noonan, S.G. 2003. Technical Challenges for Offshore Heavy Oil Field Developments. OTC15281.

Xu, W.J., Chen, Z.X., Shan, J.C. 2013. Studies and Pilot Project on Steam Stimulation with Multiple Fluids for Offshore Heavy Oil Reservoirs. SPE165383.

Zhao, D.W., Wang, J., Gates, I.D. 2014. Thermal recovery strategies for thin heavy oil reservoirs. Fuel 117: 431–441.

Zhong L.G., Sun Y.T. 2013. Cyclic steam, gas enhance Bohai heavy oil production. Oil & Gas Journal 111: 74-80.

Zhong, L., Dong, Z., Hou, J. 2013. Investigation on principles of enhanced offshore heavy oil recovery by co-injection of steam with flue gas. Paper presented at the SPE Enhanced Oil Recovery Conference, Kuala Lumpur, Malaysia, 2–4 July 2013. SPE 165231.

Discuss on application and development of bamboo and rattan materials in low carbon furniture

Jing Zhang
Tongling in Anhui Province, China; Master, Mainly Engaged in Product Development and Design

WanDong Bai
Chuxiong in Yunnan Province, China; Master, Mainly Engaged in Product Development and Design

Jia Xu
Kunming in Yunnan Province, China; Professor, Mainly Engaged in Products (Furniture) and Innovation System

WanQiu Jiang
Nanyang in Henan Province, China; Master, Mainly Engaged in Product Development and Design

ABSTRACT: The concept of low carbon life brings a new platform for furniture design, through analysing the features and characteristics of low carbon furniture, exploring the low carbon functions of bamboo and rattan materials in order to analyse the trend and development for green materials, thus finding a new direction for the future of furniture design.

1 LOW CARBON AND FURNITURE DESIGN

Since the opening of the Copenhagen Climate Change conference on 7 December 2009, this conference has been named "the most important conference ever", and "the conference that changed the destiny of planet Earth". The conference was trying to build up a frame in limiting greenhouse gas emissions, which severely evokes the deep introspection of modern human life styles and modes of production. Our country has been exploring financial patterns in energy conservation and emission reduction under this background ever since. Proposing to use sustainable development has reflected the shifting awareness to low carbon financial patterns of our country.

Low-carbon is the most natural life style, the life is truly so-called complete if we can create zero artificial temperature in our residence as well as working place, eliminating carbon emission as much as possible, using natural ventilation and day lighting instead. A low carbon life means to reduce the energy we regularly use as much as possible, especially the emission of carbon dioxide. It is a nature of life to decrease our negative influence on the environment while living healthier, a natural and safer life using a dynamic communication with nature, which may be helpful in adjusting the state of body and mind to a comfortable condition while improving our lives on earth.

Low carbon life has led and directed the development of furniture design and new materials as an advanced concept. Low carbon design would be processed on the basis of our choice, design and producing the raw materials. Aiming to use and recycle the function and influence during the whole life circle of materials in order to realize the balance and unification of the actual environment. Low carbon material is the elemental basis of low carbon furniture design, thus it is important to reduce the quantity of non-renewable materials while ensuring the quality of furniture. They will have low energy consumption, low pollution, low emission, long service life, zero hazardous substance in function duration, and can be recycled. Being simple and natural makes bamboo and rattan materials the best choices as a low carbon, environmentally friendly materials in furniture design.

2 MATERIAL FEATURES OF LOW CARBON FURNITURE

Through long term practice and development, humans have been aware that only by reducing and preventing pollution from source of the problem can the deterioration be effectively controlled. On one hand, the quality of air and earth can be improved, on the other hand, by using various raw materials the cost of design can be reduced. Among low carbon furniture design, the first rule is to make a possible plan with environmental protection interest in mind, in order to avoid negative influence to the environment during every process of production, which can enhance the competitiveness for household products. Materials of low carbon furniture shall be in features of low energy consumption, low pollution, low emission and longevity, especially if there are hazardous substances during the production. There should be several features below to realizing the low emission of low carbon furniture.

2.1 Carbon sequestration function

Low carbon furniture has a carbon sequestration function, which is the ability to eliminate carbon dioxide during the growth or forming of materials, which can reduce the greenhouse gases in atmosphere and embody the core purpose of low carbon life.

2.2 Renewability

The materials used in production of low carbon furniture are renewable, which means that they are often made using renewable resources and can be reused by humans, for instance bamboo, rattan, paper, and some other materials that are reproducible can also be included.

2.3 Low material consumption

In regular condition of low carbon furniture, the less consumption of the materials, the more saved resources, and the higher the utilisation rate.

2.4 Low pollution

There is very little pollution during the production of low carbon furniture; pollution mostly includes: water pollution, air pollution, sound pollution, radioactive contamination, and heavy metal pollution.

2.5 Non-hazardous

There would not be any hazardous substances to the consumers during the production and the use of low carbon furniture, for instance: formaldehyde, methylbenzene, or heavy metals.

2.6 Recyclable

Low carbon furniture can be recycled after discarding, and even if the materials cannot be recycled, it will cause no harm to the living environment.

3 LOW CARBON FEATURE OF BAMBOO AND RATTAN MATERIALS IN FURNITURE DESIGN

3.1 Low carbon feature of bamboo material

Bamboo forests are called "the Second Forest" in our country, producing top quantity bamboo all over the world. Bamboo is not only sustainable material but also an excellent kind of carbon plant resource. Bamboo is one of the fastest growing plants in the world, which can mature in three or four years, and can re-grow fast after being cut down. It possesses a carbon

Figure 1. Bamboo Chair.

dioxide absorbing capacity four times more than that of other trees. Its features, include the ability to be recycled, large storage and low cost. It is doubtlessly a green substitute for wood as it is the source of the constant deterioration of the 20% coverage of natural forest in our country.

Bamboo material has straight texture, high tenacity and cleavability, which makes it easy to weave whilst bringing a breeze of elegance with simplicity. Various styles of furniture are made through simple splitting, dashing, planning and grinding (refer to Fig. 1). Bamboo products are tougher than usual wooden products, thus can replace a thicker wooden material with a thinner bamboo material, in order to achieve the financial advantage. On the other hand, because of its special texture, smooth touch, and simple colour it is convenient for bleaching, dyeing and carbonising. Furthermore, it can compete with some of the broad leaf materials and is an excellent choice for integrated materials. Therefore, bamboo material is one of the most widely used materials among low carbon emission furniture.

3.2 Low carbon emission feature of rattan material

Rattan furniture is one of the oldest types of furniture in the world. In ancient India and the Philippines, rattan was used in furniture making; it was often sliced and woven into various patterns, which were used as chair backs or other ornaments, as the using time gaining, the rattan will be more and more smooth. Rattan is a plant which grows in tropical rainforests, the stem of rattan is full of pipes with brilliant breathability, while its skin has tough characteristics. (refer to Fig. 2)

Process of producing rattan furniture has contributed to its low carbon feature; rattan furniture is made using traditional crafting methods, and can be greatly diverse in style, thus can reduce the bonding

Figure 2. Rattan Furniture.

or gluing process during production. Paints used on the surface of rattan are varnish and plant lacquer. Only plant lacquer matches the surface feature of rattan material, merges with the organic feature perfectly with no chemical pollution. After stewing, drying, mould-proofing, and sterilisation, rattan can be the best low carbon material for furniture making.

Besides, the use of rattan follows the principle of energy saving. Rattans used for furniture making y are mostly cane and rattan, white and red. Rattans are mild in characteristic, namely the natural feature of warm in winter and cool in summer, which is also energy saving in both summer and winter in door.

4 SPIRITUAL CONNOTATIONS OF BAMBOO AND RATTAN FURNITURE

The cultural advantage of Bamboo-Rattan kingdom is at top of the low carbon financial world and low carbon life style; its promising prospect is manifested by touchable and visible experience which both embodies the wisdom of harmony in Chinese bamboo and rattan culture and delivers the low carbon lifestyle all over the world.

4.1 *Bamboo—simple and strong*

In Chinese culture, bamboo has being given religions idea, personality feature, aesthetic dream and emotional meaning in it is the expression of the inner spirit of the Chinese nation, and has become an important spiritual symbol in Chinese culture.

Bamboo has a long history of planting in our country, it is deeply related to the everyday life of the Chinese people; many plant features of bamboo have being given spiritual personality thus forming a cultural interest. Ancient scholars like to write, paint or sing about bamboo, which they used to express their modesty and nobility. Dongpo Su, the most famous poet ever wrote: "no prefer fresh meat, not habitat without bamboo". People have sublimated the features of bamboo into excellent personality of human spirit, such as modesty and nobility; they belong to the virtuous part of humanity, and this connotation has become a symbol of a national moral and aesthetic spirit.

4.2 *Rattan—lingering, tough and tenacious*

Rattan not only provides the basic material for craft and architecture in our daily life, but also creates a rich cultural life for people. Rattan has a long history and has become one of the classic materials for art in the Eastern world.

In traditional Chinese culture, rattan is intimately related to folk customs, but also as a spiritual symbol in many areas. Every year when the Hani people celebrate their Yeku holiday, they make swings in rattan. Because rattan is strong and durable, the Li people in Hainan province present rattan crafts when marrying off their daughters, which means the blessing for a happy life and an early arrival of a baby boy. There is art craft also, named the "Yearly Gift", which is a picture of rigorous living and colourful nature, where patterns of rattan are also included. It reflects the superb workmanship of Li people and their spirit. Some minorities may adorn themselves with rattan accessories to ward off evil.

5 MARKET TREND OF BAMBOO AND RATTAN LOW CARBON FURNITURE

The features of bamboo and rattan materials themselves inevitably have some disadvantages, because of the similarity between bamboo-rattan material and wooden material, where the competition is also obvious. The development of bamboo-rattan furniture can be described in two ways: expanding consumers' interest and expending industrial market.

5.1 *Expanding consumers' market*

So far, to reduce the cost and market price are keys to attract more consumers to purchase bamboo and rattan furniture, but the importance of reducing cost is getting less vital, while the social awareness of culture and environment are becoming major impact on the consumers' choice of green products. Bamboo and rattan products shall stress their practicability, environmental protection property and ornamental value. The consumers care about the artistic value of bamboo and rattan products, however the mildew or hemiparasites are worrying problems. Many consumers have the awareness of knowing the environmental protection property and health value of bamboo and rattan furniture, and from this point of view, they would purchase green products even if they are more expensive than plastic ones. During the development of consumers' market, features of consumers need to be considered, such as the potential influences which might occur by the differences in gender, age, education level, financial position, so that various products can be developed accordingly. Increasing varieties, enhancing quality, promoting class and aesthetics, adding cultural value, increasing commercial networks, and low carbon capacity features shall be given more publicity.

5.2 *Expanding industrial market*

Price and quality of products are major issues in expanding the market in this industry field, and to solve these problems requires technological breakthrough, for instance use of structure glue-laminated bamboo, wood-bamboo composite, sliced bamboo and to enhance the innovation in craft technology. Besides, to build an integrated bamboo-rattan network of producers distributors, consumers and all the related machines, tools, and paints with all kinds of suppliers to the relevant industries, updating advanced technological information and news and keeping enhancing the steps in developing bamboo and rattan industry as well as the development of rural areas in China. According to prediction, in the following 5 to 10 years, processing craft technology of bamboo and rattan will be revolutionary advanced, additional value of bamboo-rattan products will progress even more with the advancement of both producing efficiency and utilisation rate.

6 SUMMARY

Led by the concept of sustainable development, low carbon has constantly been an important topic for a long time, as the low carbon emission concept brings greater opportunities for furniture design, but also more challenges. Low carbon emission design may be based on the life circle function of product designs: making, using, recycling and the consequences for the environment that may occur during this process. The aim is to achieve the integration of product and the environment, designing the furniture that might be beneficial to human health and match the purpose of sustainable development. The core concept of a low carbon emission life is to fulfil human activities in unadorned simplicity, calling for lifestyles that are green, energy saving, natural, comfortable and environmentally friendly. Bamboo and rattan materials happen to match this low carbon emission feature and the big trend of sustainable development, which has very good prospects. When we promote the craft skills of processing bamboo and rattan materials, it is also important for us to embody our national spirit into expressing and branding the artistic feature of our culture. What is national is also global. Chinese bamboo and rattan furniture has a profound cultural background, which surely can prevail in the world market and become the mainstream of low carbon furniture industry.

As above, although the future development of bamboo and rattan furniture is not totally clear, but under the leading of global low carbon concept, and the deeper research in sustainable development, modern furniture market has to go in to pluralism, as green materials, the renewable material may have broader platform and become new favourites in future furniture market.

REFERENCES

Chunli Zuo, Jinfang Yue. Brief Analysis of Rattan Furniture Structure [J]. Guangxi Light Industry, 2011, 01:163–164.

Gang Huo. Featured Rattan Furniture [J]. Interior, 1986, 01:23.

Jun Liu. Technological Characteristics of Bamboo Furniture [J]. Furniture and Interior Decoration, 2001, 03:49–51.

Ruidong Wu, Zhihui Wu. Theory and Practice of Low Carbon in Furniture Industry [J]. Package Engineering, 2011, 02: 64–67.

Weiwei Sun, Dejun Li. Design Aesthetic of Traditional Chinese Bamboo and Rattan Furniture [J]. Journal of Bamboo Reseach, 2014, 01:52–58+62.

Wei Xiong. Research on Low Carbon Furniture Design [J]. Furniture and Interior Decoration, 2011, 09:13–15.

Wei Xiong. Research on Innovation of Low Carbon Furniture [J]. Package Engineering, 2012, 04:68–71.

Xingxiong Liu. Innovation of Traditional Bamboo and Rattan Furniture [J]. Artists, 2008, S1:130–131+153.

Yanping Deng, Fujian Liu, Lingchao Hai, Depei Liang. Standard Analysis of Low Carbon Furniture [J]. Standard Science, 2011, 01:36–39.

Zhongqiang Yang. Features and Applications of Bamboo and Rattan Materials [J]. Newsletter of Rattan World, 2013, 03:24–26.

Zhou Lv, Xiangdong Dai. Discuss on Low Carbon Furniture under "Two-orient Society" [J]. Hunan Social Science, 2012, 06:132–135.

Analysis of geological conditions of coal-bed methane and tight gas, and favourable areas of selection in the Upper Paleozoic of the Ordos Basin

Peng Liu, WeiFeng Wang, Lei Meng & Shuai Jiang
School of Geoscience, China University of Petroleum, Qingdao, China

ABSTRACT: The Upper Paleozoic in the Ordos Basin is favourable to form Coal-Bed Methane (CBM) and tight sandstone gas resources. The source rocks are widely developed in the whole basin with characteristics of high maturity, high intensity of gas generation and wide gas supply. The coal layer 5$^#$ of Shanxi Formation and coal layer 8$^#$ of Taiyuan Formation are the main coal reservoirs, which are mainly distributed in the eastern and western basin. The coal rank is medium-high with high adsorption capacity. The main tight reservoir intervals are Shanxi Formation and Member 8 of Shihezi Formation developing in large scale delta facies sand body with characteristics of being in multiple phases longitudinally and joining in large pieces horizontally. Based on the accumulation conditions of both CBM and tight gas, this paper selects three target areas and two favourable areas to guide the exploration of unconventional gas resources in the Ordos Basin.

1 OVERVIEW OF THE BASIN

Located in the west margin of the Huabei massif, the Ordos Basin is a multi-cycle Craton basin of 25×10^4 km^2 with a flat terrain and stable internal structures and underdeveloped faults[1]. The Permo-Carboniferous coal series and tight sandstone reservoirs are extensive in distribution as source rocks, with a high thermal evolution level indicating huge coal-bed methane and tight gas resource potential (Table 1)[2–3]. The phenomenon of CBM and tight sandstone gas coexistence in the Upper Paleozoic of the Ordos Basin is common, for example, the Sulige gas field in the north-west and the new 1×10^{12}m^3 scale reserve area[4] in the east, are abundant in both CBM and tight sandstone gas resources.

Based on the current CBM and tight sandstone gas exploration status of the Ordos Basin, this paper selected favourable exploration areas by means of analysing the gas accumulation control factors, which provides a geological basis for co-exploration of CBM and tight sandstone gas in the Ordos Basin.

Table 1. The CBM and tight sandstone gas resources potential in the Ordos Basin[2–3].

Resources	Exploration area/ 10^4 km^2	Geological resources/ 10^{12}m^3	Recoverable resources/ 10^{12}m^3
CBM	3.75	4.59	1.17
tight gas	10	6~8	3~4

2 SOURCE ROCK CONDITION

2.1 Source rock distribution

The sedimentary paleotopography was very gentle in the Ordos region in the Late Ordovician-Early Carboniferous period, and the paleo-depositional slope angle was less than 1°[4]. Controlled by the gentle ancient paleotopography, the Permo-Carboniferous is stable with respect to lithofacies and sedimentary rock thickness. The Ordos basin developed transitional facies during the period of Late Ordovician-Early Carboniferous, when the sea water retreated frequently, which formed the wide-spread, marsh-like deposits of coal series source rocks[5].

The Upper Paleozoic coal series source rocks in the Ordos Basin consists of coal and dark mudstone. The coal layer mainly developed in Benxi Formation, Taiyuan Formation and Benxi Formation, while the mudstone developed in Taiyuan Formation and Benxi Formation. The coal layer is distributed throughout almost the entire basin, its thickness mostly varies from 6 m to 20 m with the thick characteristics in the west and east, and thin characteristics in the middle. There are three thick centres of a coal layer, which are located in the Shizuishan-Yinchuan area of the north-west, Guyuan area of the south-west, and the Ordos-Shenmu area of the north-east, while the coal layer is relatively thin in the south of the basin, especially the southern area of the Pinliang-Huanxian-Jingbian-Yan'an-Hancheng line, where the thickness is below 6 m (Figure 1).

The Upper Paleozoic dark mudstone is also widely spread in the basin. It is thicker in the west, then the

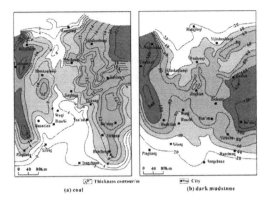

Figure 1. Thickness contour of coal and dark mudstone of the Upper Carboniferous-Lower Permian in the Ordos Basin[5].

east, and the dark mudstone in the middle basin is the thinnest, but has a stable distribution. The mudstone is more than 80 m thick in the west of the Shizuishan-Etuokeqianqi County Huan area and more than 100 m in the east of Jiaxian-Da'ning, and between them the dark mudstone is 40~90 m thick (Figure 1).

2.2 Abundance of organic matter and thermal evolution characteristics of source rock

A warm and humid paleo-climate, coastal marshes and lagoons of ancient sedimentary environments provided gainful conditions for the blooms of paleontology, leading to the formation of the Upper Paleozoic source rock, with a high organic carbon content[6]. The organic carbon content of the Upper Paleozoic coal layer is 38.31%~89.17% (the average is 67.3%), the dark mudstone is 0.05%~23.38%, with an average of 2.93%[7].

The Ordos Basin experienced strong tectonic uplift and erosion during the Early Cretaceous period, when the average thermal lithospheric thickness was down to 65 km[8], leading to the appearance of a heat flow peak. Affected by the tectonic thermal events, the organic matter evolution in the source rocks had generally entered a mature-high mature stage; the maturity of the Upper Paleozoic source rocks in most area of the Ordos Basin is relatively high with Ro > 1.3%[9]. The Ro value of the Paleozoic organic matter is highest in the south, indicating an over-mature, dry gas stage and decreases towards the north in a sequence of over-mature, dry-wet gas transition, wet gas zone and condensate oil-wet gas zone (Figure 2).

2.3 Gas generating intensity of source rock

The perfect match between high organic carbon content and a high mature stage creates beneficial conditions for gas generation of the Upper Paleozoic source rock in the Ordos Basin. The main gas generating intensity of the Upper Paleozoic source

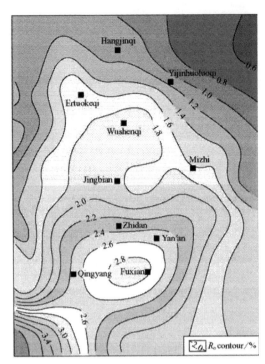

Figure 2. Ro contour of Upper Paleozoic source rocks in the Ordos Basin[9].

rock in the Ordos Basin is 8×10^8–40×10^8 m^3/km^2, the blocks with a gas generating intensity of over 10×10^8 m^3/km^2 account for 71.6% of the total basin area[5], characterized by high intensity and large area hydrocarbon generation. The gas generating intensity distribution of the Upper Paleozoic source rock is consistent with that of source rock thickness. The high values are located in the east, north-west and south-west of the basin, which are more than 24×10^8 m^3/km^2, then the Wushenqi-Jingbian-Huanxian area in the centre of the basin where the in the gas generating intensity is more than 20×10^8 m^3/km^2. Finally, in the Sulige-Etuokeqianqi-Qingtongxia area it is $(10~20) \times 10^8$ m^3/km^2, while it is less than 10×10^8 m^3/km^2 in the northern and southern areas of the basin.

3 RESERVOIRS

3.1 Coal reservoirs

The Upper Paleozoic coal reservoir is mainly located in the eastern and western basin. The coal layer 5$^#$ of Shanxi Formation and 8$^#$ of Taiyuan Formation were deposited stably in the entire basin and are the main exploration target layers[10]. Affected by thermal evolution, the coal rank of the basin is relatively high; mostly high-rank bituminous and anthracite. The coal rank decreases with a gridle pattern from the south-west towards the north-east in the sequence of dry coal, coking coal, fat coal and gas coal. It is highest around the Qingyang-Fuxian-Ningxian area where anthracite has developed, while bituminous coal has mainly developed in the other area.

The gas content of the Upper Paleozoic coal layer changes greatly, which is (0.01~23.25) m^3/t[3]. The gas content in the eastern margin of the basin is 2.46~18.36 m^3/t, in the centre it is relatively higher. In the Weibei coal field of the south basin the gas content is more than 4 m^3/t and decreases towards the east, and the highest gas content is located in the Hancheng area, where the average is 10.1 m^3/t. The western margin of the Ordos Basin has a medium gas content, which is 3.50~8.40 m^3/t.

3.2 Tight reservoirs

The Upper Paleozoic of the Ordos Basin developed many tight reservoirs, the major gas-bearing layers include the 8th Member of the Xiashihezi Formation and the Shanxi Formation[11]. During the Late Paleozoic period, the basin relief was very gentle, developing large-sized, river-delta sand bodies[5]. As that the lake level waved frequently, and the lake strandline swayed greatly and quickly, river channels of different periods migrated laterally and repeatedly, leading to multiple, vertical superimpositions and overlapping, horizontal connections of sand bodies. For instance, the major reservoir 8th Member of the Xiashihezi Formation was formed in a large area, with a width of 10~20 km, thickness of 10~30 m and extension of more than 300 km.

According to the statistics' results[11] in surface conditions, the porosities of 50.01% samples of the Upper Paleozoic of the Ordos Basin are lower than 8%, the porosities of 41.12% samples are 8%–12% and the porosities of 8.87% samples are larger than 12%; the permeabilities of 88.6% samples are lower than 0.1×10^{-3} μ, and the permeabilities of 28.4% samples are lower than 0.1×10^{-3} μ. In overburden pressure conditions, the permeabilities of 89% samples are lower than 0.1×10^{-3} μm^2, belonging to typical tight reservoirs.

4 ACCUMULATION CHARACTERISTICS

Because the source rock of CBM is the same as its reservoir, its gas accumulating time is up to the abundant gas generation time; tight sandstone gas accumulating time is controlled by the matching relationship between the gas charging time and tight reservoir forming time. The Upper Paleozoic coal series source rocks had generally entered a mature-over mature stage during the Late Jurassic-Early Cretaceous period, when considerable gas was generated[12–13], while tight sands were compacted in the Late Triassic-Middle Jurassic period[14], which means that the reservoir compaction preceded the gas migration and accumulation. Therefore, the pool formation times of CBM and tight sandstone gas were

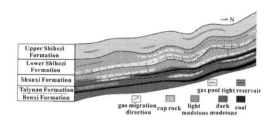

Figure 3. Accumulation model of the Upper Paleozoic coal-bed methane and tight gas in the Ordos Basin[4].

both controlled by gas generation time, that is to say their accumulation time was the Late Jurassic-Early Cretaceous period.

According to the previous study[11], the methane content of the Upper Paleozoic tight gas reservoir is apparently positive related to the Ro value; the closer between reservoirs and source rocks, the higher the gas saturation of the reservoirs, which indicates that tight gas was migrated and accumulated in short distances. The process is as follows: the gas generated from source rocks was firstly absorbed by the coal reservoirs. With the consistent generation of the gas, it gradually migrated into nearby, tight, sandstone reservoirs in a vertical plane, accumulating to form tight gas pools (Figure 3).

5 FAVOURABLE AREA SELECTION OF CBM AND TIGHT GAS

Through a comprehensive analysis of accumulation conditions of coal-bed methane and tight sandstone gas in the Upper Paleozoic of the Ordos Basin, this paper proposes favourable areas and target areas two types area for CBM and tight gas joint exploration. The former are mainly distributed in the east and west margin of the basin, characterized by a thick coal layer, high gas generating intensity and the largest exploration potential; the latter are band-like, located in the north-east and north-west of the Ordos Basin where the exploration potential is larger (Figure 4).

Based on the distribution of coal series source rock, thermal evolution, gas generating intensity and tight reservoir distribution, this paper ascertains three target areas: Block I, Block II and Block III. Block I is located in the south of Wushenzhao-Xingxian and the east of Wushenqi-Hengshan-Zizhou-Yanchang, and is the largest area among the three blocks, characterized by thick source rock (coal thickness is more than 10 m), a high mature-over mature stage and high gas generating intensity $(16~40) \times 10^8$ m^3/km^2. Besides, the tight reservoirs in the 8th Member of the Xiashihezi Formation developed stably in the area, with a thickness of 10~30 m. Above all, Block I is chosen to be the primary exploration target of CBM and tight gas. Compared with Block I, Block II and Block III have a relatively lower gas generating intensity, and are the secondary exploration targets among the target areas.

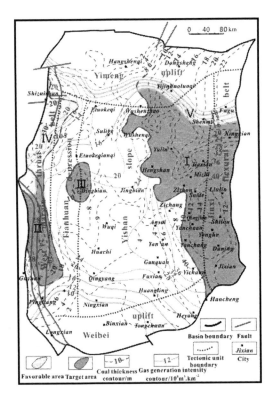

Figure 4. Favourable exploration areas selection of the Upper Paleozoic CBM and tight gas in the Ordos Basin (coal thickness contours and gas generation intensity contours are from reference [5]).

There are two favourable areas to be chosen for the Upper Paleozoic CBM and tight gas exploration in Ordos. Block IV is located in the north-west margin of the basin, with a coal thickness of 10 m∼30 m, a gas generating intensity of $(10\sim16) \times 10^8$ m^3/km^2 and tight reservoirs. Block V is around the Yijinhuoluoqi-Shenmu area, a northern basin, with the same gas accumulation conditions as Block IV, showing a large exploration potential of CBM and tight gas.

6 CONCLUSIONS

(1) The Upper Paleozoic source rocks are widely developed in the whole basin; the organic matter evolution in the source rocks has generally entered a mature-high mature stage. Maturity of the Upper Paleozoic source rocks in most areas of the Ordos Basin is relatively high with Ro > 1.3%; the main gas generating intensity of source rock is 8×10^8–40×10^8 m^3/km^2 and the blocks with gas generating intensity of over 10×10^8 m^3/km^2, account for 71.6% of the total basin area, characterized by high intensity and large area hydrocarbon generation, which provides favourable conditions for CBM and tight gas accumulation.

(2) Controlled by a sedimentary environment, the main coal reservoirs, layer 5$^{\#}$ of Shanxi Formation and layer 8$^{\#}$ of Taiyuan Formation, are distributed in the eastern and western basin. The coal rank is medium-high with high adsorption capacity. The main tight reservoir intervals are Shanxi Formation and Member 8 of the Shihezi Formation developing in large scale delta facies, a sand body with characteristics of being in multiple phases longitudinally and joining in large pieces horizontally. The porosity is less than 12% and the permeability is less than 1×10^{-3} μm^2.

(3) The Late Jurassic-Early Cretaceous period was the main pool formation time of CBM and tight sandstone gas. The CBM accumulation model is "self-generation, self-storage", while the tight gas migrated and accumulated in a short distance.

(4) The multiple vertical superimposition of the Upper Paleozoic coal layers and tight reservoirs provides perfect conditions for the unconventional gas resources joint exploration. The target exploration areas are located in the east and west of the Ordos basin, with thick coal layers, high intensity of gas generation, and tight reservoirs widely distributed.

REFERENCES

Yang Junjie 1991. The discovery of Lower Paleozoic natural gas in Shanganning Basin. *Natural Gas Industry* 11(2): 1–6.

Li Jianzhong, Zheng Min, Zhang Guosheng, et al., 2012. Potential and prospects of conventional and unconventional natural gas resource in China. *Acta Petrolei Sinica* 33(S1): 89–98.

Feng Sanli, Ye Jianping, Zhang Suian 2002. Coal bed methane resources in the Ordos Basin and its development potential. *Geological Bulletin of China* 21(10): 658–662.

Yang Hua, Liu Xinshe 2014. Progress of Paleozoic coal-derived gas exploration in the Ordos Basin, West China. *Petroleum Exploration and Development* 41(2): 129–137.

Yang Hua, Fu Jinhua, Liu Xinshe, et al., 2012. Formation conditions and exploration technology of large-scale tight sandstone gas reservoir in Sulige. *Acta Petrolei Sinica* 33(S1): 27–38.

Li Zengxue, Yu Jifeng, Guo Jianbin, et al., 2003. Analysis on coal formation under transgression events and its mechanism in an epicontinental sea basin. *Acta Sedimentologica Sinica* 21(2): 288–296.

Fu Jinhua, Wei Xinshan, Ren Junfeng 2008. Distribution and genesis of large-scale Upper Palaeozoic lithologic gas reservoirs on Yishan Slope. *Petroleum Exploration and Development* 35(6): 664–667.

Jiao Yaxian, Qiu Nansheng, Li Wenzheng, et al., 2013. The Mesozoic-Cenozoic evolution of lithosphere thickness in Ordos basin constrained by geothermal evidence. *Chinese Geophysics* 56(9): 3051–3060.

Zhang Shuichang, Mi Jingkui, Liu Liuhong, et al., 2009. Geological features and formation of coal-formed tight sandstone gas pools in China: Cases from Upper Paleozoic gas pools, Ordos Basin and Xujiahe Formation gas pools, Sichuan Basin. *Petroleum Exploration and Development* 36(3): 320–330.

Quan Haiqi, Ma Cailin, Zhang Junxue, et al., 2010. Analysis of CBM exploration prospects and law of coal accumulation in Carboniferous-Permian of Ordos Basin. *Low Permeability oil and gas Fields* 1: 23–27.

Yang Hua, Fu Jinhua, Liu Xinshe, et al., 2012. Accumulation conditions and exploration and development of tight gas in the Upper Paleozoic of the Ordos Basin. *Petroleum Exploration and Development* 39(3): 295–303.

Liu Xinshe, Zhou Lifa, Hou Yundong 2007. Study of gas charging in the Upper Paleozoic of Ordos Basin using fluid inclusion. *Acta Petrolei Sinica* 28(6): 37–42.

Dou Weitan, Liu Xinshe, Wang Tao, et al., 2010. The origin of formation water and the regularity of gas and water distribution for the Sulige gas field Ordos Basin. *Acta Petrolei Sinica* 31(5): 767–773.

Li Zhongdong, Hui Kuanyang, Li Liang, et al., 2008. Analysis of characteristics of gas migration and reservoir-forming in the Upper Paleozoic of northern Ordos Basin. *Journal of Mineralogy and Petrology* 28(3): 77–83.

Theory and practice of sustainable development

The current situation and outlook on environmental protection and sustainable development in China

LuSi Zhang
Doctor, Lecture, School of Architecture, Harbin Institute of Technology, Harbin, China

Yao Feng, GuangHao Li & XiaoGuang Liu
Harbin Institute of Technology, Harbin, China

ABSTRACT: After lengthy struggles, humans are faced with ecological damage and environmental pollution despite the great achievements in nature reconstruction, as well as social and economic development, which are posing threats to their survival and development. Therefore, it has been an urgent and arduous task for all human beings to protect and improve the ecological environment in order to realize the sustainable development of human society. Hence, environmental protection acts as the premise for social development. Besides, environmental protection, along with the realization of harmony between man and nature, serves as a precondition for further development of the economy, as well as a guarantee for the continuation of human civilization.

1 THE STATUS QUO OF ENVIRONMENTAL PROTECTION UNDER THE NEW SITUATION

1.1 Theoretical description of environmental protection

Environmental protection, just as its name implies, refers to the general term of all practical activities conducted by human beings in order to solve the existing or potential problems and maintain their personal existence as well as development. Relevant ways and means include not only engineering technology and public administration, but also laws, and the economy, as well as publicity and education. With the rapid development of the industry in recent years, environmental issues, once again, draw people's attention, for which environmental protection agencies, as well as relevant laws are further improved.

1.2 Environmental status

The environmental status in China can be summarized as: some parts have been improved, while the overall situation is deteriorating. Environmental damage pollution, along with ecological pollution, have become increasingly important constraints for economic and social development in China. In spite of the progress, there is still a severe situation concerning environmental protection in China.

2 SEVERAL ASPECTS OF ENVIRONMENTAL DESTRUCTION

2.1 The soil is destroyed, and the fresh water resource is threatened

According to the investigation of related organizations, the fertile degree of arable land is reducing due to overexploitation. Besides, forest cover is vanishing, and soil is in severe denudation. Furthermore, due to the sharp increase of the population as well as rapid development of the industry, solid waste cannot be disposed of scientifically and is being dumped and stacked constantly on the surface of soil. Moreover, harmful wastewater constantly permeates the soil; harmful gas as well as floating dust in the air also falls into the soil with the falling rain. These, to some extent will impede the normal function of the soil, which will reduce the crop yield and quality, and affect human health indirectly through food, vegetables, fruits, etc.

2.2 Climatic change, energy waste and the greenhouse effect pose a great threat to all human beings

Rise of temperature will have a strong impact on the agricultural and ecological systems. Therefore, western and developing countries should accelerate the transferring process of energy conservation technologies. Particularly, we should adopt the economic incentive method, encouraging the industrialists to develop and improve relevant technologies in order to increase the industrial resource use efficiency.

2.3 Biological diversity is decreasing, and existence is challenged

Natural areas have become smaller and smaller due to the acceleration of urbanization, agricultural development, and forest reduction, as well as environmental pollution, which leads to the loss of the original ecological environment for thousands of species. Then, extinction will be the final result. However, the extinction of some species will destroy the stability and diversity of the ecological system.

2.4 Forestry areas on the earth's surface are declining sharply

Environmental pollution will lead to the sharp decline of forestry areas, which is reflected in acid rain corrosion as well as harmful gas emission. In the event that the aforementioned gas drops into the soil as rainfall after being let into the air, the original plants will thus be affected. Serious sand storms in cities.

2.5 Deteriorating chemical pollution and air pollution

Air in most cities contains pollutants brought about by heating, transportation, and factory production. These pollutants pose a threat to the health of tens of millions of citizens. The poisonous gases includes mainly CO, SO_2, and NO_2, as well as inhalable particles. Millions of chemical compounds brought about by industry exist in the air, soil, water, plants, and animals, as well as in human bodies; even the ice cover, the last large-scale ecological system, has been polluted. All of these organic compounds, along with heavy metals and poisonous products, exist in the food chain, which will eventually threaten the health of flora and fauna, thus bringing about cancer as well as a decrease in the soil's fertility.

2.6 Chaotic urbanization

Up until the end of the present century, there will be twenty-one metropolises in the world. Living conditions will further deteriorate, for example, by the existence of crowding, polluted water, and bad sanitary conditions, as well as a lack in the sense of safety. The disorderly expansion of these metropolises also harms natural areas. Hence, unlimited urbanization will be deemed as a newly emerging disadvantage for civilization.

2.7 Overexploitation of marine resources and pollution of the seaboard

Due to overfishing, marine fishery resources are decreasing at a frightening speed. Therefore, the poor, living on the protein of marine products, are faced with the threat of hunger. Heavy metals, together with organic phosphorus compounds, etc., concentrated in fish and meat will cause great danger to the fish eaters. Besides, coastal areas will suffer huge population pressures, and 60% of the total population live less than 100 km from the ocean, which may cause a loss of balance to those fragile places.

3 INVESTIGATION ON A FEASIBLE PLAN FOR ENVIRONMENTAL PROTECTION

Environmental protection is the general term for all human activities relating to protecting and improving the environment. It aims at coordinating the relationship between mankind and the environment, and solving various problems through the theory and methods of environmental sciences. Various measures relating to science and technology are taken to utilize the natural resources reasonably and to avoid environmental pollution and destruction, so as to maintain and develop ecological balance. In this way, the development of human society is ensured. Here, the protection contains at least three meanings: protection for the natural environment; protection for people's living environment; protection for life on the earth.

3.1 Continuing to unswervingly improve local law & regulations relating to environmental protection, so as to provide a legal basis and support for environmental protection

Regarding the treatment of poisonous elements, for example, resulting from industrial production such as the "three wastes" (waste gas; wastewater; industrial residue), dust, radioactive substances, smoke, wastewater, garbage, and relevant enterprises, should be encouraged to adopt environmental protection measures. Moreover, high energy-consuming enterprises should be encouraged to install energy conservation facilities. Besides this, the "cost" resulting from the destruction of the environment should be raised in accordance with market disciplines; environmental protection laws and regulations, as well as a series of environmental evaluations, assessment, supervision, and punishment systems, shall be strictly enforced. Social productive forces should be protected by law, so as to promote the transformation of economic growth patterns as well as the implementation of a sustainable development strategy. Prevention of the environment shall be brought into the track of legalization management. Furthermore, it is of vital importance for the enterprises to develop a sense of social responsibility.

3.2 Developing people's new concept of environmental protection, and enhancing the publicity and education by laws and regulations relating to environmental protection from children

There are two aspects involved in environmental protection: the first is to protect and improve the quality of the environment and protect people's physical and mental health in order to avoid body variation and degeneration under the influence of certain types of environment; the second is to rationally utilize natural resources, and reduce or erase harmful substances (before they enter into the environment) for the benefit of people's vital movement. March 22nd is 'World Earth Day', and people are trying to use various methods to attach importance (by people) to environmental protection. Of course, children are the next generation as well as the successors for the country. Hence, it is important to cultivate good environmental protection accomplishments for the long-term development of the country.

Laws and regulations concerning environmental protection have to be incorporated into the publicity and education systems through a joint connection with the legal publicity department. Besides, these laws should be an important content of basic legal knowledge for citizens, and should be incorporated in the assessment for popularizing the legal knowledge exam. Second is to establish an educational training system for integrated decisions of the environment and development.

3.3 Strengthening the management during construction, and eradicating exploitation brought about by not considering the environment

This could be achieved by specifying the basic principles of protecting the ecological environment, establishing and improving the synthetic decision-making system for the environment and development, and avoiding pollution and destruction to the environment resulting from engineering constructions, such as large-scale hydraulic engineering, railways, arterial roads, large ports and piers, airports, and large industrial projects. Besides, great decision systems of environmental impact assessment should be developed, and environmental influences should be assessed in the event of formulating great economic and technological policies, infrastructure construction, land reclamation, etc. Those environmental quality requirements not in accordance with national and provincial standards should not be implemented. Furthermore, planning and management of engineering appear also to be important, and the development of specialists in the relevant technology is deemed as a foresight in avoiding developments in exploitation.

3.4 Managing garbage classification, and success in recycling

About 50% of garbage is biological organics, and 30% to 40% is recyclable. "Turning waste into wealth" achieves not only economic value, but also promotes the development of good environmental awareness for citizens, which is priceless. Garbage classification lowers the treatment cost, thus creating a good social atmosphere.

4 TRENDS AND FUTURE PROSPECTS IN ENVIRONMENTAL PROTECTION

Environmental protection is an arduous task. Besides the participation of all people, some official or non-governmental agencies have played irreplaceable roles in guiding environmental protection. In the future process, environmental protection agencies will be further improved, and more and more people will participate in the process to make contributions. An improved relevant environmental protection agency is: the All-China Environment Federation (ACEF), which is a national non-profit social organization formed voluntarily by people, enterprises and public institutions keen on environmental protection causes. It aims at promoting the development of environmental causes in China, thus promoting the progress of the environmental cause for all mankind. During the process, a strategy for sustainable development is implemented, the goal of the environment and development for the country is focused upon, and the public as well as social environmental rights and interests are maintained. In "Greater China, Great Environment, Great Union", the organizational advantage of ACEF is fully embodied. Furthermore, the role as a bridge and tie between the government and society should be used so as to promote the development of the environmental cause in China, thus speeding up the progress of environmental causes for all mankind.

In addition, some other international organizations have joined the effort to protect the environment. The United Nations Environment Programme (UNEP) is one of only two UN institutions with headquarters in developing countries. In the twenty-first century when the international community and all the national governments are paying close attention to the global environment as well as to the world's sustainable development, more and more importance has been attached to the UNEP, which is playing an irreplaceable role.

Whilst speaking of environmental protection, the aim of legislating for holidays is to remind people to protect the environment, as well as to show unswerving determination for the process of environmental protection. For example, the national "Land Day" sets its date as June 25th each year. According to the decision of the 83rd executive meeting of the State Council, since 1991, the date of June 25th (when the "Land Management Law" was enacted) was set as the national "Land Day". "Land Day" is the first nationwide commemorating day decided by the State

Council, and China is the first country to set a certain commemorating day for land protection.

There are similar holidays around the world. For example, February 2nd is the "World Wetlands Day". Based on the "Convention on Wetlands of International Importance Especially as Waterfowl Habitats" signed in Ramsar, Iran in 1971, "wetlands" refer to the marsh lands, peat lands, or water areas, static or flowing fresh water, brackish water or saline water, as well as those waters less than 6 metres in depth at low tide. Wetlands have played an important part in the conservation of biological diversity, especially in the lives and migration of poultry.

Under the new trend when all people are vigorously advocated to join the effort, "environmental protection" is not just a slogan. Furthermore, everyone on earth is required to take practical action to sacrifice youth and passion for the survival of the planet. Only in this way can earth be fairly treated.

5 HOW TO SOLVE ENVIRONMENTAL PROBLEMS TO REALIZE SUSTAINABLE DEVELOPMENT

The effort in reinforcing the implementation of environmental law in China is mainly reflected in the following aspects: 1), Environmental law enforcement and environmental legislation are put in equal position; environmental law enforcement is set as the key to the legal construction of the environment; 2), Strengthening publicity and education as well as personnel training for environment law, popularizing the knowledge of environment law, and improving the environmental legal sense for the whole society, especially for government officials and administrative staff; 3), Strengthening the national construction of supervision and administration systems for the environment, such as establishing and perfecting environmental management organizations for governments at all levels, improving the rank of environment administrative authorities, as well as perfecting the construction and development of environmental management and law enforcement teams; 4), The administrative scope and administration authority of environment administrative authorities is enlarging, and the responsibilities of each department relating to environmental protection are becoming clearer and reinforced, while the abilities of executive law enforcement, as well as judicial capacity, are enhanced; 5), In reality, the rights and obligations of environmental subjects of legal relations are becoming more and more explicit and specific, which is the same as for the responsibilities of pollution source units, as well as the violation of environment law. Besides, the awareness of observing the environmental law has been enhanced; 6), Definite and peremptory norms, along with prohibitive, punitive, and incentive measures relating to environmental law are increasing; 7), The institutionalization of environmental protection measures, together with operability of environmental law provisions, measures and systems have been improved, while the standardization, reutilization as well as institutionalization of environmental law enforcement has been enhanced. Moreover, the system of environmental law enforcement is improving with each passing day; 8), The coordination of the legal construction of the environment (including environmental legislation and law enforcement) and environmental management has been enhanced, so has the supervision and inspection for the environmental law; 9, Environment policemen and environment courts have been founded so as to guarantee the environmental standing for suing individuals and entities. Furthermore, environmental lawsuits (including environmental administrative proceedings, civil actions and criminal proceedings) are being actively conducted; 10), Treatment of environmental disputes is improving; in dealing with civil injury disputes thereof, liability without fault, causal relationship presumption and burden of proof transferring are implemented for actions as environmental pollution and destruction, for which longer prescribed periods for litigation are employed; 11), Among civil action and administrative proceedings, right of action is enlarged; 12), Criminal law protection for the environment has been enhanced; a legal person crime from an environmental aspect is explicitly stipulated so as to strictly crack down on potential damage offences as well as environmental crime 13), Strike on illegal criminal activities and the degree of punishment dealt have both been enhanced; for law breakers thereof, bipartite as well as multi-partite punishment systems, etc., will be adopted.

5.1 Comprehensively establishing and implementing a scientific outlook on development; building and completing the environmental-friendly, decision-making institutional system

Sticking with people first, one starts by preserving people's environmental rights and interests and improving environmental quality; urban and rural development are balanced, regional development, coordination of economic and social development, integrated and harmonious development between man and nature, and manpower development and opening up in China through maintaining people's environmental rights and interests, as well as improving environmental quality; formulating relevant laws and regulations, development strategies and plans so as to promote the harmony between man and nature, realizing the coordination between economic development and population and resources as well as the environment, thus creating production development, a rich life and a sound ecological environment. The national economic accounting method for integrated environment and development has to be studied, and resource consumption and environmental damage, as well as environmental benefit during the development process, shall be brought into the evaluation system of economic development. Furthermore, government performance examination of environmental protection during the tenure for leading cadres shall be carried out, and the tendency of pure pursuit for GDP shall be avoided.

5.2 Vigorously developing the circular economy, and taking a new road to industrialization

To develop the circular economy means to take a new road to industrialization featuring high scientific and technological content, good economic returns, low resource consumption, little environmental pollution, and a full display of advantages in human resources. It should accelerate the transformation of production and consumption methods and build a resource-saving and environment-friendly society featuring water saving, land saving, energy conservation, materials saving, etc., as well as providing environmental protection. On the basis of the principle of "reduction, reuse and recycle" and with "improving the resource use efficiency and protecting the environment" as the core concept, industrial ecologicalization should be realized through effort. Moreover, cleaner production should be positively pursued so as to renew industrial parks as well as economic and technological development zones through ecologicalization and vigorously developing ecological agriculture. Besides, developing the waste recycling and reusing industry, as well as the environmental protection industry should also be pursued so that the resource productivity and circulation utilization rate can be improved. Strictness about environmental access and raising relevant access standards should be sought; restricting and prohibiting construction projects with high energy consumption, material consumption as well as water consumption and heavy pollution should be implemented. Implement forceful elimination system, and weed out those manufacturing technique, technology, equipment and enterprises with backward technology, resource waste and environmental pollution. Carry out control system for the total amount of pollutant emission, and reduce energy material consumption as well as pollutant emission for unit output value; actively use economic means & market mechanism to encourage resource saving and pollutant discharge reduction in all sectors; continue to promote pilot projects for various circular economies.: first is to protect and improve the environmental quality, and protect people's physical and mental health in order to avoid body variation and degeneration under the influence of certain environment; second is to rationally utilize natural resources, and reduce or erase harmful substances (before they entering into the environment) for the benefit of people's vital movement. March 22nd is "World Earth Day", and people are trying to use various methods to attach importance (from people) to environmental protection. Of course, children are the next generation as well as the successors for the country. Hence, it is important to cultivate good environmental protection accomplishments for the long-term development of the country.

Laws and regulations concerning environmental protection have to be incorporated into the publicity and education system through the joint connection with legal publicity department. Besides, those laws shall be the important content relating to the popularity of basic legal knowledge for citizens, and shall be incorporated in the assessment of popularizing legal knowledge exam. Second is to establish the educational training system for integrated decision of environment and development.

5.3 Solving prominent environmental issues and maintaining social stability as well as environmental safety

First, the strictest of measures has to be taken to protect the source of drinking water. This could be achieved by speeding up the pollution prevention for key sea areas and striving for practical results. Second, during the rapid expansion of urbanization, the optimization of urban planning layout and the acceleration of infrastructure construction for environmental protection in cities is important; by constantly improving the processing rate for wastewater and garbage, and by actively protecting the natural heritages in cities, such as natural vegetation and lake water systems, as well as tidal flat wetlands. Third, by accelerating the desulphuration of coal-fired power plants, as well the air pollution treatment for metallurgy, coloured, chemical and construction industries, improvements in energy use efficiency, and striving to develop new energies and reduce acid rain, as well as atmospheric dust pollution, would be beneficial. Fourth, by strengthening environmental protection in rural areas, and focusing on transforming production and life-styles of the farmers, a comprehensive improvement of the rural environment should be developed, together with a vigorous development of ecological agriculture and organic agriculture; management the non-point source pollution as well as soil pollution for breeding industry is important so as to guarantee the safety of agricultural products. Fifth, respect for natural law and the reinforcement of ecological protection is vital; the ecological function zone should be well managed, as well as the construction and management of nature protection areas; strengthening the exploitation of mineral resources as well as environment supervision for tourism development to avoid new destruction is important Sixthly, during the development of nuclear power, the environmental security control should be enhanced and nuclear and radiation environment safety should be ensured.

5.4 Strengthening international cooperation on environment protection and sustainable development

Based on its desire to solve environment and development issues in the country, China continues to improve the survival and development environment for 1.3 billion people. Meanwhile, the international cooperation on environment protection and sustainable development shall be further promoted. On the one hand, this could be achieved by active participation in environmental conventions and relevant trading through; environmental negotiations relating to climatic change and biodiversity protection, the safeguarding of international interests, performing international obligations, and making contributions towards solving the environment and development issues facing human beings. On the other hand, this could be brought about by actively promoting "leapfrog development" for the domestic environment through; the formulation of policies, laws and regulations and environmental standards relating to breaking the green trade barrier, and avoiding pollution diversion as well as harmful species invasion, in order to promote the trade development and ensure the national environmental safety.

In order to protect the global environment and realize the globally sustainable development, joint efforts all over the world are needed. Developed countries need to actively undertake responsibilities for environmental protection, increase the environmental financial aid for developing countries, and enhance the international transmission and cooperation of environmental technology and managerial experience. Moreover, trade barriers resulting from high environmental standards should be cancelled so as to promote the joint development of environmental protection and international trade. Furthermore, the market should be opened further to ease the resource and environmental pressures in developing countries, thus promoting the rational utilization of global resources. Developing countries also need to prevent and remedy the problem of pollution during the development process, thereby protecting the ecosystem.

With the arrival of the fifth anniversary of the Millennium Summit, over 170 countries and governments will get together, twice. Whether the millennium goal can be realized will be a significant topic to be focused on, and efforts that people make for realizing the joint commitment will be tested. We will keep trying, and turn the goal into specific actions; the historic responsibility shall be shouldered, and sustainable development shall be unswervingly insisted upon, which will contribute towards the realization of the millennium development goal, as well as the great rejuvenation of the Chinese nation.

We would like to strengthen the cooperation with various countries in the world, make progress together, and co-create a wonderful future for human society. In order to realize the sustainable development of society, everyone should start small.

ACKNOWLEDGEMENT

The authors gratefully acknowledge funding from Project 51308152 and 51208135, supported by National Nature Science Foundation of China, Project QC2013C040, supported by Natural Science Foundation of Heilongjiang Province, and Project HIT.NSRIF.2013069, supported by the Fundamental Research funds for the Central Universities.

REFERENCES

Liu Nanwei, 2001. *Physical geography*, Beijing: Science Press.
Li Chunhua, 2003. *Environmental Scientific Principles* Nanjing: Nanjing University Press.
Ye Wenhu. 2003. *Introduction to Sustainable Development.* Beijing: Higher Education Press.

ial damage
Research on experimental methods of structural progressive collapse resistance

YanHua Huo & ShengRong Ding
Nanchang University College of Science and Technology

ABSTRACT: Experiments of structural progressive collapse resistance are important foundations in recognizing the performance and characteristics of structures. The characteristics of two kinds of experiment model, the plane frame structure and spacial frame structure, are analysed and compared. Some details, such as simulation of initial local damage members, loading schemes, loading processes, member damage criteria and structural collapse criteria, are discussed.

Keywords: progressive collapse, experimental research, frame structure, spacial grid structure

Because of the action of an emergency upon a building's structure, the component chain damage caused by direct local failure brings about the entire structure's, or most of the structure's, collapse, a phenomenon known as "progressive collapse". For example, in 1995, a car bomb exploded beside the Murrah Federal building in Oklahoma. The underlying three pillars were directly seriously damaged; the transfer beam had failed because of the supporting pillars' failure, resulting in the collapse of the upper column and so on. This resulted in the collapse of the whole building. Since the Ronan Point apartment collapsed in 1968, study of the progressive collapse of buildings has been continued [1–6], but in our country, the research in this area started late. Presently, many researchers focus on the analysis method of the progressive collapse. One of the most common numerical analysis methods is the alternate load path method; the initial damaged members are removed, the internal force redistribution of structure after change in the original load (gravity load) is analysed and the remaining structure's new state of equilibrium, which has occurred by the progressive collapse, is determined according to certain criteria. This analysis method can be further subdivided into linear static, linear dynamic, nonlinear static and nonlinear dynamic methods. However, the existing analysis method is based on many assumptions, is lacking in necessary testing and has certain limitations. As a result, the experimental study of progressive collapse is very necessary. Presently, the progressive collapse experimental study according to the structural system, is still rare. There are similar requirements for the design of structural progressive collapse resistance and anti-seismic design, such as the integrity of the structure, its ductility and redundancy, etc. Anti-seismic design is beneficial for progressive collapse resistance, but there are many differences between the performance of progressive collapse resistance and anti-seismic structure. The load-displacement curve, the structural failure mechanism, the plastic hinge position of structure and rotation capacity requirements and so on, are not the same; the anti-seismic design cannot replace the design of progressive collapse resistance, and the anti-seismic experiment cannot replace the experiments of structural progressive collapse resistance. The test method of structural progressive collapse resistance has its particularity in the following aspects.

1 TEST MODEL

There are two kinds of structural test models, the flat structure model and the spatial structure model. In 2007, Hunan University conducted a progressive collapse resistance experiment of a reinforced concrete plane frame model of 4 spans and 3 layers [7]; the scale ratio of the model was 1/3. The underlying middle column was damaged initially, but when the model was made, the initial damage column was replaced by a mechanical jack. The experimental model is shown in Figure 1. The conclusion from the experimental study was that the progressive collapse resistance mechanism of the frame structure was conversed from the plastic mechanism to the final suspension mechanism, and the bearing capacity of the suspension mechanism was higher than the plastic structure. Therefore, it was safe to adopt the plastic mechanism to determine the structure collapse resistance. In 2010, Nanchang University conducted a progressive collapse resistance experiment of a reinforced concrete spacial frame model of 4×2 spans and 3 layers. The scale ratio of the model was 1/3, the middle column of long side on the underlying damaged initially, the initial damage

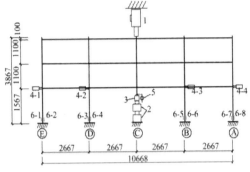

Figure 1. Plane frame test model of progressive collapse resistance.

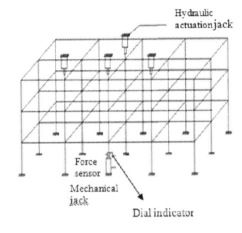

Figure 2. Spacial frame test model of progressive collapse resistance.

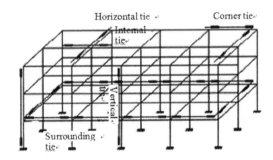

Figure 3. Tie force of spacial frame structure.

column was replaced by a mechanical jack. The experimental model is shown in Figure 2. Experimental study on the improved spatial structure model indicates that the progressive collapse resistance mechanism, plastic mechanism and catenary mechanism also play a role in the longitudinal direction of frame, while only the plastic mechanism plays a role in the transverse direction of the frame. In fact, the progressive collapse of the actual structure is spatial, under the influence of structural redundancy. The structural redundancy here contains not only the concept of an indeterminate statically number in mechanics analysis, but also the redundant members' bearing capacity[9], compared with the actual bearing capacity of members; the internal force effect produced by gravity load and design load is at lower levels. Spatial characteristics of the structure and redundancy mean that the loss of local bearing capacity, caused by initial local failure, can be compensated by the change of load path. The plane frame weakens the redundancy of the actual structure, the force is one-way, the changes in the load path are limited and the characteristics and loading process of structural progressive collapse cannot be reflected truly and effectively. In the tie force design method of structural progressive collapse resistance, tie forces include the internal tie force, the corner tie force, the horizontal tie force, the surrounding tie force and the vertical tie force, as shown in Figure 3. In the plane frame, there are only horizontal forces and vertical tie forces, as shown in Figure 4, as internal tie forces, corner tie forces and surrounding tie forces are not considered. So, the author thinks that progressive collapse resistance of frame structure should adopt a spacial frame model.

2 LOADING METHOD

2.1 *Simulation of initial local damage members*

The load method contains the simulation of initial local damage members, loading schemes and loading

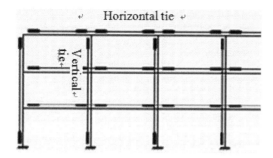

Figure 4. Tie force of plain frame structure.

Figure 5. Simulation of initial local damage column.

Figure 6. Hotel San Diego.

processes, etc. In the experimental study of progressive collapse resistance performance of a reinforced concrete frame structure[7,8], the initial damage column is replaced by a mechanical jack, when the model is making, the mechanical jack is concatenated and force sensors are adopted in advance in order to replace the underlying damage column. During the experiment, the failure of the underlying initial damage column is simulated by a mechanical jack manual classification unload, as shown in Figure 5. This is pseudo-static unloading, similar to the pseudo-static loading of the anti-seismic test. Due to the progressive collapse caused by the initial damage of the actual structure, this is generally sudden and the time is short, usually a few seconds, along with a dynamic impact effect. The above method of using pseudo-static unloading to simulate the bottom column initial damage, does not really reflect the characteristics of progressive collapses. However, given the current laboratory conditions, it is also more difficult to fully simulate a sudden destruction of the column (e.g. by an explosion, crash, etc.) of its structural entities. Presently, under laboratory conditions, for reinforced concrete frames and steel frames, the above method of adopting a mechanical jack to unload quasi-static is an acceptable method to simulate the underlying column initial damage. However, for long-span space structures such as the rack, net shell and truss, under laboratory conditions, how to simulate initial damage member is worth studying. In literature, [9] a progressive collapse resistance experiment of a space truss structure model was conducted, a device that the manual control initial damage is proposed. The device can only simulate the initial damage of the compressive bar, the set need to be installed in advance.

With the development of urban construction, there are buildings in need of demolition and blasting everywhere. Here, the author thinks that the demolition and blasting of construction entities can be used for progressive collapse resistance experiments; it is a form of recycling, model production costs may be saved and at the same time, it is more realistic to test with progressive collapse building entities. Presently, there are no reports about experimental studies on the progressive collapse of actual buildings within this country, although there are very limited studies abroad. In literature, [10] the progressive collapse resistance test of the Hotel San Diego was introduced. The Hotel San Diego was a six-layer, reinforced concrete frame structure building, which was built in 1914. The height of the bottom layer was 6.0 m, the middle layer, 3.2 m and the top layer, 5.1 m. A hollow, clay brick was adopted to fill in the walls. The building needed to be pulled down, and before dismantling, the progressive collapse resistance was tested, using the blasting method. Meanwhile, the bottom corner adjacent frame column was destroyed, as shown in Figure 6. The collapse process lasted only 1.6 seconds.

2.2 *Loading scheme*

Structure is usually vertical continuous, collapse occurs in the process of structure withstanding vertical load. In a structural progressive collapse resistance experiment, the focus should be on the vertical load simulation. In literature [7], in plane frame progressive collapse resistance tests, an electro-hydraulic servo jack was only arranged on the top of the third layer column corresponding to the initial damage column (the underlying middle column, as shown in Fig.1.); through it, vertical pressure is applied to simulate the upper vertical load bearing at the top of the pillars, whilst the other third layer columns did not have vertical pressure applied. The loading scheme does not really reflect the bearing of the bottom of the other columns. In fact, the other columns of the third floor pillars should also bear the upper vertical load at the same time, and the vertical load is transferred to the corresponding bottom column. That is to say, in the normal state of the mechanical jack instead of the initial failure column unload, the other underlying

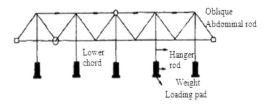

Figure 7. Load scheme of beam-truss structure.

columns should be under high pressure, are working with pressure. While the loading scheme according to literature [7], because the other columns of the third floor pillars do not apply vertical pressure pillars, the other underlying columns do not basically undertake the upper load, which is not consistent with the fact, is equivalent to increase the bearing capacity of redundancy of the other bottom columns. In the progressive collapse resistance test of space framework in the literature [8], the vertical load is applied not only on the third layer of pillars corresponding to the initial failure column, but also on the other three adjacent columns, as shown in Figure 2. The vertical load should be applied to the column originally, but restricted by the quantity of hydraulic servo jacks. Meanwhile, considering the internal force redistribution which occurs mainly on the beams and columns adjacent to the initial failure column, the distant columns are less affected, so the vertical load is not applied to the far cross-pillar, which is acceptable, and can reflect the force of the main members.

For the progressive collapse experiment of the grid, shell, truss and so on, on other large-span roof structures, the vertical load can be applied by mount mode. In literature [9], a progressive collapse test of a beam-truss structure was adopted in the mount mode to apply the vertical load, as shown in Figure 7. The devices include the boom (Φ 6 screw thread steel bar), loading pads and weights. Booms were connected to the string node in the above, and connected to the loading pad in the following location; weights were added on the loading pad to achieve progressive loading, step by step.

2.3 Loading process

During the process of structural progressive collapse, the vertical load of the structure is always being acted upon and basically remains unchanged, the vertical load is the main load which should be considered in progressive collapse. For a reinforced concrete frame and steel frame structure model, the vertical load can be simulated by applying vertical pressure on the top of the pillars by a hydraulic servo jack. In the process of the test, the hydraulic servo jack is controlled by force mode, making the vertical pressure remains the same. The mechanical jack by simulating the underlying initial damage column, can adopt a manual unload, step by step, to simulate the initial damage column out of work, the unload process uses the displacement to control, the jack-bearing reaction force can be measured by the force sensor. The total uninstall displacement needs to be estimated in advance. If a jack stroke cannot meet the requirement of the unloaded displacement, two concatenated jacks can be used; when a jack stroke is completed, another jack may continue to unload. For the grid, shell, truss and so on, on other large-span roof structures, the vertical load can be applied by derrick suspension weights, sandbags and steel blocks. It is easy to ensure that the applied vertical load remains the same.

3 PROGRESSIVE COLLAPSE FAILURE CRITERIA

In the experiment of structural progressive collapse resistance, how to judge the members' damage condition and whether the structural system occurs as a progressive collapse, are the keys to correctly evaluating the structural progressive collapse resistant capacity. Excessive deformation or rotation already means that the member or part of the frame cannot bear any more load; the member is considered to be damaged. After comprehensive analysis, United States' DoD2005 design, GSA progressive collapse damage criteria and Lu Xinzheng's (Tsinghua University) relevant research results, the author thinks that in progressive collapse resistance experiments of reinforced concrete frame structures, member failure criteria should adopt deformation failure criteria. When one of the following situations occurs, members may be determined as having failed: ①Compression concrete achieves ultimate compressive strain and crushes; ②Steel bar elongation is more than 10% and needs to be considered for breakage and failure; ③Steel bar anchorage failure; ④After losing the support of a column, the deflection of the frame beam is more than 10% of the beam span, and the beam will fail; ⑤The angle of the beam ending section is more than 12 degrees and the plastic deformation is too excessive to continue bearing and will fail. For the progressive collapse experiment of steel frame structures, with reference to the United States' DOD2005 design criteria, the deformation of a steel structure limit value is listed as follows: in a section, according to the theory of plastic, the angle is 6 degrees; in a section designed according to the theory of elasticity, the angle is 2 degrees.

For the structural system, suggestions are given to evaluate whether it has the progressive collapse resistance ability required: ①Along the planar direction, the scope of member damage should be limited to the bay which is directly connected with the initial damage member; ②Along the vertical direction, the damage member should be limited in the scope of the initial members in the layer and upper layer, as shown in Figure 8[11]. When the structure failure exceeds the above range, the risk of structural progressive collapse is bigger, the structure does not have the capacity to resist progressive collapse.

As for the difference between grid, shell, truss and other large-span roof structures and frame structures,

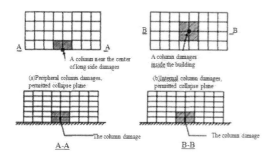

Figure 8. Permitted collapse area of frame structure.

there is an obvious difference in the force; the parameters suitable for frame structures may no longer be suitable for grid structure. There are two kinds of member damage: one, for strength failure, the stress and strain of members exceed the limit; the other, for excessive deformation, and loss of loading capacity. As for grid, shell, truss and other large-span roof structural, progressive collapse resistance experiments, when the tensile strain of a member exceeds the limit of the tensile strain of the material, the rod is considered to be a failure. If the member is controlled by compressive bar stability, and if it is unstable, the member is considered to be a failure. Grid, shell, truss and other large-span roof structures have their own structural characteristics; the rules about damage range according to the frame structure may not be suitable for these kind of large-span space structures. The problem of damage area limits needs further study.

4 CONCLUSION

(1) Experiments of structural progressive collapse resistance are the important foundations to recognize the performance and characteristics of structures. The progressive collapse resistance experiments cannot be replaced by anti-seismic experiments as the structural progressive collapse resistance experiment method has its own particularity.
(2) To really and effectively reflect the characteristics and loading processes of the structural progressive collapse, the progressive collapse experiments of frame structures should adopt space frame models.
(3) There are obvious differences in structure components and force of frame structures between the rack, net shell, truss and so on, and other large-span roof structures. In progressive collapse resistance experiments, different loading methods and failure criteria should be used according to different structure forms. Currently, research on experiments, concerning structural progressive collapse resistance, are mainly concentrated on frame structures. Experiments concerning the structural progressive collapse resistance of the rack, net shell and so on, and on other large-span structures need to be further developed.

5 AUTHOR INTRODUCTION

Yanhua Huo: female, born in 1976, Lecturer, the main research fields: Concrete Materials and Structures.

Contact method: Science and engineering department of Nanchang University College of Science and Technology, No.339, Beijing east road, Nanchang city, Jiangxi province, Zip code: 330029 Tel: 13576030962 Email: huoyanhua@hotmail.com

REFERENCES

United States General Services Administration, Progressive collapse analysis and design guidelines for new federal office buildings and major modernization projects[S]. Washington DC, GSA, 2003

Department of Defense, Unified facilities criteria: Design of Buildings to Resist Progressive Collapse[S], Washington DC DoD, 2005.

BS8110-1: 1997 Structural Use of Concrete: Part I: Code of Practice for Design and Construction[S].

European Committee for Standardization. EN1991-1-7:2006, Eurocode I: Actions on Structures. Part 1–7: General Actions-Accidental Actions[S]. Brussels, 2006.

Jiang xiaofeng, Chen yiyi. The research status of progressive collapses and control design of building structure [J], Journal of civil engineering, 2008(6):1–8

Ye lieping, Lu xinzheng, Li yi, Liang yi, Ma yifei. Progressive collapse resistance design method of concrete frame structure [J], Building structure, 2010(2): 1–7

Yi weijian, He qingfeng Xiao yan. Experimental study of progressive collapse resistance of reinforced concrete frame structure [J], Journal of building structures, 2007(5): 104–109

Wu zhaoqiang. Experimental study of progressive collapse resistance of non-seismic design frame structure [D], Nanchang University. 2010

Wang lei, Chen yiyi, Li ling, Liu hongchuan. Beam-truss structure test research of introduction of initial damage collapse [J], Journal of Tongji University (Natural science edition), 2010(5): 644–649

Mehrdad Sasani, Serkan Sagiroglu. Progressive Collapse Resistance of Hotel San Diego [J], Journal of Structural Engineering, 2008, 134(3): 478–488

Chen junling, Ma renle, He mingjuan. Progressive Collapse Resistance analysis of frame structure under accident [J], Sichuan construction scientific research, 2007(1): 65–68

Contrastive analysis on anchorage length of pre-embedded and post-embedded bars

YanHua Huo & ShengRong Ding
Nanchang University College of Science and Technology

ABSTRACT: Based on the anchorage length demands of pre-embedded and post-embedded bars in the current concrete structure design code and strengthening concrete structure code, anchorage principles, failure pattern and the influence factors of pre-embedded and post-embedded bars are analysed. The anchorage length of pre-embedded and post-embedded bars is calculated and compared. Results show that the anchorage length of post-embedded bars is overall below that of pre-embedded bars, and the anchorage length of post-embedded bars is 60–80% of that of pre-embedded bars. The anchorage length demand of the ordinary component in the current code for strengthening concrete structure is appropriate, but the anchorage length of the overhung component and the non-overhung important component may be small. It is suggested that the anchorage length of the overhung component and the non-overhung important component in the current code for strengthening concrete structure should be appropriately increased.

Keywords: post-embedded bar, pre-embedded bar, anchorage length

1 INTRODUCTION

A bar in concrete structures can bear force because it has bond anchorage action with concrete. Anchorage is the force basis of concrete structure; if the anchorage of the bar fails, the structure may lose bearing capacity and result in disastrous consequences such as collapse. To ensure this anchorage of the bar, is an important content to design concrete structure. For new reinforced concrete components, bars are usually bound and embedded before pouring the concrete. Specific provisions of anchorage length are given in the current concrete structure design code[1]. For connection of new and old concrete components in structural transformation or strengthening, a post-embedded bar is a common method and technology. The anchorage length of the post-embedded bar is specifically provided in the current strengthening concrete structure code[2]. Based on the current concrete structure design code and strengthening concrete structure code, the anchorage length of pre-embedded and post-embedded bars is analysed and compared.

2 THE ANCHORAGE PRINCIPLE AND LENGTH OF PRE-EMBEDDED BARS

The bond and anchorage effects between bar and concrete are constituted by cementation, friction, bite force and mechanical anchorage. The anchorage strength between bar and concrete is determined by

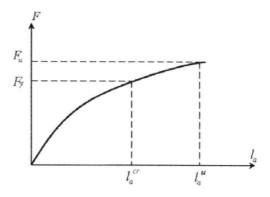

Figure 1. The relationship of anchorage length and force of bar.

an anchorage test and has many related factors: the concrete strength of wrapped layer, the appearance of the anchorage bar, the concrete cover thickness, the constraints of anchorage zone concrete, etc. Since the 1980s, the system anchorage tests have been done in our country; the anchorage length of a bar in a concrete structure is determined by the reliability calibration method. The relationship of anchorage length and force of bar in the conditions of minimum cover thickness and minimum stirrup is listed in Figure 1[3]. We can see that with the increase of anchorage length, the anchorage force is growing. When the anchorage force equals the yield force of the bar, the corresponding anchorage length is the critical anchorage length, which is the minimum length guaranteed to ensure that the bar anchorage failure could never occur before yield. When the bar yields and strengthens, the anchorage force can also increase with the anchorage length extension; when the anchorage force equals the bar tensile force, the corresponding anchorage length is the limit of the anchorage length. Obviously, the anchorage section more than limit anchorage length in the anchorage force will not work. The design anchorage length in the concrete structure design code[1] should be greater than the critical anchorage length and less than the anchorage limit length. Firstly, pull-out tests are done as the basis, then nonlinear finite element analysis is used to determine anchorage strength. The critical and anchorage limit length of different strength bars in different concrete strength levels can be deduced from the theoretical calculation, and can be verified through the verification pull-out tests. Of course, reliability calculation is needed; the bar anchorage length is decided in a certain reliability index.

According to the results of a lot of testing and study in this country, combined with reliability analysis in order to ensure the necessary reliability of anchorage forces, the calculation formula of the basic anchorage length of longitudinal tensile steel is as follows:

$$l_a = \alpha \frac{f_y}{f_t} d \qquad (1)$$

The formula is the basic anchorage length of the longitudinal tensile of a straight line, pre-embedded bar, which is provided in the current concrete structure design code[1].

3 ANCHORAGE PRINCIPLE AND LENGTH OF POST-EMBEDDED BAR

Compared with first, the conventional putting of a bar and second, pouring concrete, the anchorage performance of a post-embedded bar and concrete is through the structural glue between the bar and the concrete. The anchorage role of post-embedded bar glue is using anchorage force of its own bonding material, making the anchorage rod and the base material, anchor together effectively, producing adhesive strength and mechanical bite force. When the post-embedded bar achieves a certain anchorage depth, the post-embedded bar will have a strong resistance of pull-force and does not basically slip, so as to ensure the anchorage strength and lasting endurance strength.

According to the experimental analysis, there are four main forms of anchorage failure of post-embedded bars[4][5], which are shown in Figure 2: broken bar, bond failure, cone failure and cone-bond composite failure.

(1) Broken bar. The damage form occurs when the concrete strength and adhesive glue strength are both high, and the anchorage length is longer. The post-embedded bar failure begins with the bar yield, which

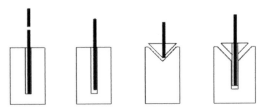

Figure 2. Four forms of anchorage failure of post-embedded bars.

has obvious signs, and it is hopeful failure form in engineering.

(2) Bond failure. When this fails, the concrete and the sub-body layer slide, the bar is pulled out together with the surrounding sub-body. The failure pattern is related to the quality of the post-embedded bar glue, the post-embedded bar depth, the concrete strength, the bar strength grade, the bar diameter, the incomplete removal of bar rust, incompletion of decontamination, the post-embedded glue has hollowing for post-embedded hole does not be fully exhausted which caused in construction process, and so on. If we use equal bond strength and ultimate tensile strength of the post-embedded bar as a limit state, we can get the minimum depth of the post-embedded bar.

(3) Concrete cone failure. When this fails, the concrete around the bar acts as a conical crack, forming a pyramidal. The failure pattern occurs when the anchorage depth is not enough or the anchorage depth is less than or equal to the minimum depth value of the embedded bar, under tensile stresses, planting-bar glue has good bond with bar, the strength of bar does not reach or just reaches the yield stage and does not exceed the ultimate tensile stress, concrete exceeds the ultimate tensile stress, the bond stress of post-embedded bar glue and bar, post-embedded bar glue and concrete and tensile strength of bar does not fully play. The failure pattern is another condition to determine the smallest depth of the post-embedded bar: the smallest depth of the post-embedded bar should meet the conditions of the concrete cone destruction strength is equal to the ultimate tensile strength of post-embedded bar, that is, bar yield and cone failure will happen at the same time.

(4) Cone-bond composite failure. The characteristic of the damage is that the cone failure happens around the bar on the concrete surface. The slipping damage occurs on the anchorage section under the cone, and the sub-body is drawn from the concrete together with the bar. When the cone-bond composite damage occurs, the limit of pull-force is composed of two parts. Part of it is the tension borne by the upper cone section of the post-embedded bar, the other part of it is the bond force between the anchorage solid and the concrete below the post-embedded bar.

In addition to some factors such as the concrete strength, the anchorage depth, the distance of the bar to the edge of the concrete and components influencing the anchorage performance of the bar, other factors such as the performance and thickness of the adhesive layer have a significant effect on the anchorage performance; the anchorage performance is more complex than the conventional anchorage bar.

The depth of the post-embedded bar should be determined by the limit state condition of the bond failure and the concrete cone failure strength equal to the tensile strength of the bar, safety reliability coefficient is introduced, the depth calculation formula of post-embedded bar is determined by the binding condition and concrete tensile condition. The calculation method of the depth of the post-embedded bar in post-embedded bar technique of the strengthening concrete structure code is:

$$l_d \geq \psi_N \psi_{ae} l_s \quad (2)$$

$$l_s = 0.2 \alpha_{asp} d \frac{f_y}{f_{bd}} \quad (3)$$

4 THE COMPARATIVE ANALYSIS OF ANCHORAGE LENGTH

Based on the calculation formula of the anchorage length of a bar in the current concrete structure design code and strengthening concrete structure code, the anchorage length of a bar is calculated in a variety of situations.

The calculation parameters of post-embedded bars of common post-embedded bar are listed as follows: the concrete strengths are C20-C40, the bar strengths are HPB235, HRB335 and HRB400, the bar diameters are from 16 to 28 mm, the concrete cover depths are 30 mm, the stirrup diameters are 8 mm or 10 mm, adhesive glue is A grade, the distance of a post-embedded bar is equal to or greater than 7 d and the side distance of a post-embedded bar is equal to or greater than 3.5 d. Due to the limited length of this paper, only two diameters, 16 and 20 mm bar anchorage length contrasts, are listed. As shown in Figures 3–5, post-embedded bars 1, 2 and 3 separately refer to the overhung components, the non-overhung important components and other components in the strengthening concrete structure code.

As we can see from Figure 3, for bar grade HPB235, the anchorage lengths of post-embedded bars are all lower than the ordinary pre-embedded ones. Specifically, as for C25 concrete, the post-embedded anchorage lengths of overhung components are a little lower than those of the ordinary pre-embedded bars, which is between 0.74~0.79 of that of the ordinary pre-embedded bars as for the C30-C40 concrete; where the post-embedded anchorage lengths of non-overhung important components are lower than those of ordinary pre-embedded anchorage lengths, the ratio is about 0.56~0.76; where the ordinary post-embedded anchorage lengths of ordinary components are lower than those of the ordinary pre-embedded bar, the ratio is about 0.50~0.66.

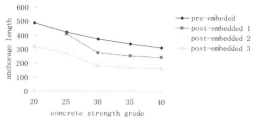

(a) diameter 16

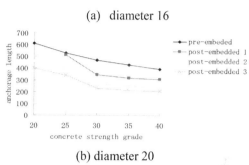

(b) diameter 20

Figure 3. Anchorage length contrasts of post-embedded and pre-embedded bars of grade HPB235.

As we can see from Figures 4 and 5, for bar grade HRB335 and HRB400, as for C25 concrete, the post-embedded anchorage lengths of overhung components are greater than those of the ordinary pre-embedded bars. In the other cases, the post-embedded anchorage lengths are lower than those of the ordinary pre-embedded anchorage lengths. Specifically, as for C25 concrete, where the anchorage lengths of overhung components are greater than those of the ordinary pre-embedded anchorage lengths, the ratio is about 1.1. As for C30-C40 concrete, where the post-embedded anchorage lengths are lower than those of the ordinary pre-embedded anchorage lengths, the ratio is about 0.84~0.90; where the post-embedded anchorage lengths of non-overhung important components are lower than those of g ordinary pre-embedded anchorage lengths, the ratio is about 0.65~0.87; where the post-embedded anchorage lengths of ordinary components are lower than those of ordinary pre-embedded anchorage lengths, the ratio is about 0.56~0.76.

If the distance and side distance of the post-embedded bars cannot reach the above requirement, for example, the distance of the post-embedded bar is 5 d and the side distance is 2.5 d (other parameters remain constant), the contrasts in the anchorage length of the diameters of the two bars are listed in Figures 6~8.

As we can see from Figure 6, for bar grade HPB235, the post-embedded anchorage lengths of overhung components are almost the same as those of the ordinary pre-embedded bars, which is only slightly less than for C30 concrete; where the post-embedded bar anchorage lengths of non-overhung important components are lower than those of ordinary pre-embedded

203

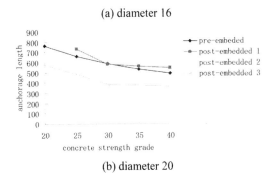

Figure 4. Anchorage length contrasts of post-embedded and pre-embedded bars of grade HRB335.

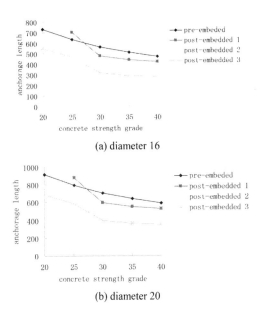

Figure 5. Anchorage length contrasts of post-embedded and pre-embedded bars of grade HRB400.

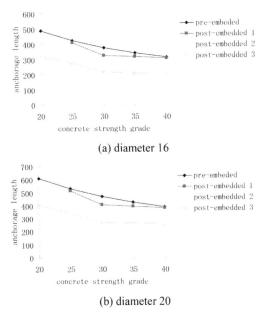

Figure 6. Anchorage length contrasts of post-embedded and pre-embedded bars of grade HPB235.

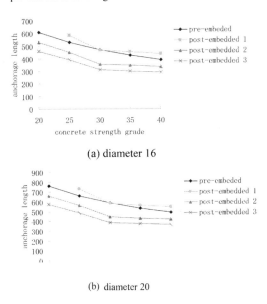

Figure 7. Anchorage length contrasts of post-embedded and pre-embedded bars of grade HRB335.

anchorage lengths, the ratio is about 0.65~0.76; where the post-embedded bar anchorage lengths of ordinary components are lower than those of ordinary pre-embedded anchorage lengths, too, the ratio is about 0.58~0.66.

As we can see from Figures 7 and 8, for bar grade HRB335 and HRB400, the post-embedded anchorage lengths of overhung components are greater than those of the ordinary pre-embedded bars, the ratio is about 1.0~1.1; the post-embedded anchorage lengths of important non-overhung components are lower than those of ordinary pre-embedded anchorage lengths, the ratio is about 0.76~0.87; the post-embedded anchorage lengths of ordinary components are lower than those of ordinary pre-embedded anchorage lengths, too, the ratio is about 0.66~0.76.

5 CONCLUSION

To analyse the above contrast results comprehensively, where the post-embedded anchorage lengths are lower than those of ordinary pre-embedded anchorage lengths, the percentage is 60~80. Especially for overhung components of C30-C40 concrete, if the distance of the post-embedded bar is greater than 7 d and the side distance is greater than 3.5 d, then the anchorage length of a post-embedded bar is 74~90 percent

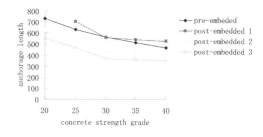

(a) diameter 16

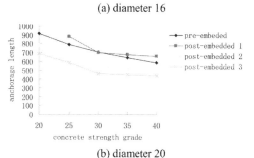

(b) diameter 20

Figure 8. Anchorage length contrasts of post-embedded and pre-embedded bars of grade HRB400.

of the length of the embedded anchorage. For non-overhung important components, where the anchorage lengths of a post-embedded bar are mostly lower than that of ordinary pre-embedded anchorage lengths, they are 56~87 percent. Considering the factors of the anchorage performance of post-embedded bars is complex. During the process of construction, bar de-rusting, decontamination, ash cleaning, exhausts not reaching the designated requirements and the creep performance of the glue structure in the long-term load, the author thinks as for overhung components and non-overhung important components, the anchorage length of pre-embedded bars is not lower than the straight line anchorage length of post-embedded bars. For the ordinary components, the anchorage length of post-embedded bars can be appropriately below that of pre-embedded bars. So, the anchorage length of ordinary components in the strengthening concrete structure code is appropriate, but for overhung components and the non-overhung important components, the depth may too small. The author suggests, for overhung components and the non-overhung important components, the depth of pre-embedded bars should be increased in the strengthening concrete structure code. The depth of pre-embedded bars not only depends on the strength of the concrete and glue, but also depends on the structural form, the using conditions and the construction quality, etc. For important structures and components, the 12 d~15 d even 8 d~10 d recommended by manufacturers or post-embedded bars company even more cannot believe in.

6 AUTHOR INTRODUCTION

Yanhua Huo: female, born in 1976, Lecturer, the main research fields: Concrete Materials and Structures

Contact method: Science and engineering department of Nanchang University College of Science and Technology, No.339, Beijing east road, Nanchang city, Jiangxi province, Zip code: 330029

Tel: 13576030962

Email: huoyanhua@hotmail.com

REFERENCES

GB50010—2002, concrete structure design code [S].
GB50367—2006, strengthening concrete structure code [S].
China building scientific research institute. Concrete structure design [M]. Beijing: China architecture & building press, 2003
Zhou Xingang, Wang Youxuan, Qu Shuying. Concrete plant steel anchorage ultimate bearing capacity analysis [J], Engineering mechanics, 2002, vol 19, No.6:82–86
Zhu Yanpeng, Liu Hui, Wang Xiuli, Zheng Jianjun. The experimental study of pre-embedded bars depth in concrete structure reinforcement [J], Journal of Lanzhou science and engineering university, 2005, vol 31, No.5:106–110

Empirical analysis of an urban development model in the western minority areas, taking Linxia City as an example of the coordination degree method

Jing Xu
School of Economics, Northwest University for Nationalities, Lanzhou, China

ABSTRACT: In this paper, urbanization motivation factors of Linxia City are analysed by application of improved exponential function and the coordination degree analysis method. The rules of variation with time of the coordination degree between industrial structure development and urbanization rate is analysed so as to summarize urbanization development regularity features of Linxia City, and analyse the factors of these features from their essence, providing theoretical support for an urbanization development strategy for Linxia City.

Keywords: urbanization, coordination degree, Linxia

1 FOREWORD

Given the conditions that China's rapid economic growth is facing shrinking international demand, hot market competition and domestic resources, and environmental constraints, urbanization is not only the main task and key point for development of the western regions but also an important means and condition for expanding domestic demand and stimulating economic growth. Meanwhile, urbanization is also a social and economic phenomenon. It is influenced not only by economic factors but also by institutional factors and socio-cultural factors. Thus, urbanization dynamic mechanism is complex.[1][2] Specifically, for each region, urbanization motives are comprised of different characteristics. Thus, urbanization needs to be promoted in a focused, coordinated and orderly manner depending on natural resources, cultural traditions, industrial foundation and other conditions.

Urbanization development motivation is always a hot issue in the economic field. In terms of research of the urbanization development model and motivation, Li Liping and Guo Baohua have analysed the economic motivation of urbanization formation mechanism[3]; Chen Lijun and Wang Keqiang have performed empirical analysis of the relationship between China's urbanization development and industrial structure[4]; Ma Chengwen and Wei Wenhua have analysed the process of urbanization in Anhui Province and performed an in-depth study of the relationship between urbanization and industrial structure upgrading[5]; Lu Dadao et al., have pointed out the negative impact of our government-led rapid urbanization on environment and resources, and also discussed pressure on employment, regional spatial control and the principles for developing urbanization process in line with China's national conditions[6].

In the study of urbanization of underdeveloped areas in western China, Li Yanxi has analysed the economic motivation of the urbanization process in undeveloped western regions, summed up the geographical features of undeveloped western regions and the impact on the urbanization process, and put forward concrete measures for promoting urbanization in such regions[7]. Wu Xiao and Zhang Pei have analysed the factors affecting lagging of urbanization of western regions based on an overview of urbanization development in the western regions[8].

In this paper, Linxia City, Linxia Hui Autonomous Prefecture, Gansu Province, located in a minority area in the northwest of China, is the research object. Through analysis of the historical process of urbanization of Linxia City, focusing on research of the relationship between industrial structure and urbanization of Linxia City, as well as analysis of the urbanization development mode and the fundamental motivation of Linxia City, a policy basis is proposed for the adjustment of the industrial structure of Linxia City.

2 MATHEMATICAL MODEL AND ALGORITHM

Based on the calculation of the coordination degree of power function of different industries and urbanization rate, a coordination relation between industry and urbanization development is obtained so as to analyse the urbanization development motivation of Linxia City.

2.1 *Power functions*

In the multi-index evaluation system of power function, there are many types of power functions, wherein the common ones include linear power function methods (also known as traditional power function methods), exponential power function methods, logarithmic power function methods and power function power function methods, which mainly differ in function form. In this paper, the improved exponential power function is adopted, wherein the model formula is [9]:

$$u = Ae^{B\frac{x-x^s}{x^h-x^s}} \quad (1)$$

wherein, u is for the evaluation value of a single evaluation index (ie efficacy score); x is actual value of a single index; x^s is intolerable value (or unallowable value); x^h is satisfaction value (or just permissible value); A and B are positive parameters to be determined.

Improved exponential power function has the following properties: (1) monotonicity; (2) convexity; (3) forward form and reverse form have a unified power function form; (4) index value can exceed satisfaction value and unallowable value for ease of historical comparison, making up for the shortcomings of the power function power function.

2.2 Measuring of coupling degrees

A coupling degree is calculated according to the capacity coupling concept and capacity coupling coefficient model.[10] A coupling degree is used to calculate the coordination degree[11] and create a coupling degree model of multiple systems (or elements)[12], namely the capacity coupling concept and capacity coupling coefficient model in physics, wherein variables $u_i (i = 1, 2, \ldots, m)$, $u_j (j = 1, 2, \ldots, n)$, respectively represent the systems; the coupling degree model with a plurality of systems can be expressed as:

$$C_n = n \left\{ \frac{u_1 \cdot u_2 \ldots u_n}{\prod (u_i + u_j)} \right\}^{\frac{1}{n}}$$

When there are only two systems, for ease of analysis, the coupling degree function can be obtained directly and expressed as:

$$C_2 = 2 \left\{ \frac{u_1 \cdot u_2}{(u_1 + u_2)(u_1 + u_2)} \right\}^{\frac{1}{2}} \quad (2)$$

Obviously, the coupling degree value C_2 is between 0 and 1. When C_2 tends to be 1, it indicates a high coupling degree between the systems, which means good resonant coupling between the two systems, and the systems will tend to new ordered structure; when C_2 tends to be 0, it indicates a low coupling degree between the systems, which means the two systems are in an independent state and the systems will undergo disorder development.

2.3 Coupling coordination degrees

In order to more accurately reflect the coordination between systems, coupling coordination degree measurement is adopted and the model is as follows:

$$D = (C \cdot T)^V \quad (3)$$
$$T = \alpha u_1 + \beta u_2$$

wherein, D is the coupling coordination degree; C is the coordination degree; T is the comprehensive evaluation index, reflecting the overall benefit or level of the two; α and β are parameters to be determined, and generally $v = 0.5$.

3 COORDINATION DEGREE CALCULATION AND RESULT ANALYSIS

At the core of urbanization lies the transformation process of employment structure and economic industry structure, and the change process of the urban and rural space structure. The essential features of urbanization are mainly reflected in three aspects: first, conversion of the rural population in space; second, gathering of the non-agricultural industry to urban areas; third, transfer of the agricultural labour force to a non-agricultural labour force. The urbanization motivation of Linxia City is analysed through calculation of coordination between industrial structure change and urbanization of the population. The urbanization rate of the population is a symbol indicator of the urbanization process and the data source is Gansu Province Statistical Yearbook. An analysis was made about the impact of industrial development and industrial restructuring from 2001 to 2011 on the development of urbanization of Linxia City and the laws.

3.1 Relationship between industrial structure development and urbanization process

Data from the yearbook was selected to respectively analyse the coordination degree of the annual output of the primary, secondary and tertiary industry of Linxia City as a separate system, and the proportion of the non-agricultural population (urbanization rate) of Linxia City. According to the formulae (1), (2) and (3), the urbanization coordination degree between the population of Linxia City and three industries was obtained, as shown in Figure 1.

Seen from Figure 1, before 2007, the coordination degree between the primary industry and urbanization rate of Linxia City was relatively high in three major industries, and after 2007 the coordination degree between the tertiary industry and the urbanization rate of Linxia City was the highest; the coordination degree between the secondary industry and urbanization rate of Linxia City was slightly higher than that of the tertiary industry before 2004, and the coordination degree between the secondary industry and urbanization rate of Linxia City was the lowest among the three industries after 2004.

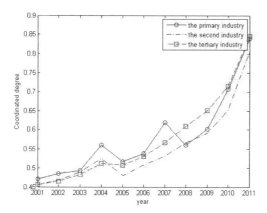

Figure 1. Comparison chart of the coordination degree between the industrial structure and urbanization development of Linxia City.

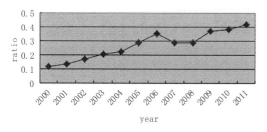

Figure 2. Proportion of the construction industry output value accounting for the secondary industry.

3.2 The function of the construction industry in the economic development of Linxia City

The secondary industry of Linxia City is relatively weak. Under the guidance of government-led and investment-led ideas, as other cities across the country, Linxia City is also going the same way of stimulating the economy through infrastructure projects. Figure 2 shows the change of the proportion of the construction industry output value accounting for the secondary industry of Linxia City since 2000.

Figure 2 shows that after 2000, after the construction industry of Linxia City witnessed rapid development, the construction industry accounted for a higher and higher proportion of the secondary industry output, increasing from 0.11 in 2001 to 0.42 in 2011, but the rapid development of the construction industry had no obvious impact on improving the urbanization rate of Linxia City.

3.3 The coordination degree among the construction industry and the wholesale and retail industry, and the urbanization rate development

For analysing the motivation to promote the process of urbanization in Linxia City, the coupling degree method is applied to compare the coordination degrees

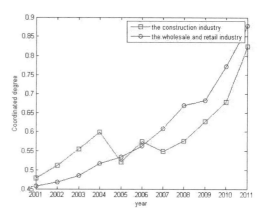

Figure 3. The coordination degree between the construction and retail trade industry and the urbanization development.

amongst the development of the Linxia City construction and wholesale and retail industry, and the development of the urbanization rate, as shown in Figure 3.

Figure 3 shows that before 2004 the coordination degree between the construction industry and urbanization development was higher than that between the wholesale and retail industry and urbanization development, which witnessed stable growth after 2006. However, the coordination degree between the construction industry and urbanization development was greatly reduced, significantly lower than that between the wholesale and retail industry and urbanization development, which was obviously matched with the proportion of the construction industry output in the sedentary industry, indicating that the strategy of promoting urbanization by the construction industry was facing a bottleneck.

4 CONCLUSION AND DISCUSSION

Based on the perspective of the coordination degree, the relationship between the level of urbanization development and industrial development since 2000 is analysed in this paper. The following conclusions are obtained: First, the industrial structure of Linxia City has obvious regional characteristics, which are mainly reflected in a relatively low coordination degree between the secondary industry development and urbanization development, and is not the main motivation for urbanization development of this city; Second, due to national and historical reasons, the tertiary industry has always been the main motivation for urbanization development of this city; Third, with urbanization development, the primary industry is also simultaneously developed, which is due to the fact that urbanization development needs to promote upgrading of the agricultural industry so that the output value of the unit land is enhanced; Fourth, the promotion function of the construction industry promoted by investment has a limited contribution to urbanization, so that it is difficult to become the continued motivation for promoting urbanization development.

The key to the urbanization of Linxia City should be the development of commercial markets and a great development of the secondary industry to enhance the city's gathering ability. Meanwhile, efforts should be made to avoid the one-sided, government-led economic model led by investment in the construction industry. This practice in the last decade has proved that after a certain stage this model has a limited role in promoting urbanization. Thus, we should insist in the orderly urbanization development way of promoting the city by industry.

ACKNOWLEDGEMENT

This work was supported by the Fundamental Research Funds for the Central Universities of Northwest University for Nationalities (Grant No: 31920130098)

REFERENCES

Zhou Yixing; *Thinking about China's Urbanization Rate [J]; City Planning Review*; the 30th Volume Supp. 2006.

Xu Yida; *Motivation, Mechanism and Others of Modern Western Urbanization Process [J]. New Architecture*; the 5th issue, 2000.

Li Liping, Guo Baohua; *Economic Analysis of Urbanization Formation Mechanism [J]; Academic Journal of Zhongzhou*; 2006, (05)

Chen Lijun, Wang Keqiang; *Empirical Analysis of the Relationship between Chinese Urbanization Development and Industrial Structure [J]. China Population, Resources and Environment;* 2010, (02)

Ma Chengwen, Wei Wenhua; *Analysis of the Relationship between Urbanization and Industrial Structure Upgrading of Anhui Province [J]. Special Zone Economy*; June 2012

Lu Dadao, Yao Shimou, Li Guoping, Liu Hui, Gao Xiaolu; *Comprehensive Analysis of Urbanization Process Based on China's National Conditions [J]; Economic Geography.* The 6th issue, the 27th Volume, November 2007

Li Yanxi; *Economic Motivation Analysis of Urbanization Process in Undeveloped Western Regions [J]. Territorial Economy*; the 1st issue, 2002

Wu Xiao, Zhang Pei; *Study of Western Region Urbanization Impact Mechanism and Power Systems [J]. Journal of Anhui Agricultural Sciences*; 2012, 40 (3):1802–1805, 1833

Peng Fei, Yuan Wei, Hui Zhengqin; *Discussion of Improvement of Exponential Power Function of Comprehensive Evaluation Methods [J]; Statistical Research;* the 12th Issue, the 24th Volume, December 2007

Liu Yaobin, Li Rendong, Song Xuefeng; *Analysis of Coupling Degree Between China's Urbanization and Ecological Environment [J]. Journal of Natural Resources;* 2005.20 (1):106–111.

Valerie Ming worth . *The Penguin Dictionary of Physics [M]. Beijing: Beijing Foreign Language Press*, 1996.

Liu Yaobin, Li Rendong, Zhang Shouzhong; *Research of Urbanization and Ecological Environment Coordination Standards and Evaluation Models [J]. China Soft Science*, 2005, (5):140–148.

Empirical analysis of main industrial capacity utilization in Gansu Province, focusing on an overcapacity situation

Jing Xu
School of Economics, Northwest University for Nationalities, Lanzhou, China

ABSTRACT: The oil and gas exploration industry, the ferrous metal mining industry, the non-ferrous metal mining industry and the oil processing and coking industry of Gansu Province are those researched in this paper. Peak value analysis methods and non-parametric frontier methods are used to measure the capacity utilization and input efficiency. The results are as follows: the four sectors witnessed severe overcapacity in capacity utilization levels during 2007 and 2008, a moderate average annual capacity or capacity reuse during 2008 and 2010, and an insufficient average annual capacity during 2010 and 2012.

Keywords: overcapacity, peak value method, non-parametric frontier, major industries in Gansu Province

1 DEFINITIONS OF THE ISSUES TO BE RESEARCHED

1.1 Overcapacity

Capacity refers to the production capacity within a certain period, namely the product yield after putting the various production factors into consideration. Overcapacity refers to the phenomenon of the actual or theoretical production capacity of the producer being greater than the market demand because of changes in the producer, investor, government, market and so on. There are a lot of theoretical formation mechanisms of overcapacity. Overcapacity is featured in relativity, differences and structural characteristics. There are differences among the standards for overcapacity in different industries, markets, regions, period.

1.2 Definition of research objects

The major industrial sectors of Gansu Province refer to the important ones with a high economic proportion in the economic structure of Gansu Province. According to the industrial production and industrial input data from the Statistical Yearbook of Gansu Province, the oil and gas exploration industry, the ferrous metal mining industry, the non-ferrous metal mining industry and the oil processing and coking industry of Gansu Province are those researched in this paper. The output of such four industries reaches about 56.5%, far higher than several dozens of other industries.

2 RESEARCH METHODS

Considering the difficulty in data acquisition and applicability, applicability of model methods and linkage among the model methods and other factors, peak value analysis methods and non-parametric frontier methods are used in this paper.

2.1 Peak value analysis method

Advantages of the peak value analysis method lies in its low requirements for data, but its limitation lies in only considering the impact of technological progress factors on the capacity while ignoring economic fluctuation, changes in economic structure and other factors. Within a certain period, the proportion of the maximum of the peak value higher than the actual output is the excess capacity. Assumptions include: T1: unit output coefficients must have an input-output relationship from the data and the technical conditions at the time, the input in the peak value area corresponds to 100% of the output; T2: a capacity utilization change in the peak value area is caused by a technical progress or technical change.

Model formula of peak value method:

$$Q = AMT \quad (1)$$

$$Q_t = AM_t T_t$$

$$M = M_t / M_{t-1} \quad (2)$$

Q_t is the output in period t, A is the output coefficient proportional constant, M is the composite index for the input and T technical progress (constant returns to scale). According to the above, only an increase in all input can get the output in proportion.

The average change rate of productivity represents a technical level rate under different peak values, namely that T can be calculated according to Q and M:

$$T_t = T_{t-x} + \left[\frac{\frac{Q_{t+x}}{M_{t-x}} - \frac{Q_{t-y}}{M_{t-y}}}{\frac{y+x}{y}} \right] \quad (3)$$

Y is the duration of the previous peak value and X is the duration of the next peak value. T_t is the technical level within the range of the previous peak value and is also known as the average productivity of the period. $\left[\frac{\frac{Q_{t+x}}{M_{t-x}} - \frac{Q_{t-y}}{M_{t-y}}}{\frac{y+x}{y}} \right]$ is the cumulative technical level change. The technical level indicators of the next peak period can be get by plus the two. T3: $A = 0$, and it is expressed by the capacity utilization level of technology utilization as $T_t = \frac{Q_{1t}}{M_t}$, wherein Q_{1t} is the output level. It can get the following:

$$Q_1 = M_t \times T_t \quad (4)$$

It can finally get:

CP (capacity utilization rate) $= Q_1/Q$ (5)

CG (overcapacity rate)=1-CP (6)

According to the model of a peak value analysis method, it can be obtained: output $Q =$ industrial output; composite index of the input is:

$M =$ balance of the fixed assets in the current year/balance of the fixed assets in the previous year. The technical progress index T is calculated according to the formula.

According to evaluation criteria for capacity utilization rates in European countries, America and Hong Kong: the normal capacity utilization rate is generally between 79%–83%. The Board of Governors of the Federal Reserve System believes that the capacity utilization rate is as follows: more than 90% is insufficient production capacity, 85%–90% is capacity reuse, and 75% or less is serious excess capacity.

2.2 Non-parametric frontier method

Input efficiency is proposed in this paper based on non-parametric frontiers. Expansion of output under determined input, and compression of input under determined assumed output condition, are two basic ways to improve the efficiency of resource allocation (or economic efficiency). In this paper, the production technology efficiency function based on input is used for description of efficiency. It is technology efficiency under the assumption of a given output by considering the maximum compression degree of existing input levels to a feasible extent.

2.2.1 Model assumption

T1: Assuming for every kind of industry, $Q = 1, \ldots, Q$ enterprises; T2: Q enterprises subject $F = 1, \ldots, F$ types of input; T3: Q enterprises' output $R = 1, \ldots, R$ types of volume of output; T4: C represents the amount of input of type F of the Qth enterprise; H_{QR} represents the amount of input of type R of the Qth enterprise. For convenience of description, the set is used, namely, $A = T_{QF}, S = H_{QR}$.

Firstly, the input set with a constant returns to scale and free dispose of input is considered, and it is referred to as (C, S) input set:

$$L(u|C,S) = \{x : u \leq zA, zA \leq x, zS \leq x, z \in R_+^Q\}, \quad (7)$$

$$u \in R_+^A$$

For $\lambda > 0$, (C, S) input set meets:

$$L(\lambda u|C,S) = \{x : \lambda u \leq zA, zS \leq x, z \in R_+^Q\}$$
$$= \lambda \left\{ \frac{x}{\lambda} : u \leq \left(\frac{z}{\lambda}\right)A, \left(\frac{z}{\lambda}\right)S \leq \frac{x}{\lambda}, \left(\frac{z}{\lambda}\right) \in (1/\lambda)R_+^Q \right\} \quad (8)$$
$$= \lambda L(u|C,S)$$

It is assumed that (x^Q, u^Q) is the input-output vector of the Qth enterprise. The (C, S) input-based technical efficiency measurement function is as follows:

$$F_i\{x^Q, u^Q|C,S\} = \min\{\lambda : \lambda x^Q \in L(u^Q|C,S)\}, \quad (9)$$
$$Q = 1, \cdots, Q, \lambda > 0$$

Secondly, if the technology set restriction is enhanced once again, the variable input set returns to scale and the strong input disposability is called (V, S) input set and is expressed as:

$$L(u|C,S) = \left\{ \begin{array}{l} x : u \leq zA, zS \leq x, z \in R_+^Q, \\ \sum_{Q=1}^{Q} z_Q = 1 \end{array} \right\}, u \in R_+^Q \quad (10)$$

The (V, S) input set-based technical efficiency measurement function is as follows:

$$F_i\{u^Q, x^Q|V,S\} = \min\left\{\lambda : \lambda^{x^Q \in L(u^Q|V,S)}\right\}, \quad (11)$$
$$Q = 1, \cdots, Q$$

Finally, if all inputs are weakened, the variable weakened set of returns to scale is (V, W) set, which can be expressed as:

$$L(u|V,W) = \left\{ \begin{array}{l} x : u \leq zA, zS = \sigma x, 0 < \sigma < 1, \\ z \in R_+^Q, \sum_{Q=1}^{Q} z_Q = 1 \end{array} \right\}, u \in R_+^Q \quad (12)$$

The (V, W) input set-based technical efficiency measurement function is as follows:

$$F_i\{x^Q, u^Q|V,W\}$$
$$= \min\{\lambda : \lambda x^Q \in L(u^Q|V,W)\}, \quad (13)$$
$$Q = 1, \cdots, Q, \lambda > 0$$

The comprehensive technical efficiency of the industry is:

$$\begin{Bmatrix} F_i\{u^Q, x^Q | C, S\} \times F_i\{u^Q, x^Q | V, S\} \\ \times F_i\{u^Q, x^Q | V, W\} \end{Bmatrix} \quad (14)$$

2.2.2 Decomposition of production resource allocation efficiency, based on the inputs

The above three cases only describe the connotation and quantity relationship of technical efficiency under different gain modes, and the contents below further decompose it into scale efficiency, disposal degree of the input factors and pure technical efficiency. Returns to scale function is:

$$S_i\{u^Q, x^Q\} = F_i\{u^Q, x^Q | C, S\} = F_i\{u^Q, x^Q | V, S\}, Q = 1, \cdots, Q \quad (15)$$

Input factor disposal degree function:

$$C_i = F_i\{u^Q, x^Q | V, S\} = F_i\{u^Q, x^Q | V, W\}, \quad Q = 1, \cdots, Q \quad (16)$$

The disposable technical efficiency of variable input elements of returns to scale:

$$F_i\{u^Q, x^Q | V, W\} \quad (17)$$

After excluding the impact of the variable scale and the disposal ability, the pure technical efficiency is:

$$F_i\{u^Q, x^Q | C, S\} = F_i\{u^Q, x^Q | V, W\} \times S_i(u^Q, x^Q) \times C_i = F_i\{u^Q, x^Q | V, S\} \quad (18)$$

3 EMPIRICAL ANALYSIS OF OVERCAPACITY

3.1 Measurement results

According to the aforementioned peak value analysis steps, the balance of fixed assets is introduced to the formula (2) to get the composite index of input M; the total industrial output value and composite index of input M are introduced into the formula (3) to obtain technical progress index T; the composite index of input M and technical progress index T are introduced into formula (4) to get the actual output level Q_1; the actual output level Q_1 and total industrial output value are introduced into the formula (5) to obtain the capacity utilization rate CP; and finally, the capacity utilization rate CP is introduced into the formula (6) to obtain the overcapacity rate CG.

According to the aforementioned non-parametric frontier analysis steps, the number of enterprises and balance of fixed assets are selected as the input volume, and the total industrial output and gross profit are selected as the output volume. The input volume and output volume are input into Deap 2.0 software, which will achieve a calculation of formula (8) to formula (18) to obtain the input efficiency.

These initial calculations above are from the total industrial output (price in current year), the annual average balance of net fixed assets and the total profit of four industrial sectors in 2000 to 2012, found in the *Statistical Yearbook of Development of Gansu Province*. The results are shown in Table 1.

Table 1. Capacity utilization rate and input efficiency measurement results of four sectors.

	Oil and gas exploration industry		Ferrous metal mining industry	
Year	Input efficiency	Capacity utilization rate	Input efficiency	Capacity utilization rate
2000	1	1.0006	0.394	1.0732
2001	1	0.6937	0.216	0.6982
2002	0.568	0.5736	0.256	1.0000
2003	0.516	1.0000	0.107	0.1497
2004	0.344	0.6408	0.229	0.0852
2005	0.566	0.4726	0.555	0.0643
2006	0.775	0.4470	0.593	0.0480
2007	0.775	0.5051	0.633	0.0408
2008	0.983	0.4893	1	1.0000
2009	0.364	0.7610	0.453	1.0000
2010	0.428	0.7505	0.606	1.0000
2011	0.563	0.7264	1	1.0000
2012	1	0.6326	1	1.0000

	Non-ferrous metal mining industry		Oil processing and coking industry	
Year	Input efficiency	Capacity utilization rate	Input efficiency	Capacity utilization rate
2000	0.499	1.1208	0.517	1.0070
2001	0.466	0.7225	0.346	1.0000
2002	0.47	0.4946	0.347	0.5116
2003	0.393	0.3902	0.47	0.3158
2004	0.683	0.3092	0.586	0.2361
2005	0.593	0.2438	0.654	0.1786
2006	1	0.2133	0.797	0.1588
2007	1	1.0000	0.797	0.1610
2008	0.672	1.0000	1	1.0000
2009	0.4	1.0000	0.648	0.9434
2010	0.652	1.0000	0.711	0.7186
2011	1	1.0000	1	1.0000
2012	1	1.0000	1	1.0000

3.2 Sector analysis

The input efficiency capacity utilization rate comparison graph of each section is drafted according to data in Table 1.

3.2.1 Oil and gas exploration industry

Figure 1 shows frequent fluctuation in the capacity utilization level in the industry, but the trend of the

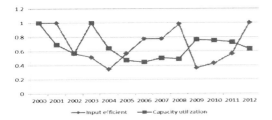

Figure 1. Input efficiency capacity utilization rate comparison graph of the oil and gas exploration industry.

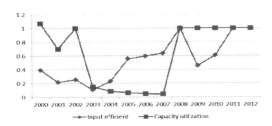

Figure 2. Input efficiency capacity utilization rate comparison graph of the ferrous metal mining industry.

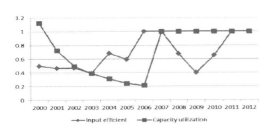

Figure 3. Input efficiency capacity utilization rate comparison graph of the non-ferrous metal mining industry.

two indexes is basically synchronous, with the average capacity utilization level of 68.3%, belonging to serious overcapacity. According to statistical data from the *Statistical Yearbook of Development of Gansu Province*, it is inferred that the industry has problems such as reconstruction, over-investment, idle production equipment, insufficient utilization or a serious impact of some kind, of some link in the entire production system.

3.2.2 Ferrous metal mining industry

Figure 2 shows two indexes of the industry in capacity utilization level that were not simultaneous before 2008, and were basically the same after 2008 but were anisomerous. It is determined that the industry witnessed low levels of capacity utilization and serious excess capacity before 2008, and with an average capacity utilization level of 90% after 2008, a high level of capacity utilization and mild insufficient capacity

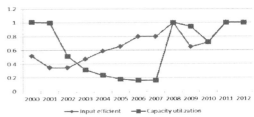

Figure 4. Input efficiency capacity utilization rate comparison graph of the oil processing and coking industry.

3.2.3 Non-ferrous metal mining industry

As can be seen from Figure 3, the two indexes of the capacity utilization level of the non-ferrous metal mining industry are basically the same in change trend, and featured in insufficient capacity utilization and high fluctuation of the capacity utilization level. According to statistics data from the *Statistical Yearbook of Development of Gansu Province*, it is inferred that the industry suffered production or operational problems in 2010, but still witnessed the efficient use of capacity.

3.2.4 Oil processing and coking industry

As can be seen from Figure 4, the capacity utilization level of this industry fluctuated greatly, falling from 100% in 2000 to 17.86% in 2007, year by year, with low levels of capacity utilization. After 2008, it rose to more than 85% or above, and there may be reuse of capacity or mild insufficient capacity.

4 CONCLUSION

Before 2007 or 2008, the capacity utilization levels of the four sectors were in severe shortage state (75%); during 2008 and 2010 they were in a moderate state (75–90%); and after 2010 they were more than 90% and in an insufficient capacity state. This shows that 2007 to 2008 was a point of inflection from low to high for the capacity utilization levels of the four major industrial sectors in Gansu Province.

REFERENCES

Tian Yanfang; Barriers to Exit and Overcapacity [D]; Dongbei University of Finance and Economics; 2010 (11), 1–16

Lu Nanlin; Research of Correlation between Overcapacity and Market Structure [D]; Jilin University; 2007(3), 6–50

Zhu Minxian; Research of Formation Mechanism of Domestic Overcapacity Under the Transitional Regime [D]; Zhejiang University; 2012(5), 3–6

Wang Liguo, Gao Yueqing; Research of Overcapacity Based on the Perspective of Technical Progress[J]; Research on Financial and Economic Issues; 2012(2)2, 2–7

Bao Jian; Research of China's Total Demand Structure Adjustment and Balance of Payments [D]; Fudan University; 2011(4), 2–20

Li Peng; Research of China's Coal Chemical Industry Overcapacity [D]; Dongbei University of Finance and Economics; 2011(11), 7–45

Na Yimei; *Research of Coal Industry Development Model Based on the Production Function and Cost Function* [D]; Fudan University; 2011(4), 1–55

Zhao Ying; *Quantitative Measure of Overcapacity and Its Correlation with Macro-economy* [D]; Anhui University; 20011(4), 10–45

Sun Wei; *Non-parametric Analysis of Efficiency and Productivity* [M]; Beijing: Social Sciences Academic Press; 2010, 23–77

Du Zhonghua; *Research of Overcapacity in China's Iron and Steel Industry- Taking 22 Domestic Steel Companies as Example* [D]; Dongbei University of Finance and Economics; 2011(11), 6–49

Research on business model innovation for China's environmental industry

Y.F. Li
Economics and Management Institute, Southeast University, Nanjing, Jiangsu, China
Business Institute, Anhui University of Finance and Economics, Bengbu, Anhui, China

D. Li
Economics and Management Institute, Southeast University, Nanjing, Jiangsu, China

ABSTRACT: Currently the criticism of environmental pollution in China has increased at home and abroad. Energy efficiency and emissions reduction have become the driving force of the environment industry which involves air pollution treatment, water pollution treatment, solid waste pollution treatment and ecological restoration, etc. A good business model can shape and reshape the environment industry and drive spectacular growth. Business model innovations for China's environmental industry include creating a new Customer Value proposition, optimizing the design of Profit models and improving Operation systems.

1 INTRODUCTION

China has paid a high price for its rapid development over the past 30 years, as nowadays the environmental problem is becoming more and more serious. The traditional economic development pattern which means high energy consumption and heavy pollution but low efficiency, is not fit for the needs of enterprises' subsistence and development. Against this background, energy efficiency and emissions reduction are the inevitable choice for the social and economic development for today's China, and have become the driving force of the environment industry which involves air pollution treatment, water pollution treatment, solid waste pollution treatment and ecological restoration, etc. If the environmental problem remains unresolved for a long time, it is not enough to only consider techniques; emphasis needs to be placed upon bringing forth new ideas and business models. Once the environment industry finds out and applies a good business model, a huge market will open up and enjoy enormous growth potential. Therefore, combining technical innovation with business model innovation is a must to the development of the environmental industry in China.

2 BUSINESS MODEL

2.1 *Definition and components*

Despite lineage going back to when societies began engaging in barter exchange, business models have only been explicitly catapulted into public consciousness during the last decade or so (Teece, 2010). Later in the 1990s, the term "business model" was widely used in management speak and the business press (Feng et al. 2001). This surge is levelled with the advent of the Internet in the business world and the steep rise of the NASDAQ stock market for technology-emphasized companies (Osterwalder et al., 2005). Strategy has been the primary building block of competitiveness over the past three decades, but in the future, the quest for sustainable advantage may well begin with the business model (Casadesus-Masanell & Ricart 2011). The problem is that most people know that business models matter, but not every business model suits. Only a good business model remains essential to every successful organization, whether it's a new venture or an established player (Magretta, 2002). While it has become uncontroversial to argue that managers must have a good understanding of how business models work with effect if their organizations are to thrive, the academic community has only offered early insights into the issue. To date, there is (as yet) no agreement on the distinctive features of superior business models (Casadesus-Masanell & Ricart, 2010). Because the same idea or technology applied to the market by two different business models will yield two different economic outcomes, it makes good sense for companies to develop the capability to innovate their business models (Chesbrough, 2010).

A business model is a story that explains how an enterprise works (Magretta, 2002), a mediating construct capturing value from early-stage technology (Chesbrough & Rosenbloom, 2002), a method of doing business (Rappa, 2004), a blueprint of how a company does business (Osterwalder et al., 2005), and a reflection of the firm's realized strategy (Casadesus-Masanell & Ricart, 2010). Teece (2010) argues that a business model articulates the logic and provides data and other evidence that demonstrates how a business creates and delivers value to customers, and also outlines the architecture of revenues, costs, and profits associated with the business enterprise delivering that value. Osterwalder & Pigneur (2010) conceptualize a business model as a description of the rationale of how an organization creates, delivers, and captures value. Amit & Zott (2012) define a business model as a system of interconnected and interdependent activities which are performed by the focal firm and by its partners and the mechanisms that link these activities to each other, and thus it determines the way the focal firm "does business" with its customers, partners and vendors.

Johnson et al. (2008) hold that a business model consists of four interlocking elements (namely, Customer Value proposition, Profit formula, Key resources and Key processes) which, taken together, create and deliver value. Baden-Fuller & Mangematin (2013) contend that a business model is composed of four elements: Customers, Customer engagement, Monetization and Value chain and linkages. Bocken et al. (2014) put forward a viewpoint that three components constitute a business model, that is, Value proposition, Value creation and delivery, and Value capture.

In this paper, a business model represents a way the focal firm do business with the stakeholders to capitalize on market opportunities, and includes three major components: (a) a Customer Value proposition which describes the nature of targeted customers, the offerings, as well as the value elements incorporated within the offerings; (b) a Profit model which describes how the focal firm captures the values which are generated through the value proposition, and how all kinds of resources are monetized in the value constellation; and (c) an Operation system which describes how those values are created and delivered.

2.2 Classification

Baden-Fuller & Morgan (2010) point out that the functions of business models as models are embodied in three forms: to provide means to describe and classify businesses; to operate as sites for scientific investigation; and to act as recipes for creative managers. At present, business models of the environmental industry in China can be subdivided into several types as follows.

A BOT (build–operate–transfer) model, is essentially, a way of investment, construction and business operation for the public infrastructure. The public administration and an environmental protection company come to a concession agreement that the latter is authorized to design and build infrastructure, to operate and maintain the facility for a certain period, and then to transfer the facility to the former at the end of the concession agreement. The main reason that the BOT model is widely employed in the municipal public utility industry is that many governments are short of money.

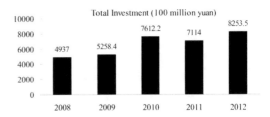

Figure 1. China's total investment in the treatment of environmental pollution.
Data sources: China Statistical Yearbook 2013.

A BTO (build–transfer–operate) model, is where the environmental protection company is responsible for the public facility construction, and the facility is required to immediately transfer to the public administration when it is completed, and then the company will receive the franchising license from the public administration over a long period of time.

A BOO (build–own–operate) model, is where, in accordance with government concessions, contractors finance, build, and operate the project, without need of giving it back to the public.

A DBOT (design–build–operate–transfer) model, is where the environmental protection company is responsible for designing and building the pollution-treatment infrastructure, and ensures that the operation meets the contract requirements under the terms of the contract, and after the expiration of the contract, the property ownership must be transferred to the owner.

An EPC (engineering–procurement–construction) model, is where the environmental protection company, as the general contractor, assumes complete responsibility for project engineering, procurement, construction and commissioning, and the focal firm really consider handing the project over to the owner only when those jobs are properly done.

A PBCES (performance-based contract for environmental services) model, is where payment for environmental services is only made when a customer gains a given environmental performance.

3 CHINA'S ENVIRONMENTAL INDUSTRY

3.1 Current situation

China's total investment in the treatment of environmental pollution was 8253.5 billion RMB yuan in 2012, which is 1.67 times the amount of 2008, as shown in Figure 1.

According to the Twelfth Five-Year (2011 to 2015) Plan for China's Energy Saving and Environmental Protection Industry, the gross product has already reached 2 trillion RMB yuan since 2010, and by 2015, the total output value will add up to 4.5 trillion RMB yuan with an average increase of more than 15%. It is a task of this stage's development to breed and cultivate a crop of large enterprise groups with international competitive ability in energy saving and environmental protection.

Table 1. Nine listed companies in 2013.

Stock code	Company name (abbreviation)	Revenues Billion RMB yuan	Weighted average ROE %
000826	SOUND ENVIRONMENTAL	26.84	14.16
002549	KAIMEITE GASES	2.32	8.30
002573	SPC	7.65	8.3
002672	DONGJIANG ENV	15.83	9.61
300070	BOW	31.33	19.35
300172	CEEP	5.41	8.21
300187	YONKER	6.40	6.53
300190	WELLE	2.78	7.52
600292	CPIYD	32.51	7.3

Source: 2013 Annual Report of four companies.
Note: ROE means Rate of Return on Common Stockholders' Equity.

3.2 Listed companies

According to the China Securities Regulatory Commission industry classification standard (the first quarter of 2014), there are nine listed companies in the industry of ecological conservation and environmental clean-up (industry classification code: 77), including three enterprises listed on the Small and Medium-sized Enterprise Board whose stock code begins with the numbers "002", while another four enterprises are listed on the Growth Enterprise Market, whose stock code begins with the number "3", as shown in Table 1.

Sound environmental resources co., ltd., or Sound Environmental for short, is a pioneer in China's solid waste disposal industry, and a comprehensive environmental service provider. Hunan kaimeite gases co., ltd., or Kaimeite gases for short, has become the largest provider of food-grade liquid carbon dioxide based on the tail gases emitted from the production process of chemical enterprises, and is also a specialist to recover the tail gas. Beijing spc environment protection tech co., ltd., or Spc for short, is a comprehensive service provider mainly engaged in atmospheric environmental governance. Dongjiang environmental co., ltd., or Dongjiang env for short, is a highly technological environmental protection enterprise which specializes in industrial waste treatment and municipal waste treatment. Beijing originwater technology co., ltd., or Bow for short, is committed to solving three main problems in the field of water environment, that is, "polluted water", "water scarcity", and "water insecurity", whose main business area covers the entire water industrial chain. Nanjing cec environmental protection co., ltd., or Ceep for short, mainly provides comprehensive environmental business, including wastewater treatment of the key national industry, seawater desalination, municipal wastewater treatment etc. Yonker environmental protection co., ltd., or Yonker for short, is a force of the soil restoration business in China's Xiangjiang river basin. Jiangsu welle environmental co., ltd., or Welle for short, is a pioneer specializing in municipal solid waste and waste leachate treatment. CPI Yuanda environmental protection co., ltd., or Cpiyd for short, is involved in flue gas desulfurization, the manufacture of denitration catalysts, and other businesses.

From Table 1, we can see that the nine listed companies differed in their revenues and profit abilities in 2013. BOW, listed on the Growth Enterprise Market, has grown into the most profitable environmental protection company, while CPIYD, listed on the Main Board Market, remains the company with the largest annual revenues. Only two companies' weighted average ROE was above 10% in 2013.

4 CASE STUDY: BUSINESS MODELS OF KITCHEN WASTE TREATMENT

4.1 Market background

The total amount of China's kitchen waste is far beyond huge. China now generates around 50,000,000 to 60,000,000 million tons of food waste per year, accounting for more than 37% of municipal solid waste. According to the "Twelfth Five-Year (2011 to 2015) Construction Plan for National Urban Household Garbage Harmless Disposal Facilities, until 2015, China will comprehensively push forward a garbage classification pilot, and will have initially executed food waste collection, classification, transportation and treatment in more than 50% of cities with sub-districts, and strive to achieve the capacity of 3 0,000 tons per day. Even so, the rate of China's garbage classification and treatment will still be less than 20% by 2015.

The improper disposal of food waste does damage to both the environment and human health. Food waste is high in organic content, high in moisture content, and is extremely apt to rot and breed flies in the short term. Thus, problems of secondary pollution exist extensively because of its improper treatment. For the same reason, food waste is a sort of organic garbage, which is hard to be disposed of in the traditional manner, such as through landfills or incineration. Moreover, illegal cooking oil which is usually made from discarded kitchen waste, may flow back onto people's dinner tables in China.

However, kitchen waste could also be viewed as a useful resource. Kitchen waste can be converted into bio-diesel, bio-gas, organic fertilizer, high protein feed and other burning sticks etc., through different technological pathways.

4.2 Yantian model

Yantian District in Shenzhen City, China, is made up of four sub-district offices and had about 213, 900 residents within a total area of 72.63 square kilometres at the end of 2013. According to Yantian's website, a project of kitchen waste harmless treatment and resource utilization has been undertaken since 2012, and it has become one of the most important projects related to people's well-being for the Yantian government in 2013. Nowadays, Yantian basically achieves a harmless treatment of kitchen waste and has not yet increased the amount of refuse burning. In fact, the real value of the Yantian model lies in its finding a sound business logic in a 'win-win' situation, as shown in Figure 2.

An environmental protection company (expressed in Franchisee A) obtains the franchise right to food waste treatment in Yantian District through public bidding. Franchisee A can turn food waste into bio-diesel and fuel rods, and the process does not require incineration. Thus, Franchisee A actually improves and beautifies the environment, and prevents the usage of illegal cooking oil. With an integrated process design, the food waste can be discarded into sortable garbage in a service centre of waste reduction easily and conveniently, and can be classified by the residents themselves because of the high number of centres. The residents can also put their food waste into the specified locations twice a day, in the morning and evening. The highly-involved residents will receive a small prize provided by Franchisee A. The dishwashers are recruited to classify the food waste of their restaurants by Franchisee A. The restaurants' food waste achieves an 85% reduction and harmless disposal through solid-liquid separation and bio-chemical treatment in the integration treatment station nearby. Franchisee A also pays a fee (30 RMB yuan per ton) to their restaurants for carting away food waste. Franchisee A will get subsidies from the local government (200 RMB yuan per ton). The benefits for all stakeholders, which include the local government, the franchisee, the restaurant owners, the residents and the dishwashers, are taken into account with the utmost level of detail.

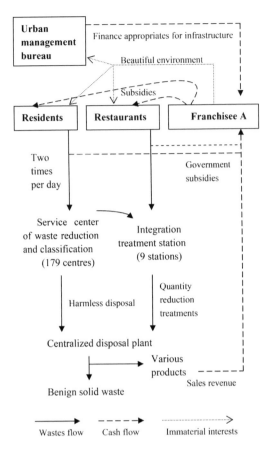

Figure 2. Yantian model.

4.3 Nanshan model

Nanshan District also located in Shenzhen City, China, had more than 1,100 thousand residents within a total area of 510.95 square kilometres, which is about 2.8 times as large as its land area, at the end of 2013. Nanshan is the pioneer in using the concession system which established the concession company (expressed in Franchisee B) for food waste disposal in Shenzhen City. But now Franchisee B, by contrast, is caught up in trouble because of local restaurants' unwillingness to sign up to it.

Because Franchisee B converts food waste into feed, it demands a higher level of garbage sorting, which increases its business and operational costs. If it does not, the quality of feed may deteriorate and could not be sold easily. For example, Franchisee A of Yantain can turn the food residue with small amounts of plastic bags into bio-fuels, while Franchisee B cannot transform the food residue into feed for livestock or poultry because the animals will get sick or die by eating the feed.

In addition, some restaurants have established long-term and stable business relations with the companies that make illegal cooking oil from discarded kitchen waste. The waste can be refined, and the illegal oil came out. Once used, the restaurants can gain more profit rather than gain subsidies from Franchisee B. However, local government has failed to combat illegal kitchen waste treatment effectively.

So the main reason why specification processing of kitchen waste in Nanshan has been caught in the dilemma as "the castle in the air", lies in its failing to handle the interests of the stakeholders.

4.4 Lanzhou model

Lanzhou is the capital city of Gansu Province, China. Lanzhou city started a new programme for the comprehensive collection and disposal of food residue from the winter of 2011 with the BOT model, appointing a bio-energy company (expressed in Franchisee C) as sole agent of dealing with food waste in the city. Based on the food waste, Franchisee C produces bio-diesel, bio-gas, organic fertilizer, and electricity. The programme, described by the media as the "Lanzhou model" which bans the usage of illegal cooking oil and turns the waste into wealth, has been caught in an awkward position since 2013.

Lanzhou charges for the disposal of food residue from the restaurants on the basis of management areas, at 2 RMB yuan per m^2, and pays Franchisee C. The restaurants are definitely not willing to pay for waste disposal. Of course local businesses may have some doubts about Franchisee C who already enjoys governmental financial incentives and even seeks a large treatment fee from the restaurants in general.

5 CONCLUSIONS AND FUTURE WORK

The protection of the environment is one of the basic state policies in China. Currently, the criticism of environmental pollution in China is increasing at home and abroad. Therefore, China has strong environmental needs. But there is a great gap between environmental needs and the environmental industry due to a lack of effective business models. As we can see from the above cases, a good business model is a necessary condition in order to make technology work effectually.

5.1 Creating a new Customer Value proposition

Seen from the Customer Value proposition, the focal firm (Franchisee) needs to identify different customers' jobs and balance distinct sets of stakeholders' interest. The focal firm might have several target consumer groups according to the heterogeneity of consumers' demands or jobs. The term "Job" here means a fundamental problem that the customers need to solve in a specific situation (Johnson et al. 2008). A customer problem refers to the gap between the expected state and the actual state of a certain customer job. It is worth noting when the situation is of critical importance. In order to adapt to a new situation, a customer will almost always need a new solution. A complete solution describes a certain customer's job process and setting. The focal firm (Franchisee) is expected to understand and help to finish the jobs of local government, residents and restaurants in the environmental industry. For example, the reason why disposing of waste through landfill and incineration is often accompanied by some dispute, especially when a new tipping site is planned, is that by using this approach, is it is hard to avoid repeated pollution.

5.2 Optimizing the design of the Profit model

Seen from the Profit model, the focal firm needs to create and select profit points. A profit point, also called a revenue generator, is where the focal firm expects to make its money. The focal firm should design and then decide what products, processes, functions or property rights are chargeable and what ones are free in the output system, as shown in Figure 3.

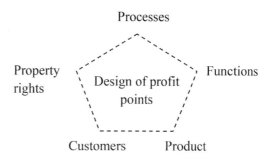

Figure 3. The design of "profit points".

The ability to add revenue sources is mainly manifested in five aspects: (a), Distinguish products. Not every product and service has to make money, just some for a fee, and some for free. Free items should contribute to increasing the number of customers for fee-based products and services; (b), Distinguish processes. The Profit point is not always dependent upon the existence of a specific product and service. If the payments process is completed ahead of shopping, for example, the focal firm may get additional income; some large department stores get a lot of advance payment by issuing shopping cards. In the same respect, the stores can also retain a large amount of cash income by late payments to their suppliers; (c), Distinguish customers. Not every customer has been charged for the same business. The most common internet companies do not charge the netizens, but advertisers; (d), Distinguish property rights. Some commercial and real estate companies gain higher revenues by a sale-leaseback which separates ownership and uses the right of house property; and (e), Distinguish functions. The different functions of a product or service could have different charging modes. The ignition function of a lighter may charge the end-consumer, while its advertising function may charge a certain restaurant that imprints a message (the address and the telephone) on it. Thus, on the one hand, the focal firm decreases operating costs through a modernized management system, whilst on the other, it promotes economic performance by improving the utility level and looking for better ways of resource recycling and innoxious treatment and disposal.

5.3 Improving the Operation system

Seen from the Operation system, the focal firm needs to take advantages of network effect and technology innovation. The Operation system is the basic material conditions for realizing the Customer Value proposition and Profit model. This case study shows a business model can be on the high class way only by building the system of food waste collection and transportation and disposition. Some individuals or firms illegally collect, cart away and dispose of food waste for potential business benefits. The focal firm needs to depend on government power to halt trafficking. Forming an industrial chain is necessary to bring environmental pollution under control, and is the key to China's environmental protection industry. It is important to recognize that technical innovation may bring a new Customer Value proposition and changes in the Operation system to create more value for the stakeholders. Therefore, one goal for the ecologicalization of the industry and industrialization of ecological environment protection, is to establish a virtuous circle of business ecosystems.

Environmental concerns have shifted from the water to the air and the soil, and to the whole of China. So, future research agendas are to focus on four dimensions: (a), Preventing and controlling air pollution needs not only institutional guarantees at a legal level, but also a viable business model at the economic level; (b), Soil remediation as a new industry is in its infancy, with imperfect and impertinent laws and regulations, uneven quality of techniques, and hard-to-find financing; (c), The urban water supply and wastewater treatment market is developed well, but the water problems of the countryside and small towns are far from being resolved in a systematic way because of this lack of a business model; and (d), The Environmental service industry is an important component of the environmental industry. Except for the municipal industry, the industrial water treatment industry, and the garbage incineration industry, its business model remains largely unclear in other areas.

ACKNOWLEDGEMENTS

Financially supported by the 2014 Jiangsu Provincial Innovation Program for Colleges and Universities Graduate Students (Grant No.: KYZZ_0079) and National Natural Fund Project of China (Grant No.: 71372196).

REFERENCES

Amit R. & Zott C. 2012. Creating value through business model innovation [J]. *MIT Sloan Management Review*, 53(3): 41–49.

Afuah A. & Tucci C. L. 2000. Internet business models and strategies: Text and cases [M]. *McGraw-Hill Higher Education*.

Baden-Fuller C. & Morgan M. S. 2010. Business models as models [J]. *Long Range Planning*, 43(2): 156–171.

Baden-Fuller C. Mangematin V. 2013. Business models: A challenging agenda [J]. *Strategic Organization*, 11(4) 418–427.

Casadesus-Masanell R. & Ricart J. E. 2010. From strategy to business models and onto tactics [J]. *Long Range Planning*, 43(2): 195–215.

Casadesus-Masanell R. & Ricart J. E. 2011. How to design a winning business model [J]. *Harvard Business Review*, 89(1–2): 100–107.

Chesbrough H. Business model innovation: opportunities and barriers [J]. *Long range planning*, 2010, 43(2): 354–363.

Chesbrough H. & Rosenbloom R. S. 2002. The role of the business model in capturing value from innovation: evidence from Xerox Corporation's technology spin-off companies [J]. *Industrial and corporate change*, 11(3): 529–555.

Feng H. et al. 2001. A new business model? The capital market and the new economy [J]. *Economy and Society*, 30(4): 467–503.

Johnson M. W. et al. 2008. Reinventing your business model [J]. *Harvard business review*, 86(12): 57–68.

Magretta J. 2002. Why Business Models Matter [J]. *Harvard Business Review*, 80(5): 86–92.

Osterwalder A. & Pigneur Y. 2010. Business model generation: a handbook for visionaries, game changers, and challengers [M]. *Wiley. Com*.

Osterwalder A. et al. 2005. Clarifying business models: Origins, present, and future of the concept [J]. *Communications of the association for Information Systems*, 16(1): 1–25.

Teece D. J. 2010. Business models, business strategy and innovation [J]. *Long range planning*, 43(2): 172–194.

Zott C. & Amit R. 2013. The business model: A theoretically anchored robust construct for strategic analysis [J]. *Strategic Organization*, 11(4) 403–411.

Bocken N.M.P. et al., 2014. A literature and practice review to develop sustainable business model archetypes [J]. *Journal of Cleaner Production*, 65: 42–56.

The establishment of a public participation system in regulatory detailed planning—a study based on the interactive model of government and NGOs

Dan Li
Graduate Student, School of Architecture and Urban Planning, Huazhong University of Science and Technology, China

JinFu Chen & Qian Chen
Urban and Rural Panning Policy Research Centre, Huazhong University of Science and Technology, China

ABSTRACT: Public participation in regulatory detailed planning is a guarantee of community planning implementation. Faced with the current situation of superficial and false public participation, tried to establish a public participation system that is suitable for China's current conditions, using the cooperation theory of "government – non-governmental organizations". The basic logic is: in urban and rural planning and management, a set of operation systems in which the government is the main body and is responsible for planning approval, implementation and supervision, and a set of systems in which non-governmental organizations are the main body and are responsible for the whole public participation process, are implemented. Second, based on the game theory, analysed the internal mechanism for how non-government organizations affect public participation, and concluded that public participation is a game process between the public, development groups, experts and scholars. In the public participation system which is based on an interactive mode of the government and non-government organizations, there are many participants and games, which ensure the effectiveness of public participation.

1 BACKGROUND

At present, public participation in our country is still at the primary stage, the level of public participation is low, and participation strength is weak. The current system of public participation has some problems. On the one hand, the government and its subordinate departments are exclusively responsible for urban and rural planning, from planning to implementation and supervision, that is to say the government is both "referee" and "athlete". In recent years, the government has begun to introduce public participation in planning and management processes. The common practice is that experts' points of view replace public opinion; the interaction with other participation parties is small. On the other hand, the public participation system is not in place; there are only some rules in some laws and regulations. These general rules cannot guide the public to participate in planning and management, and government departments at all levels have various methods to do these things. Information disclosure, public feedback and feedback mechanisms are lacking, which leads to a complete lack of participation or superficial participation. The establishment of a public participation system in regulatory detailed planning.

1.1 The new mode of public participation—the interaction between governmental and non-governmental organizations

Non-governmental organizations are non-profit and welfare-oriented organizations which are independent of the government and enterprise market system, they have increasingly become an important force in social governance, and play a bridge role in social relationships. Non-governmental organizations are of vital significance in helping the public understand the government's planning and decision-making, realizing public participation, and protecting public interests, as well as alleviating social conflicts.

Benjaimin Gidron, Ralph Kramer, and Lester M.Salmon founded the classic theory about the relationship between government and non-government organizations. They used fundraising and authorization, and service distribution as the core variables, and they proposed the four basic models for the relationship between government and non-government organizations[1]: government dominant model, non-government organizations dominant model, dual mode and cooperation mode. At present, Western countries, especially America, adopt the "cooperation model";

they think this kind of cooperation relationship is greatly conducive to social democracy, and non-government organizations have large autonomy and a say in public service management.

Faced with China's unique social and political conditions, completely copying the Western approach is not feasible. But, if we want to change the present situation of superficial and false public participation, we need to introduce a new body which is independent of the government and is responsible for public participation, and promote the smooth progress of public participation through the establishment of an interactive cooperation relationship. The author hopes to establish a public participation system within the Chinese planning administrative system which is suitable for China's current conditions, using the cooperation relationship theory of "government – non-government organization".

1.2 Design of a public participation system in regulatory detailed planning

Assuming there are non-government organizations which are protected by laws and responsible for public participation in regulatory detailed planning, an interactive mode can be established in urban planning and management. In one system, the government is the main body and is responsible for controlled planning, approval, implementation, supervision and operation; in another system, non-governmental organizations are the main body and are responsible for the whole process of public participation. The two systems interact with each other and achieve coordination development, so as to ensure the effectiveness of public participation. For the government, its administrative system and work process have not changed, although non-governmental organizations are responsible for public participation, from the perspective of long-term development, and its work efficiency is improved significantly. The specific steps and processes are shown below. Table 1 shows the interactive model of government and NGOs in public participation, and can be found at the end of this paper.

1.3 Description of the public participation system

1.3.1 Legal documents and guidance documents involving different public objects

The main content of detailed planning control are mainly land use control, environmental capacity control, building construction and control, city design guidance, city municipal engineering facilities, public service facilities, traffic control and regulations of environmental protection, as well as qualitative and quantitative control, and guidance on different plots, different projects and different development processes according to indexes and provisions. Each area of our country has slightly different preparation methods for regulatory detailed planning but, overall, there are two levels, a control planning unit and a regulatory management unit, which form two planning systems, regulatory detailed planning guidelines and regulatory detailed planning rules. Guidelines and rules are divided into two parts, legal documents and guidance documents. In the process of examination and approval, the legal documents are the fundamental basis for urban and rural planning and management, and they must be publicly disclosed and reported to the Municipal Government for approval. Guideline documents are the basis for planning design and management, and must be submitted to the competent department of urban and rural planning for approval. In the implementation process, the government planning department should propose conditions for regulatory detailed planning, and guiding opinions for land use and construction projects. The fulfillment of these conditions is necessary for obtaining the "two certificates".

Decision-making and implementation of different public affairs have different influences on the public, and the degree of public participation also varies. Generally speaking, an open degree of public participation is closely related to the correlativity of interest. More direct stakeholders have more power in public participation. At the same time, we need to notice that public participation has its own disadvantages, for example, the opinions from the public may be related to their own interests and be lacking in overall consideration. And, development groups pursue their own profit growth, and are likely to infringe upon public interests. Therefore, we should not introduce public participation into every link of the planning process; participation objects and forms should be based on planning content and key points.

The legal documents of regulatory detailed planning mainly include function positioning, public facilities, etc. This content strengthens the control on public service facilities, transportation, water, green land, infrastructure, cultural relics, and other public resources which are subject to legal protection. In the implementation process, all the content in the legal documents will directly enter the planning conditions. In the process of examination and approval, the regulatory legal documents are very important, and involve the interests of numerous citizen groups and enterprises. Therefore, compared with guidance documents, the public participation degree of the legal documents should be of less importance.

Because non-governmental organizations are folk, non-profit making, autonomous, and non-political, they can determine public participation objects and carry a certain power. In public participation of legal documents approval, non-governmental organizations can build a group that offers consulting services and advice for planning decision-making. Through this group, related experts and personnel from departments related to urban construction can take up the vast majority of the work, and representatives from enterprises and the public can take up a smaller proportion. In order to be distinguished from other groups, this group is known as an "advisory" group. The "advisory" group can examine legal documents, put forward opinions and suggestions, but the final decision is made by the government. After the legal document is approved, the design unit can improve the plan according to recommendations from the "advisory" group, and adopt those advices according to specific conditions. The roles of non-governmental organizations are: (1), carrying out publication before approval of statutory documents, collecting public opinions through multiple channels; (2), choosing appropriate members for the "advisory" group; (3), organizing "advisory" group discussion, making the group members to offer mature advice according to public opinion; (4), collecting suggestion materials and submitting to the government departments; (5), publishing recommendations.

Guidance documents mainly include population scales, land use control, construction strength index control, city design requirements, etc. The core issue at this stage is the optimal allocation of space resources, including service facilities, land arrangement for large-scale enterprises, important nodes and landscapes in the city, etc. These involve direct interests of different groups, and public participation is the discussion and game between representatives from these different groups. The degree and intensity of public participation will increase, and the key point for public participation is the coordination of multi-stakeholders; government planning departments must consider whether the plan will make all relevant stakeholders satisfied with the process of examination. NGOs can determine candidates of the public participation objects and build a temporary group, which is mainly composed of experts and scholars, business representatives and public representatives; stakeholders that are related to urban construction will only take up a small proportion. In order to be distinguished from other groups, the group is called the "intervention" group. The roles of non-governmental organizations are: (1), carrying out publication before approval of guidance documents, and collecting public opinions through multi-channels; (2), choosing proper "intervention" group members; (3), organizing "intervention" group discussion, and listening to requirements from various interested parties; (4), promoting negotiations between conflicting parties and resolving conflicts through various effective means, with the use of professional and popular descriptions; (5), reaching a final amendment with which all parties are satisfied; (6), submitting the amendment to the government planning department. "Intervention" group members hope that their interests are well respected, and that the modified opinion will reflect the needs of the stakeholders. The Government planning department must seriously consider the modified opinion and integrate it into the construction plan.

In the proposition of the planning conditions for the implementation process, the government planning department should propose planning conditions about the location of land, the construction strength (building density, building height, volume rate, green rate), the required infrastructure and public facilities, the population carrying capacity, the building form, etc. At this stage, non-governmental organizations need to use detailed information of the construction land, redetermine "intervention" group members, and make them discuss and propose planning conditions with personnel from the planning department. Due to the certainty of the construction land in the implementation process, the number of direct stakeholders will greatly increase, public participation should bet thorough. The proportion of stakeholders in the "intervention" group should be increased.

1.3.2 *Methods of public participation that can be used in non-governmental organizations*

For the implementation of public participation, non-governmental organizations complete each process through choosing proper methods. In terms of information disclosure, newspapers, television media, Micro messages, websites, micro-blogs, brochures, web query systems, and community bulletin boards are used for disclosing information about public projects. Easy to understand descriptions are also designed to explain planning intent and professional indexes. In public opinion collecting, interviews, questionnaire surveys, public hearings, citizen forums, email, micro messages, micro-blogs, and other means are adopted. In terms of publicity, planning knowledge lectures, planning exhibitions, planners' hotlines, and on the spot Q&As are used to stimulate the public's enthusiasm; in public opinion feedback, public opinion books, public Q&A meetings, and rewards for active participants can be used.

1.3.3 *Guarantees for non-governmental organizations*

Non-governmental organizations are very effective in promoting public participation, but they have not achieved great developments in China. The reasons are mainly as follows: (1), the government is the main body involved in allocating public interests. Its traditional decision-making may not be able to quickly accept public participation, public participation is inevitably in conflict with traditional governance method; (2), many non-governmental organizations depend on the government for their funding and personnel, and their main task is to follow the command of the government and to complete work; they cannot represent the public in public decision-making. (3), because of the influence of planned economy systems, public consciousness is not open, growing environments, management functions and organizational structures of non-governmental organizations are not standardized, and their development is slow.

The balance of power is the basis of resource allocation, and the precondition of social fairness (S. S. Fainstein, 2000). Effective protection of non-governmental organizations' roles depend on their abilities to obtain social resources, and their development needs effort in the following four aspects: legal guarantees, financial support, great ability to resolving social contradictions, and strong public support.

Table 1. The interactive model of government and NGOs in public participation.

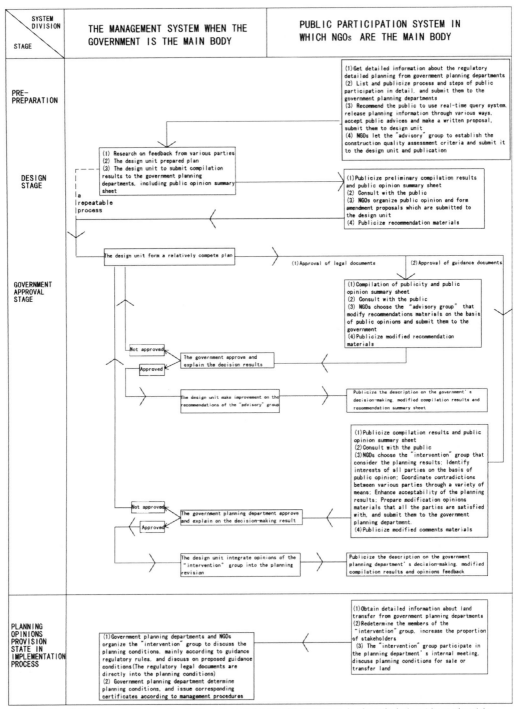

Table 2. Multi-subjects and games in interactive modes of government and non-government organizations.

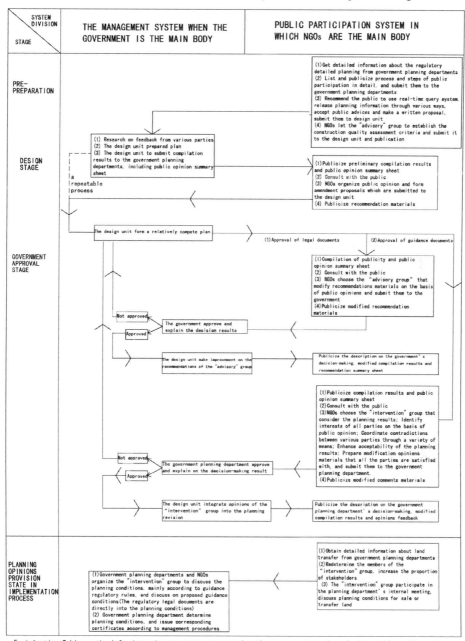

2 INTERNAL MECHANISMS FOR NON-GOVERNMENT ORGANIZATIONS AFFECTING PUBLIC PARTICIPATION IN REGULATORY DETAILED PLANNING

The main parties of public participation can be divided into four categories[3]: (1). The public. The public is an important and extensive element for stakeholders in regulatory detailed planning. In a modern society, the public's democratic quality and participation awareness gradually improve and increase. The public who want to take part in planning, consist of two types of people. The first type participate because certain planning decisions have actual impacts on their own interests, and they want to express their desires in public participation. The second type do so because of their master consciousness; although some planning decisions do not have a direct effect on their own legitimate rights and interests, as the ultimate owners of the city, they are bound to be affected, therefore, they want to express their views; (2). Development groups. Land developers or investor enterprises are the most active subjects in city construction, they have a say in the selection of the project site, the nature of land use, and the development intensity. They hope to express their intentions in planning decisions and implementation, so as to maximize their own interests; (3). Experts and scholars. Experts and scholars are technical personnel who have professional knowledge and occupation abilities related to urban planning, their responsibilities are to tell the truth and impart knowledge to the public, thereby protecting public interests and vulnerable groups; (4). The government. The government is the biggest provider and protector of public interests. The government can maintain open channels of communication with the public, listen to public feedback, promote social justice, and ensure smooth implementation of the plan through public participation. But, the government also has the pressure of survival, and a reform of the system of tax distribution has made the government pay great attention to the requirements of development groups, in order to maintain the city's effective operation.

Analysis of the effectiveness of public participation, Public participation can be regarded as a game process between the public, development groups, experts and scholars, and the government. Game objects are lacking and game times are few. The first game is between the development group and the government, and it happens at the beginning of the planning preparation. The development group plays a game with the government, with the intention of gaining more profits. Since the government's public policies need to be realized through economic means, they are inevitably restricted by economic factors.

The development group changes and even controls planning, and does not consider the city's overall benefit and the social public interest. It tends to infringe on the interests of other parties. The government is faced with a dilemma ("fairness" versus "efficiency") in the game, but in general, it needs to reduce the negative impact of the development group. The second game is between the government and experts and scholars; it happens at the experts' review meeting before the approval. Experts and scholars will propose positive amendments, due to their planning education and moral constraints, and the government will selectively accept the experts' opinions, at the same time taking financial conditions and the investment group's requirements into account.

The public participation system, which is based on interactive modes between government and non-government organizations, provides multi-subjects and games. The specific conditions are shown as follows. Table 2 shows multi-subjects and games in interactive modes of government and non-government organizations, and can be found at the end of this paper.

3 CONCLUSION

Public participation is an eternal and important content in urban and rural planning. The author has tried to establish a public participation system that is suitable for China's current conditions, using the cooperation theory of "government – non-governmental organizations". The public participation system is still in the stage of theoretical research; the author hopes to bring about more academic contend with this paper.

REFERENCES

Benjaimin Gidron, Ralph Kramer, Lester M. Salmon. Government and The Third sector, San Francisco, Josser-Bass Publishers, 1992, pp. 18.

Feng Ling. Improve the "light", the anti-corruption in depth [N]. Legal Daily, 2008-01-22 (3).

Xixin Wang. The public participation in the administrative system in the process of practice [M]. Chinese legal publishing press, 2008, pp. 205–207.

Zhiqiang Wu. Huade LI. Principle of city planning (Fourth Edition) [M]. BeiJing: Chinese Building Industry Press. 2010, pp. 300.

Other related topics

Input-output analysis of CO₂ emissions embodied in international trade and the analysis of geopolinomic structure implications

QinNeng Tang
Institute of Policy and Management, Chinese Academy of Science, Beijing, China

Zheng Wang
Institute of Policy and Management, Chinese Academy of Science, Beijing, China
East China Normal University, Key Laboratory of Geographical Information Science,
Ministry of State Education of China, Shanghai, China

JunBo Xue
Institute of Policy and Management, Chinese Academy of Science, Beijing, China

XiaoNan Cong
Institute of Policy and Management, Chinese Academy of Science, Beijing, China
Institute for Urban and Environmental Studies, Chinese Academy of Social Sciences, Beijing, China

ABSTRACT: In this paper, based on the latest GTAP V8 database, a multi-regional input-output (MRIO) model is used to calculate the CO_2 embodied in international trade according to the Consumption-based Accounting Principle. Results show that there is a significant regional difference in CO_2 embodiments in international trade. Such regional differences affect the responsibility allocation for climate change mitigation, resulting in a certain geopolinomic structure on negotiation for allocating responsibility for emissions reductions.

Keywords: Multi-Regional Input-Output analysis, CO_2, Embodied emissions, International trade, Consumption-based Accounting Principle, Geopolinomic Structure

1 INTRODUCTION

It has been widely accepted that human beings are facing increasingly severe global warming issues and the global temperature rise is mainly due to the anthropogenic greenhouse gas emissions since the Industrial Revolution. Since the 1992 Rio summit, countries of the world have been discussing intensively how to stabilize the level of the greenhouse gases (GHGs), especially the CO_2 emissions from fossil fuel combustion. The signing of the Kyoto Protocol which has come into force shows that the world's major industrial countries are committed to reducing GHG emissions, but in the post-Kyoto Protocol era, the world has not yet reached an agreement on long-term emissions reduction responsibility. According to the research, about 50% of the anthropogenic greenhouse gases accumulated in the atmosphere can be attributed to developed countries up until the 2005 (Höhne et al., 2011). In recent years, however, the rapid economic growth in developing countries, especially in China and India, has caused a significant increase of global CO_2 emissions. Consequently, the share of the CO_2 emissions in developing countries has been expected to rise rapidly.

For that reason, developing countries also need to be incorporated into the global GHG emissions reduction framework, and consequently the corresponding responsibility should be reassigned across the world. However, the level of economic development and real GHG emissions need to be considered when allocating the responsibility for emissions reduction. Therefore, a scientific and rational approach is required to calculate the national GHG emissions.

Currently, the annual national greenhouse gas emissions reported by the United Framework Convention on Climate Change (UNFCCC 2005) are calculated based on the Production-based Accounting Principle (Munksgaard and Pedersen 2001). The Production-based Accounting Principle refers to the GHGs emitting during the process of production within one national territory and offshore area over which countries have jurisdiction (Dong et al., 2010). In the context of globalization, a rapid increase of international trade volume, and a large-scale flow of goods worldwide, is consequently causing a geographic separation between the production and consumption of goods. Goods produced in one country are not only for domestic consumption but also for exporting to foreign countries. Obviously, the Production-based Accounting Principle ignores the final destination of products and assigns the emissions of all the products to the producing countries rather than to the consuming countries, which will distort the true emissions of countries to some extent. As a result, greenhouse gas emissions calculated based on the Consumption-based Accounting Principle (Munksgaard and Pedersen 2001) receive much more attention. From the perspective of the consumers, this principle calculates the CO_2 emissions in the final use of goods and services, regardless of whether they are domestically produced or imported from foreign countries, and a large number of studies have used this principle to calculate the CO_2 emissions up to now (Lenzen et al. 2007, Rodrigues and Domingos 2008, Vetőné Mózner 2013). Meanwhile, large amounts of embodiment of CO_2 emissions, along with international trade, has given rise to great concern about the reallocation of the emissions reduction responsibilities. The calculation of the embodiment of CO_2 emissions in international trade reflects the responsibilities of GHG emissions requiring a shift from producer to consumer, which may lead to a new global emissions reduction responsibility programme. In fact, a Chinese delegation has already put forward the embodiment of CO_2 emissions issues during the United Nations Climate Change Conference held in 2007 in Bali.

In the future, international climate change negotiations need to pay more attention to the impact of international trade on GHG emissions. In addition, negotiation on climate change mitigation is a typical geopolinomic issue (here, the terminology 'geopolinomic' refers to the comprehensive summary of international, spatial, political and economic relations which was first proposed by George in 1988). To receive political benefits or reduce the abatement cost, it is possible for different countries to ally together to achieve emissions reduction commitments. Researchers have suggested that the international coalition is considered to be a promising and effective strategy to achieve the global abatement goal (Peters and Hertwich 2008, Chen et al. 2010). Therefore, this paper firstly uses the Consumption-based Accounting Principle to calculate the CO_2 emissions embodied in international trade, and then analyses the possible alliance in global climate change negotiations from the geopolinomic perspective. The remainder of this paper is organized as follows: Section 2 briefly reviews the existing literature on calculating the embodiment of CO_2 emissions in international trade; Section 3 describes data sources and the methodology; Section 4 analyses the results which are calculated in Section 3. Finally, a discussion of geopolinomic structure is given in Section 5, d followed by conclusions in Section 6.

2 LITERATURE REVIEW

There are two main approaches to calculate the CO_2 emissions embodied in international trade. One is based on the import and export volume of each product and emissions' coefficients of the products. This approach can be easily used, but it does not take into account the intermediate inputs in the process of production, and ignores the CO_2 emissions in the supply chain no matter if it is calculated on unit emissions' coefficients of exporters or importers. The other approach is based on the input-output (IO) table. It traces all the intermediate inputs consumed in the production process, and calculates the CO_2 emissions of all intermediate inputs which are recognized as the embodiment CO_2 emissions in the trade. This approach fully considers correlations among industries and tracks the total energy used and CO_2 emissions, from the primary inputs to the final product, consequently leading to a more accurate result.

The single-region input-output (SRIO) model and the multi-region (MRIO) model are frequently used to estimate the pollution content of trade. The SRIO model, because of its simple structure and the consistency of data processing, is widely used to estimate the GHG emissions (Ghertner and Fripp 2007, Li and Hewitt 2008). The SRIO model adopted in most researches, assumes that a single region is considered as a closed economy, and imported goods and services are produced with the same technology as the domestic technology in the same sector, and have the same carbon emissions coefficient (Weber et al., 2008, Pan et al., 2008, Su and Ang, 2012). In general, the imports of one country come from a number of different countries with different production technologies. Each of these regions also places import demands on foreign economies, thus the embodied production factors may drill up the entire international supply chain. The path of supply and production is not reflected in a SRIO model, and therefore the amount of CO_2 emissions embodied in international trade deriving from this model will be far from the actual situation. The MRIO model, however, is able to distinguish the differences of production technologies between domestic and imported goods from different regions, which will significantly improve the statistical precision. In addition, the MRIO model is able to calculate the CO_2 emissions embodied in trade between any two regions, thus based on this model, it is more rigorous and comprehensive. As a result, the MRIO model becomes an important technique for calculating the embodiment of CO_2 emissions in national or supranational scale trade. (Wiedmann 2009) systematically reviews the research of embodiment CO_2 emissions in trade based on the IO models. In addition, other research based on the MRIO model includes: analysing the CO_2 emissions influenced by international trade between Japan and South Korea (Rhee and Chung 2006), and analysing the embodied CO_2 emissions induced by fossil fuel combustion, with the world divided into three supranational coalitions, i.e., G7, BRIC, and the rest of the world (ROW) (Chen and Chen 2011). For a more reasonable estimate of the CO_2 embodied in international trade, we employ the MRIO model in this study.

3 DATA SOURCE AND METHODOLGY

3.1 Data source

The international input-output data used in this study was obtained from the Global Trade Analysis Project (GTAP) database. The GTAP database provides global multi-regional input-output data, bilateral trade statistics, trade protection data, energy consumption data and CO_2 emissions data, and so on, and it is a superset of the social accounting matrix. Users can easily construct the interrelated global input-output table according to the research purpose. The GTAP database, version 8, used in this study integrates the global economic data which boasts dual reference to 2004 and 2007, and it has already expanded to 129 regions, which also keeping the previous version containing 57 production sectors. The GTAP database has reached a state of equilibrium and no balancing procedures are processed when it is used.

The GHG emissions calculated in this paper only refer to CO_2 emitting from fossil fuel combustion, and the emissions' coefficients of the fossil fuel refer to the GTAP-E database. The GTAP-E database involves five types of fossil energy; 'gas' refers to the merger of 'gas' and 'gdt' in the GTAP source data, therefore we set the same emissions coefficient for the 'gas' and 'gdt'. Coal, petroleum and gas generate carbon emissions and their CO_2 emissions coefficients' values are shown in Table 1 (Nijkamp et al. 2005). The reason why crude oil has a zero CO_2 emissions coefficient is that crude oil is mainly used as a material input for petroleum refining. The zero coefficient of electricity is used to avoid double-counting since electricity is produced by other primary fuels with non-zero CO_2 emissions' coefficients.

3.2 Methodology for calculating embodiment of CO_2 emissions

The input-output analysis framework is often used to study CO_2 emissions embodied in international trade, which can track the direct and indirect energy consumption and CO_2 emissions during the process of production. The total output of a region can be expressed as:

$$V = C + G + I + VI + \left(\sum_i X_i - M\right) + T \quad (1)$$

where the variables are n-dimensional column vector, n is the number of sectors. V, C, G, I, VI, X_i and M denotes total output, private demand, government demand, investment, intermediate input, exports to the i-th region and total imports respectively. It is different from the general IO table, since the GTAP model adds a virtual global transportation sector which serves international trade; the total output should add up to the value that exports to the virtual global transportation sector T. Because the GTAP database provides a non-competitive IO table, equation (1) can be decomposed into two parts which represent the domestic supply and foreign supply. We obtain:

$$V = (C_D + C_F) + (G_D + G_F) + (I_D + I_F) \\ + (VI_D + VI_F) + \left(\sum_i X_i - M\right) + T \quad (2)$$

where subscripts D and F denote the domestic and imported goods respectively. Eliminating the imports component in equation (2), we obtain:

$$V = C_D + G_D + I_D + VI_D + \sum_i X_i + T \quad (3)$$

where i represents all the other countries. According to equation (3), the total output of a region consists of private household consumption, government purchases, investment, intermediate input and the exports to foreign countries and virtual global transportation sector. According to the relationship that domestic intermediate inputs are equal to the product of domestic direct consumption coefficients and the total output, namely the equation $VI_D = A_D V$, equation (3) can be rewritten as:

$$V = (E - A_D)^{-1}\left(C_D + G_D + I_D + \sum_i X_i + T\right) \quad (4)$$

where E denotes the unit matrix, A_D denotes the input-output coefficient matrix of using the domestic products in the production process and $(E - A_D)^{-1}$ denotes the corresponding Leontief inverse matrix. By left multiplying on both sides of equation (4) by the unit output emissions, we will obtain the total emissions of any sector in the production process:

$$eV = e(E - A_D)^{-1}\left(C_D + G_D + I_D + \sum_i X_i + T\right) \quad (5)$$

where e is a diagonal matrix, its diagonal element $e_{j,j}$ represents the emissions per unit output of the j-th

Table 1. CO_2 emissions' coefficients in the GTAP-E model.

Energy products	Coal	Patroleum	Crude oil	Gas	Electricity
CO_2 coéfficient	3.8107	2.7638	0	1.8844	0

In tons of CO_2 per tonne of oil equivalent (toe).
Source: Nijkamp et al. (2005, p.964).

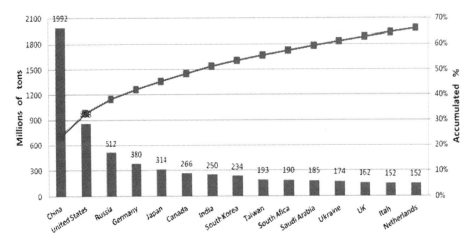

Figure 1. Top 15 regions of CO_2 outflows embodied in global trade (in millions of tons).

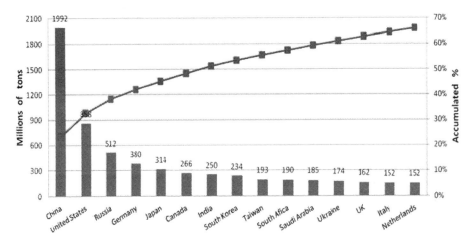

Figure 2. Top 15 regions of CO_2 inflows embodied in global trade (in millions of tons).

sector. Then we will obtain the CO_2 emissions embodiments that exports to the *i-th* region through the various commodities:

$$EV_i = e(E - A_D)^{-1} X_i \qquad (6)$$

where X_i and EV_i are n-dimensional column vectors. The total CO_2 emissions embodiment exports to the *i-th* region are:

$$TEV_i = \sum_{j=1}^{n} EV_{i,j} \qquad (7)$$

4 RESULTS

4.1 *CO_2 flows embodied in imports and exports*

Based on the equations (1)–(7) given in Section 3, we use the non-competitive MRIO data provided by GTAP 8 database to calculate the CO_2 emissions embodied in international trade flows for 129 regions around the world in 2007. The results show that the total amount of CO_2 emissions embodied in global trade have reached 8909.13MtCO$_2$, accounting for 29.43% of the world CO_2 emissions. The huge amount of the CO_2 flows embodied in global trade shows that international trade has an important influence on the global CO_2 emissions and consequently affects the allocation of responsibility for global emissions reductions.

Despite the huge amount of CO_2 emissions embodied in global trade, there are significant regional differences of CO_2 flows embodied in exports and imports. With respect to the CO_2 flows embodied in exports, the top 15 countries (or regions) account for 67.51% of the CO_2 emissions embodied in global trade, as shown in Figure 1. China, the United States and Russia are the top three countries, especially China, whose amount of CO_2 flows embodied in exports is up to 1991.83MtCO$_2$, accounting for 22.36% of the total CO_2 emissions embodied in global trade, and 29.73% of the total emissions of China in that year.

Table 2. Net outflow and inflow volumes of CO_2 emissions embodied in global trade of the top15 regions (in millions of tons of CO_2).

Regions	Net inflow Volume	Regions	Net Outflow Volume
United States	420.11	China	−1465.26
United Kingdom	225.76	Russia	−329.54
Germany	182.27	South Africa	−139.36
France	171.37	Saudi Arabia	−111.84
Japan	167.91	Ukraine	−102.32
Italy	141.34	India	−68.90
Belgium	104.29	Taiwan	−43.43
Hong Kong	81.61	Rest of Former Soviet Union	−34.92
Spain	80.67	Australia	−34.72
Switzerland	78.89	Iran	−30.95
Turkey	74.46	Argentina	−29.27
United Arab Emirates	72.81	Venezuela	−27.04
Singapore	63.68	Malaysia	−21.80
South Korea	59.13	Thailand	−18.88
Austria	44.42	Kazakhstan	−18.81

With respect to the CO_2 flows embodied in imports, the top 15 regions covering 61.9% of the CO_2 emissions inflows embodied in global trade, are shown in Figure 2. It illustrates that the main regions of CO_2 emissions inflows are found in developed and developing countries which mostly engage in manufacturing. The largest amount of inflow countries are the United States, Germany and China.

In this paper, we define net flows of regional CO_2 emissions embodied in global trade as the difference between outflows and inflows. A positive value of net flows indicates that there are net inflows of CO_2 emissions embodiments in one country. By contrast, a negative value indicates that there are net outflows of CO_2 emissions embodiments. The larger value of the net flows represents the greater distortion on the real CO_2 emissions calculated by the Production-based Accounting Principle. We keep the value of net flows in order from and the result shows that most regions' net flows are located around zero. Most regions are balancing on their CO_2 emissions embodied in exports and imports, and only a few regions deviate far from the equilibrium position.

Table 2 lists the top 15 regions of net outflow and inflow volumes of CO_2 emissions embodied in global trade. Developed regions show a high value on net inflow volumes of CO_2 emissions embodiments in global trade. It is worth mentioning that the United States, the United Kingdom and Germany are the top three destinations of inflow of CO_2 emissions embodiments. The United States is the largest CO_2 emissions embodied country in the world and the amount reaches $420.11 MtCO_2$. However, in contrast to the net inflow of CO_2 emissions, the net outflow countries are mainly from the developing regions, i.e., BRICs countries (except for Brazil) and the former Soviet Union countries. Here, we must pay more attention to China which has the largest amount of net outflow of CO_2 emissions with the value of 1465.26 $MtCO_2$, which surpasses the sum of the volume of other countries with net outflow, and accounts for 21.87% of total emissions of that year in China. Thus, CO_2 emissions calculations based on the Production-based Accounting Principle will greatly exaggerate the responsibility for emissions reduction in developing countries, especially for China.

In general, as seen clearly in Figure 3, developed countries are the main destinations of the net inflow of CO_2 emissions embodiments. But due to the differences in their industrial structure as well as the status of resources, different developed countries have different import and export structures, which consequently lead to differences in CO_2 emissions embodiments. For example, the resource-rich countries such as Australia and Canada are the net outflow countries of CO_2 emissions embodiments. Likewise, not all the developing countries exhibit net outflow of CO_2 emissions embodiments. For example, low-level developing countries such as sub-Saharan Africa show net inflow of CO_2 emissions embodiments, and due to the impact of a slightly unfavourable trade balance and the structure of imports and exports, some of the Association of South East Asian Nations (ASEAN), e.g., Vietnam, Cambodia, Myanmar and Laos, also show net inflow of CO_2 emissions embodiments to some extent.

4.2 CO_2 emissions embodied in units of trade

Due to the differences in the level of production technologies and the structures of import and export products, different CO_2 emissions are embodied in unit import and export in a country. We define the indicator as the ratio of the CO_2 emissions embodied in unit export and unit import:

$$\rho = \left(\frac{TEVX}{X}\right)\bigg/\left(\frac{TEVM}{M}\right) \qquad (8)$$

where TEVX and TEVM represent the total outflow and inflow of CO_2 embodiments in one region, respectively. X and M respectively, denote one region's total

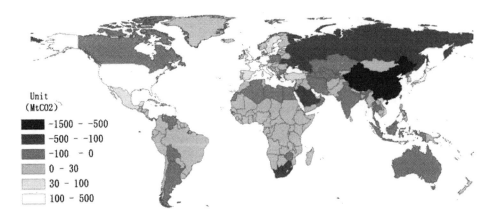

Figure 3. Regional distribution of net flow of CO_2 embodied in international trade.

Table 3. CO_2 embodied in unit exports & imports and the ρ value of the top 15 regions.

Region	Exports (Unit: tCO$_2$/US$)	Region	Imports (Unit: tCO$_2$/US$)	Region	ρ value
Ukraine	0.00312	Botswana	0.00226	South Africa	3.75
South Africa	0.00224	Namibia	0.00197	Ukraine	3.31
Rest of Former Soviet Union	0.00220	Mozambique	0.00196	China	3.03
Panama	0.00219	Mongolia	0.00165	Panama	2.80
Greece	0.00183	Kyrgyzstan	0.00160	Russia	2.39
China	0.00171	Rest of Eastern Europe	0.00154	Greece	2.07
Rest of East Asia	0.00168	Zimbabwe	0.00149	Egypt	1.88
Russia	0.00166	Rest of Former Soviet Union	0.00144	India	1.79
Zimbabwe	0.00161	Malawi	0.00143	Bulgaria	1.70
Egypt	0.00144	Rest of East Asia	0.00138	Venezuela	1.59
Mongolia	0.00138	Nepal	0.00128	Rest of Europe	1.57
Kazakhstan	0.00136	Ethiopia	0.00127	Rest of Former Soviet Union	1.52
Belarus	0.00123	Armenia	0.00125	Poland	1.40
Kyrgyzstan	0.00119	Rest of South Asia	0.00122	Argentina	1.38
India	0.00118	Vietnam	0.00115	Romania	1.36

exports and imports. So, the higher ρ value means that a region has a higher CO_2 embodiments content in exports compared to the imports; when $\rho = 1$ indicates that a region has equal $CO_{2embodiments}$ content in unit exports and imports and $\rho > 1$ or $\rho < 1$ respectively, indicates that a region has greater or lower CO_2 embodiments content in unit exports and imports.

As shown in Table 3, regions with high value are mainly concentrated in the manufacturing regions. The top three regions are South Africa, Ukraine and China, followed by Panama, Russia and Greece, as their values are all greater than 2. It is worth noting that the high values in Panama and Greece are because of their large portion of the sea transportation sector in exports, which accounts for 34.11% and 34.32% of their total exports respectively; the sea transportation sector is a carbon-intensive sector.

Furthermore, ρ values of BRICs countries (except for Brazil) are at the forefront. By contrast with the other BRICs countries, Brazil's ρ value is just 0.67. Moreover, the ρ values of North American countries are close to 1, among them, the United States and Canada with a value of 1.04 and 1.01 respectively, which shows that they have almost equivalent carbon contents embodied in exports and imports.

As shown in Figure 4, regions with low ρ values are mainly concentrated in two regions, Europe and Sub-Saharan Africa. The ρ value of Mozambique is the lowest in the world and just 0.086. From the perspective of an economic development level, the two regions represent two extremes.

On the one hand, Europe has low ρ values mainly because their majority of exports are services and low-carbon, high-tech products. On the other hand, Sub-Saharan Africa mainly imports deep-processing products and exports raw materials and primary processed products. Since the shorter supply chain of exports is relative to imports, those countries have a small ρ value.

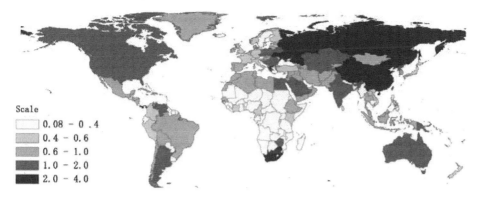

Figure 4. Regional distribution of the ratio of CO_2 embodied in unit export and unit import (ρ value).

5 DISCUSSION ON THE GEOPOLINOMIC STRUCTURE

Regional specialization in the global economy has an important impact on the CO_2 emissions embodied in the international trade which shows a typical geopolinomic structure. From the results of Section 4, on the issue of CO_2 emissions embodied in international trade, we have found that the international community does not simply divide into the North and the South, not entirely in accordance with Park and Labys 1994. However, it does divide into three types of countries or regional groups: financial and high-tech industrial-based countries, manufacturing-led countries and resource-based countries. In addition, resource-based countries are themselves divided into two subtypes: resource-based high-income countries and resource-based low-income countries.

Firstly, the financial and high-tech industrial-based countries such as the United States, the European Union countries and Japan, export a large amount of financial services products and low-carbon high-tech products, and import high-energy-consuming industrial products. Thus, these countries have a smaller value and larger net inflow of CO_2 embodiments. Secondly, manufacturing-led countries such as the BRICs countries export a great amount of high-energy-consuming industrial products. These industrial products have a relatively longer supply chain, and consume more energy and resources. Thus, these countries have a higher value and a significant net outflow of CO_2 embodiments. Consequently, manufacturing-led countries are the countries or regional groups who suffer the most serious damage with the assignment of responsibility for emissions reductions under the Production-based Accounting Principle. Thirdly, low-income developing countries such as Sub-Saharan Africa and some ASEAN countries, such as Vietnam, Cambodia, Myanmar and Laos have low production technology but relatively rich resources. These countries mainly export raw materials and primary industrial products, and import manufactured goods from the other two types of countries. Therefore, these countries have a relatively low unit export of embodied carbon with a high unit import of embodied carbon, and consequently, a smaller value and net inflow of CO_2 embodiments. Rich in mineral resources, developed countries such as Australia and Canada also show a net outflow of emissions embodiments, and their value is greater than 1.

CO_2 embodied in global trade manifests a spatial economic and political relationship, which is the so-called "geopolinomic structure". It has directive significance for the international negotiations on the assignment of responsibility for emissions mitigation. Represented by China, manufacturing-led countries are inclined to re-estimate the actual emissions of each country, using the Consumption-based Accounting Principle instead of the Production-based Accounting Principle. By comprehensively considering both the regional net flow of CO_2 emissions embodied in global trade, and the structure of CO_2 emissions embodied in units of trade, China's main negotiation opponents are developed countries such as the European Union countries and Japan, while the BRICs (except Brazil) countries have similar situations concerning the embodied CO_2 emissions issue, and so they will be China's main negotiation partners. For China, it is difficult to keep a long-term negotiation partner relationship with some low-income developing countries, e.g., Sub-Saharan Africa and some ASEAN member countries, owing to the different situation on the issue of CO_2 embodiment emissions. Additionally, China can seek the opportunity to cooperate with the United States on the aspect of emissions reduction technologies. Although the United States has a larger net inflow of CO_2 emissions embodied in global trade, CO_2 emissions embodied in unit import and export is almost equal. Lastly, the developed countries with rich mineral resources such as Australia and Canada are also the negotiation partners that China needs to win over.

6 CONCLUSIONS

In this research, we used the MRIO model to calculate the CO_2 emissions embodied in international trade based on the Consumption-based Accounting Principle. The calculation results show that the total amount of CO_2 embodied in the international trade has reached 8909.13 $MtCO_2$, accounting for 29.43% of the global CO_2 emissions. Such a huge amount of CO_2 embodied in international trade surely leads to a brand new scheme for the allocation of responsibility for climate mitigation.

In addition, the results also show that there exist big differences between embodied CO_2 emissions in different groups of countries. Most of the countries' net flow of embodied CO_2 emissions are around zero. It is worth mentioning that China has the largest net outflow of embodied CO_2 emissions in the world; other countries with large outflow mainly include the BRICs countries (except for Brazil), the former Soviet Union as well as high and medium-income resource-based countries such as Saudi Arabia, Iran and Venezuela. On the contrary, the United States, the European Union and Japan have relatively large net inflow of embodied CO_2 emissions; other countries with relatively large amounts of net inflow of embodied emissions include Sub-Saharan Africa and some ASEAN member countries. Furthermore, due to the difference in product structure between imports and exports, there are significant differences in CO_2 emissions embodied in unit export and import among different countries. South Africa, Russia and China have similar embodied CO_2 structure, and their CO_2 embodied in unit export is significantly higher than import. Countries with low values are concentrated in Europe and Sub-Saharan Africa, which indicates that both the highest and lowest levels of economic development areas, possess a lower value. National net flow of CO_2 emissions embodied in international trade show a similar regional structure with value. Finally, the regional differences in embodied CO_2 affect the responsibility allocation for emissions mitigation, resulting in a certain geopolinomic structure on negotiation for allocating responsibility on climate change mitigation. According to the analysis of the geopolinomic structure, the world has differentiated into three types of countries or regional groups: financial and high-tech industrial-based countries, manufacturing-led countries and resource-based countries. Manufacturing-led countries exhibit large scales of embodied CO_2 outflow. Financial and high-tech industrial-based countries and resource-based countries with low technology levels have significant net inflow of CO_2 embodiments and possess relatively lower value.

Therefore, we reach the conclusion that the negotiation partners of China on the issue of global emissions reductions are the countries which primarily engage in manufacturing, while resource-based but low-tech developing countries are not the most appropriate negotiation partners.

REFERENCES

Chen, Z. M. & Chen, G. Q. 2011. 'Embodied carbon dioxide emission at supra-national scale: A coalition analysis for G7, BRIC, and the rest of the world.' Energy Policy, 39:5, 2899–909.

Chen, Z. M., Chen, G. Q. & Chen, B. 2010. 'Embodied Carbon Dioxide Emissions of the World Economy: A Systems Input-Output Simulation for 2004.' Procedia Environmental Sciences, 2:0, 1827–40.

Dong, Y., Ishikawa, M., Liu, X. & Wang, C. 2010. 'An analysis of the driving forces of CO_2 emissions embodied in Japan–China trade.' Energy Policy, 38:11, 6784–92.

George, D. J. 1988. 'Geography beyond the Ivory Tower.' Annals of Association of American Geographers, 78:4, 575-79.

Ghertner, D. A. & Fripp, M. 2007. 'Trading away damage: Quantifying environmental leakage through consumption-based, life-cycle analysis.' Ecological Economics, 63:2–3, 563–77.

Höhne, N., Blum, H., Fuglestvedt, J., Skeie, R. B., Kurosawa, A., Hu, G., Lowe, J., Gohar, L., Matthews, B. & de Salles, A. C. N. 2011. 'Contributions of individual countries' emissions to climate change and their uncertainty.' Climatic change, 106:3, 359–91.

Lenzen, M., Murray, J., Sack, F. & Wiedmann, T. 2007. 'Shared producer and consumer responsibility — Theory and practice.' Ecological Economics, 61:1, 27–42.

Li, Y. & Hewitt, C. N. 2008. 'The effect of trade between China and the UK on national and global carbon dioxide emissions.' Energy Policy, 36:6, 1907–14.

Munksgaard, J. & Pedersen, K. A. 2001. 'CO_2 accounts for open economies: producer or consumer responsibility.' Energy Policy, 29:4, 327–34.

Nijkamp, P., Wang, S. & Kremers, H. 2005. 'Modeling the impacts of international climate change policies in a CGE context: The use of the GTAP-E model.' Economic Modelling, 22:6, 955–74.

Pan, J., Phillips, J. & Chen, Y. 2008. 'China's balance of emissions embodied in trade: approaches to measurement and allocating international responsibility.' Oxford Review of Economic Policy, 24:2, 354–76.

Park, S.-H. & Labys, W. C. 1994. 'Divergences in manufacturing energy consumption between the North and the South.' Energy Policy, 22:6, 455–69.

Peters, G. P. & Hertwich, E. G. 2008. 'CO_2 Embodied in International Trade with Implications for Global Climate Policy.' Environmental Science & Technology, 42:5, 1401–07.

Rhee, H.-C. & Chung, H.-S. 2006. 'Change in CO_2 emission and its transmissions between Korea and Japan using international input–output analysis.' Ecological Economics, 58:4, 788–800.

Rodrigues, J. & Domingos, T. 2008. 'Consumer and producer environmental responsibility: Comparing two approaches.' Ecological Economics, 66:2–3, 533-=46.

Su, B. & Ang, B. 2012. 'Structural decomposition analysis applied to energy and emissions: Some methodological developments.' Energy Economics, 34:1, 177–88.

UNFCCC 2005. 'Key GHG Data: Greenhouse Gas (GHG) Emissions Data for 1990—2003 submitted to the UNFCCC.' Bonn, Germany: United Nations Framework Convention on Climate Change.

Vetöné Mózner, Z. 2013. 'A consumption-based approach to carbon emission accounting – sectoral differences and environmental benefits.' Journal of Cleaner Production, 42:0, 83–95.

Weber, C. L., Peters, G. P., Guan, D. & Hubacek, K. 2008. 'The contribution of Chinese exports to climate change.' Energy Policy, 36:9, 3572–77.

Wiedmann, T. 2009. 'A review of recent multi-region input–output models used for consumption-based emission and resource accounting.' Ecological Economics, 69:2, 211–22.

Energy and Environmental Engineering – Wu (Ed.)
© 2015 Taylor & Francis Group, London, ISBN 978-1-138-02665-0

Dynamic simulation and experimental investigation on the performance of a refrigeration system under frosting conditions at low temperatures

HaiJie Qin, WeiZhong Li & Bo Dong
School of Energy and Power Engineering, Dalian University of Technology, Dalian, Liaoning, P. R. China

WeiYing Zhu
Dalian Sanyo Compressor Co. Ltd., Dalian, Liaoning, P. R. China

ABSTRACT: In the present study, a dynamic simulation model is developed to simulate a refrigeration system by considering the fan characteristics and the dynamic response of the variations of condensing temperature (T_e) and evaporating temperature (T_c) due to frosting to the system. In the model, the mean frost thickness of the overall control volumes in the evaporator is taken as a variable to calculate the air volume flow rate, and the mean frost thickness of the air tunnel is taken as a variable to consider the air maldistribution at windward side. A series of experiments have been carried out to validate the proposed model by means of the microphotograph method. The deviations for T_e and T_c were limited to 1–2°C, respectively. Both the simulation and experimental results demonstrate that the refrigerating capacity, the T_c, and T_e of the refrigeration system are increased slightly at the early stage of frosting, and then are gradually decreased; finally they are reduced sharply at the late stage of frosting.

1 INTRODUCTION

When moist air flows over the surface of an evaporator whose temperature is lower than its dew point temperature and the freezing temperature of water, the frosting phenomenon may occur on its surface. The frosting on the surface of an evaporator in a refrigeration system results in an increase of thermal resistance or the reduction of heat exchange efficiency. The reduction of heat exchange efficiency means that the evaporating temperature is lowered, and as a result, the throttle device cannot work in the normal state; even more liquid slugging may be taken place, which leads to damage of the compressor in some serious cases.

Frosting is a coupled heat transfer, mass transfer, and fluid dynamic phenomenon. It is difficult to estimate the performance of a heat exchanger working under frosting conditions because the rate of frost growth varies from inlet to outlet of the heat exchanger circuits. In order to improve the efficiency of the heat exchanger under frosting conditions, a series of dynamic simulations and experiments have been conducted. Kondepudi et al. (1987) established a numerical frosting model for a finned tube heat exchanger, and calculated the frost growth, variations of the heat transfer coefficient and the heat flux during the frosting/defrosting process, respectively. In their studies, experimental investigations were carried out to validate the proposed model. Lenic et al. (2009) proposed a transient two-dimensional mathematical model of frost formation on a finned tube heat exchanger and compared numerical and experimental analyses of heat and mass transfer during frost formation. Padhmanabhan et al. (2011) proposed a semi-empirical model that was able to predict non-uniform frost growth on a heat exchanger. In their model, redistribution of the refrigerant due to non-uniformities in the geometry was handled with the air redistribution.

Based on the above analysis, the purpose of this paper is to establish a dynamic simulation model for a refrigeration system under frosting conditions with a direct expansion evaporator at low temperature, and conduct the experiments to validate the proposed method.

2 DYNAMIC SIMULATION OF A REFRIGERATION SYSTEM

The compressor, condenser, evaporator and throttle device are the main components in a typical vapour compression refrigeration system. In order to simulate a refrigeration system, each component model should be combined into an overall model based on the relationship of parameters between the components. This paper only focuses on the dynamic response of operating parameters for a refrigeration system under frosting conditions in a steady state. In this section, the model for each component and the solution procedure of the system cycle are elaborated, respectively.

2.1 Compressor model

A map-based model for the compressor is adopted in this study. Although it is proposed for steady operation, it can also be used to simulate the dynamic performance of the compressor (W.E.Murphy & V.M. Goldschmidt, 1986). The mass flow rate $\dot{m}_{com}$ or heat transfer rate Q_{com} can be calculated (ARI540, 2004). This model calculates the refrigerant mass flow rate and the power consumption based on the compressor suction and discharge saturation temperatures:

$$X = C1 + C2 \cdot T_e + C3 \cdot T_c + C4 \cdot T_e^2 + C5 \cdot T_c \cdot T_e + C6 \cdot T_c^2 \\ + C7 \cdot T_e^3 + C8 \cdot T_c \cdot T_e^2 + C9 \cdot T_c^2 \cdot T_e + C10 \cdot T_c^3 \quad (1)$$

where X represents $\dot{m}_{com}$ or Q_{com}. The compressor in this study is a scroll compressor and the coefficients C1–C10 are provided by the manufacturer.

The enthalpy at the outlet of the compressor is expressed by:

$$h_{o,com} = h_{in,com} + \frac{Q_{com}}{\dot{m}_{com}} \quad (2)$$

Formula 1 and 2 are derived from the compressor performance data under specific rating conditions. Mullen et al. (1998) found that the mass flow rate and power consumption for off rating conditions can be corrected using the density and isentropic power ratios:

$$\dot{m}_{com,new} = \left[1 + F_v(\frac{\rho_{com,new}}{\rho_{com}} - 1)\right] \cdot \dot{m}_{com} \quad (3)$$

$$Q_{com,new} = [0.00144 \times (t_{sh} - 11.1) + 1] \cdot Q_{com} \quad (4)$$

where $F_v = 0.62$ (Dabiri, 1981).

2.2 Throttle device model

In the present study, the throttle device is an electronic expansion valve (EEV). The opening degree of the EEV can be set dynamically as a function of the refrigeration system operating condition. Since the EEV responds quickly, the effects of its operation process on the dynamics of the refrigeration system can be ignored, so the steady-state model can be used.

The enthalpy at the inlet and outlet of the EEV is equal, namely: $h_{in,exp} = h_{o,exp}$

The throttling mechanism of the EEV is identical to that of the TEV (thermal expansion valve), so the EEV model of the mass flow characteristic can take the form of TEV:

$$\dot{m}_{exp} = C_d A_{exp} \sqrt{\rho_{in,exp}(p_{in,exp} - p_{o,exp})} \quad (5)$$

The parameters of C_d and A_{exp} were explained in Xue Zhifang et al. (2008).

2.3 Evaporator model

In order to set up a dynamic model for an evaporator, the following assumptions should be made:

(1) The refrigerant flow through the heat exchanger tube is one-dimensional.
(2) The frosting process is a quasi-steady-state.
(3) Conductivity of metals is so large that its thermal resistance can be ignored.
(4) The air between tube rows is not mixed, and the frost thickness within any control volume is uniform.
(5) The air in contact with the frost is considered as saturated moist air during the frosting process.
(6) The heat conduction coefficient of the frost is only dependent on its density.

Following the above assumptions, the governing equations for the refrigerant, the frost and the air can be written as shown below.

2.3.1 Governing equations in the refrigerant side

As a direct expansion evaporator, the pressure losses and the phase change of the refrigerant must be calculated in detail. The refrigerant state in the evaporator at normal is changed from the two-phase state to the superheated state.

The mass conservation equation is:

$$\frac{\partial \rho}{\partial t} + \frac{\partial (\rho u)}{\partial x} = 0 \quad (6)$$

The momentum conservation equation is:

$$\frac{\partial (\rho u)}{\partial t} + \frac{\partial (\rho u^2)}{\partial x} = -\frac{\partial p}{\partial x} - F_{pw} \quad (7)$$

The energy conservation equation is:

$$\frac{\partial (\rho h)}{\partial t} + \frac{\partial (\rho u h)}{\partial x} = \frac{4}{d_{in}} \dot{q} + \frac{\partial p}{\partial t} + u \frac{\partial p}{\partial x} \quad (8)$$

with

$$\dot{q} = \alpha_{ref}(T_w - T_e) \quad (9)$$

The heat transfer coefficient α_{ref} of the refrigerant in two-phase state is calculated by the experimental fitting formula from Kandlikar (1999).

The heat transfer coefficient of the refrigerant at superheated state is calculated by the formula given by Incopera (1990):

$$\alpha_{ref} = \frac{\lambda_v}{d_{in}} \times \frac{(f/8) \mathrm{Re}_v \mathrm{Pr}_v}{1.07 + 12.7(f/8)^{0.5}\left(\mathrm{Pr}^{\frac{2}{3}} - 1\right)} \quad (10)$$

where $f = (3.64 \log_{10} \mathrm{Re}_l - 3.28)^{-2}$

Calculations of the pressure loss of the refrigerant side are from ASHRAE Fundamentals Handbook 2009.

2.3.2 Governing equations in the air side

The energy conservation equation in the air side without frosting is:

$$\frac{\dot{m}_a dh_a}{dy} = \frac{\alpha_a A}{c_{p,a}}(h_a - h_w) \quad (11)$$

The energy conservation equation under frosting conditions is:

$$\frac{\dot{m}_a dh_a}{dy} = \frac{\alpha_a A}{c_{p,a}}(h_a - h_{fr,s}) \quad (12)$$

where $h_{fr,s}$ is the enthalpy of saturated moist air on the frost surface.

The mass conservation equation under frosting conditions is:

$$\frac{\dot{m}_a dX_a}{dy} = \alpha_m A(X_a - X_{fr,s})i_{sv} \quad (13)$$

where $X_{fr,s}$ is the absolute humidity of saturated moist air on the frost surface, and $\dot{m}_a$ is the air mass flow rate which is determined jointly by the performance characteristics of the fan and the pressure drop across the finned tube coil.

The pressure drop within the heat exchanger given by Kays & London (1984) consists of four contributions: an entrance effect, a momentum effect, core friction, and an exit effect. The pressure drop can be expressed as Eq.14.

In general, the core frictional pressure drop is the most significant contribution to the pressure drop. The entrance pressure loss is partially recovered by the exit effect:

$$\Delta p_a = \frac{\dot{m}_a^2}{2 \cdot \rho_{a,in}}\left[\left(1 - \sigma^2 + K_c\right) + 2\left(\frac{\rho_{a,in}}{\rho_{a,o}} - 1\right)\right.$$
$$\left. + \left(f_a \frac{A}{A_{\min}} \frac{\rho_{a,in}}{\rho_{a,m}}\right) - \left(1 - \sigma^2 - K_e\right)\frac{\rho_{a,in}}{\rho_{a,o}}\right] \quad (14)$$

where $f_a = 0.152 \cdot Re_a^{-0.164} \cdot \xi^{-0.331}$ and
$$\sigma = \frac{(P_f - \delta_{fr})(S_2 - d_o)}{P_f \cdot S_2}$$

Define function of resistance characteristic function:

$$fs(\delta_{fr}) = \frac{1}{2 \cdot \rho_{a,in}}\left[\left(1 - \sigma^2 + K_c\right) + 2\left(\frac{\rho_{a,in}}{\rho_{a,o}} - 1\right) + \right.$$
$$\left.\left(f_a \frac{A}{A_{\min}} \frac{\rho_{a,in}}{\rho_{a,m}}\right) - \left(1 - \sigma^2 - K_e\right)\frac{\rho_{a,in}}{\rho_{a,o}}\right] \quad (15)$$

$$\Delta p_a = fs(\delta_{fr})\dot{m}_a^2 \quad (16)$$

The mass transfer coefficient α_m is related to the heat transfer coefficient by Lewis number (Seker D et al. 2004), and can be calculated by:

$$\alpha_m = \frac{\alpha_a}{Le \cdot c_{p,a}} \quad (17)$$

In order to determine the heat transfer coefficient in the air side in dry conditions, a dimensionless number X^* is introduced based on a large number of tests in an enthalpy difference test lab, which is defined as:

$$X^* = \frac{n_r \cdot S_2}{Re_a Pr_a d_e} \quad (18)$$

and

$$Nu_a = \frac{\alpha_a d_e}{\lambda_a} = a + b \cdot (X^*)^c \quad (19)$$

$a = -2.13; \quad b = 2.77; \quad c = -2.31\ln(n_r) - 0.249$

According to Gao et al. (2011), the frost initial condition is given by:

$T_{fr} = T_w, \quad \delta_{fr} = 2 \times 10^{-2} mm, \quad \rho_{fr} = 25 kg/m^3$

The air flow rate at a given time can be determined from the intersection between the pressure drop curve and the fan. Here, the fan's performance curve is given by the manufacturer, and the pressure drop is calculated according to the frost layer thickness. Huee-Youl Ye et al. (2013) calculated the pressure drop by applying experimental data, but they did not consider the air flow maldistribution on the windward side. To study effects of the frost growth on air maldistribution, the air side pressure loss is calculated by use of the average frost thickness of the overall control volumes, and the air flow rate is determined by the air side pressure loss combined with the fan curve. The air distribution bases on the resistance characteristic at one air flow tunnel, the main processes are listed as follows:

(1) The static pressure value $\Delta P_{a,0}$ is determined by the mean frost thickness $\bar{\delta}_{fr}$ over all control volumes in the evaporator, and then under the above conditions, the air volume rate V in the windward side of the evaporator and $fs(\bar{\delta}_{fr})$ can be determined. The static pressure value $\Delta P_{a,0}$ is calculated by Eq.14.

$$\bar{\delta}_{fr} = \frac{1}{N_{cv}}\sum_{i=1}^{i=N_{cv}}\delta_{fr,i} \quad (20)$$

(2) The calculations of each $fs(\bar{\delta}_{fr,i})$ in each air flow tunnel are performed by utilizing the mean frost thickness of the air flow tunnel.
(3) The air volume flow rate for each air flow tunnel is calculated by:

$$V:V_1:\cdots:V_N = \frac{1}{\sqrt{fs(\bar{\delta}_{fr,0})}} : \frac{1}{\sqrt{fs(\bar{\delta}_{fr,1})}} : \cdots : \frac{1}{\sqrt{fs(\bar{\delta}_{fr,N})}} \quad (21)$$
$$V_1 + V_2 + \cdots + V_N = V$$

Thus, the air volume flow rate distribution in each air flow tunnel is completed.

2.3.3 The frost model

The frost side governing equations are given by Tso, C. P. et al. (2006).

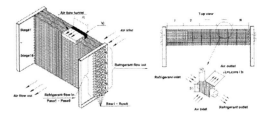

Figure 1. Schematic diagram of grid generation for the evaporator.

2.3.4 Calculation procedures

The governing equations for the heat transfer and mass transfer processes of the evaporator are solved by the finite volume method. The evaporator is divided into a number of continuous non-overlapping control volumes of equal length along the tube as shown in Figure 1. The width of a control volume is the column spacing (S2), and the height of a control volume is the tube spacing (S1).

Each heat exchanger control volume is calculated one by one along the refrigerant flow path from the coil inlet to outlet. The inlet condition of each control volume is equal to the outlet condition of the previous control volume. For the control volumes located at the coil inlet, the inlet condition is equal to the coil inlet condition. On the air side, the inlet condition varies depending on the location of the control volume. If the control volume is in the front row of the heat exchanger, the inlet condition is the same as the coil inlet condition. If the control volume is not at the front row, the inlet condition is the outlet condition of the nearest upstream control volume (Jiang et al. 2006).

The flow chart of the numerical algorithm is shown in Figure 2.

2.4 Condenser model

The condenser is a finned tube heat exchanger with air cooling. Compared with the evaporator, there is no frosting process in the condenser, so the calculation process is relatively simple, and the model for the condenser is the Tube-by-Tube Model (Domanski, 2003).

2.5 Solution procedure of the system cycle

In the present study, the refrigerant mass flow rate was selected as the key variable to connect the components of the refrigeration system. The solution procedure of the refrigeration system is shown in Figure 3.

In the simulation, the structure parameters of the evaporator and the condenser are included. The operation conditions include T_e and T_c. The refrigerant flow circuit is cut from the suction point of the compressor and set the initial iteration variables. Following the flow direction of the refrigerant, the compressor

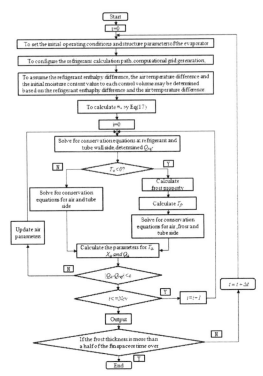

Figure 2. The flow chart of the numerical algorithm for the evaporator.

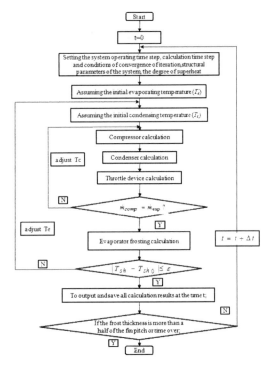

Figure 3. The solution procedure of the refrigeration system.

Table 1. Operating condition parameters.

Parameter	Units	Value
$T_{a,in}$ of the evaporator	°C	6
The degree of superheat	K	5
$T_{a,in}$ of the condenser	°C	32
Suction pressure loss corresponding saturation temperature drop	K	0.2
RH at the evaporator inlet	%	85
The degree of sub cooling	K	3
RH at the condenser inlet	%	60
Discharge pressure loss corresponding	K	0.3

Table 2. The structure parameters of the evaporator.

Parameter	Units	Value
Tube outer diameter	mm	9.53
Row Number	–	6
Pass Number	–	9
S_1	mm	25
Fin pitch	mm	10
Tube inner diameter	mm	8.8
Width	mm	1080
S2	mm	21.6
Fin thickness	mm	0.18
Fin Type	–	wave

model, the condenser model, the electronic expansion valve model and the evaporator model were calculated successively. As shown in Figure 5, by adjusting T_e and T_c through the bisection method, the refrigeration system model can be solved. The calculation period is 120 minutes, the time step is 5 minutes. The initial T_e is set as 5 K lower than the inlet air temperature of the evaporator, h and the initial T_c is set 5 K higher than the inlet air temperature of the condenser.

2.6 Experimental investigation

A series of experiments were carried out to validate the proposed dynamic simulation model. The refrigerant is HCFC22. Table 1 and Table 2 present the parameters of operating conditions and the structure of the evaporator prototype. The experimental equipments and procedure were introduced by Haijie Q et al. (2014).

3 RESULTS AND ANALYSIS

3.1 Variation of frost layer thickness

Figure 4 presents the variation of the simulated mean frost thickness as a function of time for various tube columns in the evaporator. Figure 5 demonstrates the comparison of the frost thickness between simulation results and the experimental data in the observed region. Since it takes about 3 minutes to get steady operation conditions for the refrigeration system, so the chart coordinates of the refrigeration system operating time is started at 5 minutes. It can be found that

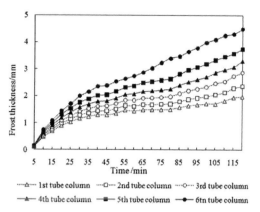

Figure 4. Variation of the simulated mean frost thickness for various tube columns in the evaporator.

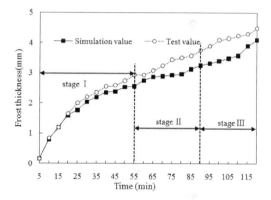

Figure 5. Comparison of the frost thickness between simulation results and the experimental data in the observed region.

the frost growth tendencies are consistent. Both the experiment and simulation indicate that more frost is formed at the air inlet than that at the air outlet, and the frost thickness growth rate at the early stage is higher than that at the late stage in the same tube column. The observed images of frosting thickness at different moments are shown in Figure 6.

The main causes for the occurrence of such trends lie in the three aspects as follows:

(1) During the frosting process, at the early stage (stage I in Figure 5), the relatively large temperature difference between the moist air and the fins indicates the big driving force for frosting, and as a result, the frosting rate is fast. As the frost layer grows, a thermal resistance is formed between the moist air and finned tubes which hinders the heat and mass transfer, and therefore, the frosting rate is decreased.

(2) At the middle stage (stage II in Figure 5), as reported by Hayashi et al. (1977) and Tao et al. (1993), defrosting phenomenon occurs on the frost

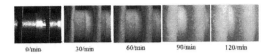

Figure 6. The variation monitoring observed images of frosting thickness at different moments.

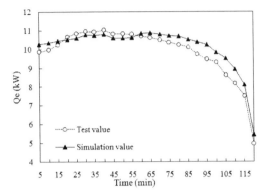

Figure 7. Comparison of the refrigerating capacity of the simulation and the test data.

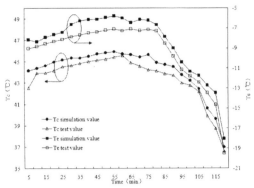

Figure 8. Comparison of T_c & T_e for the simulation and the test data.

surface, so the growth rate of frost layer is lowered remarkably.
(3) At the late stage (stage III in Figure 5), T_e is reduced significantly due to the increasing of frost layer, so that the temperature on the frost surface is decreased quickly. Consequently, the temperature difference between moist air and the frost surface becomes large again, and then the driving force for frosting increases, so the growth rate of frost layer is increased.

3.2 Variation of refrigerating capacity

The variation of the refrigerating capacity in the refrigeration system is shown in Figure 7. It is found that the refrigerating capacity of the refrigeration system is increased slightly at the early stage of frosting, and then it is decreased gradually. Finally, a sharp reduction of refrigeration capacity is observed at the late stage of frosting. It is because that at the beginning of frosting, the heat transfer area is enlarged due to the attachment of the frost particles to the tube surface, and at a result, the heat transfer is enhanced. As the frost layer's thickness increases, the increase of the frost thermal resistance leads to the reduction of the heat transfer coefficient, and the increase of the air flowing resistance results in the decreasing of the air volume flow rate, thus, the refrigerating capacity is decreased. At the late stage of frosting, the fast increase of frost density makes the thermal resistance become large sharply, and at the same time, the air flow resistance is increased. The heat transfer is weakened significantly; as a result, the refrigerating capacity is decreased.

In addition, it can be seen that the deviation between the calculation and experimental data for the refrigerating capacity is less than 10%.

3.3 Effects of operating parameters

Comparisons between the simulation and experimental data in terms of T_c and T_e of the refrigeration system are shown in Figure 8. The simulation and experimental results present the same tendency. Both T_c and T_e are increased slightly at the early stage of frosting, and then they are decreased gradually. At the late stage of frosting, they are reduced sharply. The reasons for this phenomenon are that at the early stage of frosting, the attached frost particles enhance the heat transfer, so the refrigerant volume flow rate is increased. As a result, T_c and T_e are increased slightly. As the frost layer grows, the thermal resistance of the frost layer degrades the heat transfer, and the refrigerant flow rate is reduced. T_c and T_e are decreased accordingly. At the late stage of frosting, the dual effects of increasing the frost thermal resistance and decreasing the air volume flow rate result in the rapid deterioration of the heat transfer, so the refrigerant flow rate is decreased significantly and the evaporating pressure is also decreased sharply. Consequently, T_c is reduced.

4 CONCLUSIONS

A dynamic simulation model for a refrigeration system under frosting conditions has been proposed. The visual experiment has been carried out to validate the proposed model by utilizing the microphotograph method. The frost growth and the variations of refrigerating capacity, and the evaporating temperature and condensing temperature have been studied by the proposed dynamic simulation model. Moreover, the mean frost thickness of the overall control volumes in the evaporator is taken as a variable to calculate the air volume flow rate, and the mean frost thickness of the air tunnel is taken as a variable to consider the air maldistribution on the windward side. The conclusions can be summarized as follows:

(1) As air maldistribution on the windward side is considered, the model simulation accuracy can be improved significantly and the deviation between the simulation and experimental data of the frost layer thickness for the observed region is less than 10%.
(2) The frost at the air inlet is thicker than that at the air outlet. As for the same tube column, the frost grows faster at the early stage of frosting than that at the late stage of frosting. The refrigerating capacity is increased slightly at the early stage of frosting, and then it is gradually decreased. Finally, it is reduced sharply at the late stage of frosting. The evaporating temperature and condensing temperature vary with the same tendency. They are increased slightly at the early stage of frosting, and then they are decreased gradually. Finally, they are reduced sharply at the late stage of frosting.
(3) The dynamic simulation method can reflect the dynamic variations of operating parameters of the refrigeration system under frosting conditions, and the deviations between the simulation and experimental data for the evaporating and condensing temperature of the system are smaller than 1–2°C. Consequently, the proposed dynamic simulation model can be used to guide the system design effectively.

REFERENCES

Ashrae F. Fundamentals Handbook. SI Edition, 2009.
De Monte, F., 2002. Calculation of thermodynamic properties of R407C and R410A by the Martin–Hou equation of state—part I: theoretical development [J]. International journal of refrigeration, 25(3): 306–313.
Domanski, P.A., 2003, EVAP–COND, Simulation models for finned tube heat exchangers. National Institute of Standards and Technology Building and Fire Research Laboratory Gaithersburg, MD, USA
Gao, T., Gong, J., 2011. Modeling the airside dynamic behavior of a heat exchanger under frosting conditions. Journal of mechanical science and technology, 25(10), 2719–2728.
Haijie Q, Weizhong L, Bo D, et al, 2014. Experimental study of the characteristic of frosting on low-temperature air cooler. Experimental Thermal and Fluid Science, 55: 106–114.
Hayashi, Y., Aoki, A., Adachi, S., Hori, K., 1977. Study of frost properties correlating with frost formation types. Journal of heat transfer 99: 239–253.
Huee-Youl Ye, Lee K S., 2013. Performance prediction of a fin-and-tube heat exchanger considering air-flow reduction due to the frost accumulation. International Journal of Heat and Mass Transfer, 67: 225–233.
Incopera, F. D., D. P. Dewitt, 1990. Fundamentals of Heat and Mass Transfer, 3rd. John Wiley and Sons, New York.
Jiang, H.; Aute, V.; and Radermacher, R., 2006.CoilDesigner: A general-purpose simulation and design tool for air-to-refrigerant heat exchangers. International journal of refrigeration, 29, pp. 601–610.
Kandlikar, S.G., ed. 1999. Handbook of phase change: Boiling and condensation. Taylor and Francis, Philadelphia.
Kays W.M. & London, A.L. 1984. Compact Heat Exchangers, third ed., McGraw-Hill, New York.
Kondepudi, S., O'Neal, D., 1987. The effects of frost growth on extended surface heat exchanger performance: a review. ASHRAE Transactions, 93: 258–274.
Lee, Kwan-Soo, Sung Jhee, Dong-Keun Yang, 2003. Prediction of the frost formation on a cold flat surface. International journal of heat and mass transfer 46.20: 3789–3796.
Lenic, K., Trp, A., Frankovic, B., 2009. Prediction of an effective cooling output of the fin-and-tube heat exchanger under frosting conditions. Applied Thermal Engineering 29:2534–2543.
Murphy, W.E. & Goldschmidt, V.M. 1986, Cycling characteristics of a residential air conditioner-modeling of shutdown transients. ASHRAE Transactions 92(1A):186–202.
Padhmanabhan S K, Fisher D E, Cremaschi L, et al. Modeling non-uniform frost growth on a fin-and-tube heat exchanger [J]. International Journal of Refrigeration, 2011, 34(8): 2018–2030.
Standard, A.R.I. 540, 2004. Performance Rating of Positive Displacement Refrigerant Compressors and Compressor Units. Air-conditioning and Refrigeration Institute.
Tao, Y.-X., Besant, R., Mao, Y., 1993. Characteristics of frost growth on a flat plate during the early growth period. Transactions-American Society of Heating Refrigerating and Air Conditioning Engineers 99:746–753.
Tso, C. P., Cheng, Y. C., & Lai, A. C. K. 2006, Dynamic behavior of a direct expansion evaporator under frosting condition. Part I. Distributed model. International journal of refrigeration, 29(4), 611–623.
Zhifang, X., Lin, S., & Hongfei, O., 2008. Refrigerant flow characteristics of electronic expansion valve based on thermodynamic analysis and experiment. Applied Thermal Engineering, 28(2), 238–243.

NOMENCLATURE

Roman symbols

A = Area, m^2
A_{min} = Minimum flow area, m^2
Bo = Boiling number, –
C_d = Electronic expansion valve flow coefficient,
d = Diameter, m
F_{PW} = Flow resistance, N
Fr = Froude number, –
G_{ref} = Refrigerant flow rate, kg/(m^{2}s)
h = Enthalpy, kJ/kg
i_{sv} = Water-ice latent heat of phase change, kJ/kg
K_c = Contraction loss coefficient at entrance
K_e = Expansion loss coefficient for flow at exit
K_p = Proportional constant, –
Le = Lewis number, –
m = Mass, kg
$\dot{m}$ = Mass flow rate, kg/s
N = Control volumes of heat exchanger width, –
N_{cv} = Numbers of control volumes, –
n_d = Number of tube stages, –
n_r = Number of tube rows, –
p = Pressure, Pa

P_f = Fin spacing, m
Pr = Prandtl number, –
$\dot{q}$ = Heat flux, W/m^2
Q = Heat transfer rate, kW
R = Gas constant 8.314472, J·/(K·mol)
Re = Reynolds number, –
S_1 = Tube pitch, m
S_2 = Column pitch, m
T = Temperature, K
t_f = Fin thickness, m
u = Velocity, m/s
X = Moisture content, kg/kg$_a$
X^* = Extraction coefficient, –
x = Dryness, –

Greek symbols

α = Heat transfer coefficient, W/(m^2·K)
δ = Thickness, m
ε = Convergence coefficient, 1E-4
θ_m = Logarithmic mean temperature difference, K
λ = Heat conduction coefficient, W/(m·K)
Δ = Difference, –
ρ = Density, kg/m^3

Subscripts

a = Air
c = Condenser
com = Compressor
e = Evaporating/equivalent
exp = Expansion device
fr = Frost
i = The ith zone
ice = Ice
in = Inlet/inner side
l = Refrigerant liquid phase
lat = Latent
m = Mean
new = New value
o = Outlet/outer side
ref = Refrigerant
s = Saturation
sh = Super heat
v = Refrigerant vapor
w = Tube wall
0 = Initial value

A study of the physical properties of soil under different ecological restoration measures in degraded red soil

YanYan Li
Research Institute of Ecology & Environmental Sciences, Nanchang Institute of Technology, Nanchang, Jiangxi, China

ABSTRACT: By taking bare ground and strong interfered *Pinus massoniana* as a control, the physical property of the soil of *Pinus massoniana*, standing under different ecological restoration measures in degraded red soil areas, was studied. The results indicated that bulk density in the layer of 0–20 cm was far lower under different restoration measures than bare ground, and was the least in check-dam *Pinus massoniana* stand. Soil water was greatly improved in grass-bamboo-burl-groove *Pinus massoniana*, in which the maximum water-holding capacity, the capillary water-holding and the soil water-holding was 44.27%, 48.24% and 46.28%, respectively.

1 INTRODUCTION

There is 20.31 billon km² red soil in southern China, accounting for 21% of the total land. Because of its topography and unique natural conditions such as the rainfall intensity and concentration, and prominent contradiction between people and land, the destruction of natural vegetation and soil and water loss in this area is serious. In the governance process over the years, according to its climate conditions such as high temperatures, rain, no vegetation and strong restoration of natural ability, a recovery mode has formed, based on artificial vegetation and forest ecological reconstruction in the red soil degradation area.

The different types of forest reconstruction have increased the content of the soil's organic matter and improved the ventilation conditions, microbial biomass and enzyme activity. With natural restoration, the quality of forest soil is better than artificial tending to the forest, which has an important significance for ecological restoration of red soil areas. In this study, we selected four ecological restoration forests, and combined with engineering measures (horizontal ditches etc.), we studied the physical and chemical properties of the soil in order to explore the improvement effect of different restoration patterns on soil fertility.

2 SUMMARY OF STUDY SITE

The study site is located in the Tiger Mountain watershed, east longitude 114°52′ ∼ 114°54, 26°50′∼26°51 north latitude, and belongs to a subtropical monsoon climate, with an average rainfall of 1,363 mm. The frost-free period is 288 days, the average temperature is 18.6°C, the accumulated temperature of 5 918°C more than 10°C. Extreme maximum is 40 temperature of 4°C and the extreme minimum temperature is −6°C. Tiger Mountain watershed is a plain and hillygully region, 80∼200 m above sea level, with a slope of around 5°. The soil has developed from the quaternary red clay soil and the thickness is commonly 3∼40 m.

3 METHODS

3.1 Sample site design and experiment

The experiment took bare ground (no vegetation restoration) as its control (expressed in CK). We adopted different modes of ecological restoration in the degraded red soil region to governance. The artificial restoration patterns included six kinds: mode A: strong interfered *Pinus massoniana* (restoration based

Table 1. Mechanical composition in five forests.

Forest type	<0.25	0.25–0.5 mm	0.5–1 mm	1–2 mm	2–5 mm	>5 mm
CK	6.9	7.1	12.9	7.6	34.5	31.0
Mode A	3.2	4.5	11.1	7.8	35.5	37.9
Mode B	7.0	12.1	18.9	10.0	34.2	17.8
Mode C	4.4	9.5	19.6	9.7	31.4	25.4
Mode D	3.4	10.9	18.1	10.8	36.0	20.8

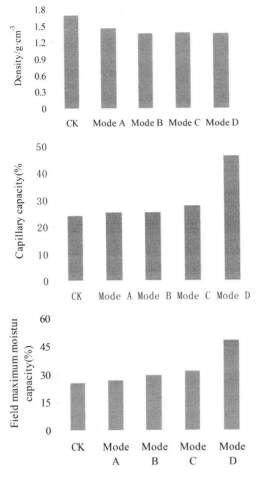

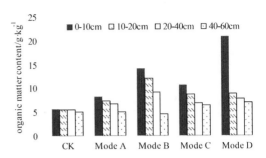

Figure 2. Organic matter content in the 0–60 cm soil layer in different restoration forests.

Figure 1. Field maximum moisture, density and water capillary capacity in different restoration forests.

on bare ground treatment, playing the pine, pine needles, no tending or management measures), mode B: no-disturbance *Pinus massoniana*; mode C: bamboo-burl-groove *Pinus massoniana* forest, and ModeD: grass-bamboo-burl-groove *Pinus massoniana* forest (ribbon grass and excavation horizontal corrugated channel).

3.2 *Results*

From Figure 1 we can see that the bulk density in the 0–20 cm soil layer reached $1.67 g \cdot m^{-3}$ in the bare surface soil without vegetation protection and with long periods of rainfall. After artificial restoration, the bulk density in the 0–20 cm soil layer decreased, respectively to 1.46, 1.36, 1.37 and $1.36 g \cdot m^{-3}$, in the strong interference of the Masson pine forest, non-disturbed Masson pine forest, bamboo-burl-groove Masson pine forest and grass-bamboo-burl-groove Masson pine forest. This showed that the soil bulk density in the *Pinus massoniana* forest, in different slope positions, decreased.

In the red soil degradation area, water content of soil plays an important role in vegetation restoration and it affects the nutrient transport in soil and the vegetation growth and development. In addition to no-disturbance *Pinus massonina*, the rest of the standing (0 ~ 20 cm) depth of the maximum water capacity and capillary water content was higher than bare land. The water content condition in the grass-bamboo-burl-groove pine was the best, and the 0–20 cm soil maximum moisture capacity and capillary water content were 92.40%, 93.28% higher than the control, which showed that through the process of restoration of forest degradation in the red soil region of grass, an obvious increase in soil water content can be found; this has a very important role for the forest succession. This is because the check-dam raises the base level of erosion and reduces the water loss and soil erosion, creating maximum water.

Table one lists the mechanical composition of all the forest. From the table, we can see that the proportion of larger than 1 mm soil particles in the strong interference Masson pine forest, was 81.2%, higher than those of the closed pine and bare land. There were more particles in the check-dam forest than in the bare land. The proportion of the range from 0.25 to 1 mm particles, increased in the bamboo-grass-groove pine, while the fine gravel, less than 0.25 mm, was reduced.

As can be seen from graph two, the content of organic matter in all stands gradually reduced with the increase of soil depth.

The organic matter content in the soil above 10 cm from high to low was: grass-bamboo-ditch Masson pine forest, no-disturbance Masson pine forest, strong interference Masson pine forest and bamboo-burl-groove Masson pine forest. Due to the lack of vegetation coverage on the bare ground and in the greatly disturbed Masson pine forest, the soil layers between 0 and 10 cm and below, had no significant difference. The most obvious difference was in the grass-bamboo-ditch Masson pine forest, where the average organic matter content was between 0-10cm; the soil layer was more than $12.14 g \cdot kg^{-1}$ below 10 cm soil. So, we can conclude that in the red soil areas of Bahia grass, the

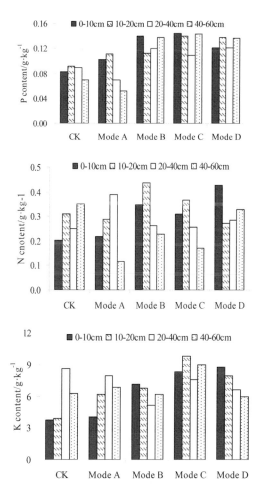

Figure 3. N, P and K content in the 0-60cm soil layer in different restoration forests.

types and quantity of microorganisms in the soil can be increased, in order to increase the content of organic matter in the soil.

There are three kinds of nutrient content of the average value of the order of K>N>P according to the average nutrient value; the nutrient content in different restoration forests was greater than the bare ground and the greatly disturbed Masson pine forest. The P and K content were greatly improved in the bamboo-ditch Masson pine forest. In the grass-bamboo-ditch Masson pine forest, the N content of the soil was the highest, while P and K were lower than the bamboo-ditch Masson pine forest, probably because herbaceous plants demanded more P and K greater demand, but the less for N during the growth.

The P and K content were greatly improved in the bamboo-ditch Masson pine forest. In the grass-bamboo-ditch Masson pine forest, the N content of the soil was the highest, while P and K were lower than the bamboo-ditch Masson pine forest, probably because herbaceous plants demanded more P and K greater demand, but the less for N during the growth.

As can be seen from the graph, the nutrient content of the soil surface layer increased in all artificial repair stands, but the vertical distribution of nutrients is more complex. The nitrogen content in the 0–10 cm soil layer was higher than that of the below 10 cm soil layer in the bare ground and grass-bamboo-ditch Masson pine forest, but it was the opposite in the other stand. In the bare ground, soil was eroded by rain for a long time and lost many nutrients, so the nutrient content of the 0–10 cm soil layer was lower than that of strong interference Masson Pine forest.

There was long-term disturbance by humans in the bamboo-ditch Masson pine forest and strong interference Masson Pine forest, so the N and K element contents in the surface soil was lower than that of the 10–20 cm soil layer. In the non-disturbed Masson pine forest with no human interference, the N and K element contents of the 10–20 cm soil layer was lower than that of the surface soil. The N and K content of the surface soil had a better improvement on the grass-bamboo-ditch Masson pine forest because Bahia grass was planted; it was significantly higher than the N and K content of the 10–20 cm soil layer.

4 CONCLUSION

Soil structure and water content were improved by combining a natural recovery stand and by taking seriously the soil and water conservation project in the southern degraded red soil region. Of the forests on the slope, the grass-bamboo-ditch Masson pine forest had the best effect on the soil's physical structure. In the 0-20 cm layer of the soil, its bulk density, water holding capacity, capillary water content and soil water were $1.36 \text{ g} \cdot \text{m}^{-3}$, 44.27%, 48.24% and 46.28%, respectively. The physical properties of the holding capacity, capillary water content and soil water of the check-dam Masson pine forest at the bottom of the slope, were improved, too. Its soil bulk density and water holding content were 1.21 gm^{-3}, 29.96%, 44.76% and 40.23%, respectively in the 0–20 cm layer.

From the view of soil nutrients, the organic matter content, N, P and K was significantly higher than that of bare land and strong interference woodland, which improved the result of the grass-bamboo-ditch Masson pine forest with regards to organic matter. The bamboo-ditch Masson pine forest on the improvement of K, had the best effect. In the study of litterfall in the project, the annual litterfall and nutrient return were at a maximum in the greatly disturbed, bamboo-ditch Masson pine forest, the grass-bamboo-ditch Masson pine forest and the non-disturbed Masson pine forest. Zhou Likai pointed out that, soil enzyme activity correlated largely soil organic matter and other fertility factors, so in the following research work, we should further explore in the course of recovery, whether there are fall-related content decomposition and soil enzyme characteristics of degraded vegetation in red soil.

ACKNOWLEDGEMENTS

The project is supported by Youth Fund of the Education Department of Jiangxi Province (GJJ13749)

REFERENCES

Chen, G.Q., Huang, D.Y., Su, Y.R., etc. Effect of soil organic matter in hilly red soil from Mid-subtropics region under various utilization patterns .Jour of Agro-Environment Science, Vol24 (2005), p. 256–260.

Huang, S.Y. Effects of Different Ecological Restoration Measures on Soil Fertility in Red Soil Eroded Degradation Land Research of Soil and Water Conservation Vol 16(2009),p38–42.

Li, F., Huang, R.Z., Fan, H.B. etc. Study on species diversity and community stability of rehabilitating forest restoration on degraded red soil 27 years later[J]. Journal of soil and water conservation. Vol. 25(2011), p. 189–193.

Liu, Y.Q., Yang, G.P., Du, T.Z. etc. A Study on Ecological Benefit of Reforestation Soil in Seriously Degraded Red Soil Regions. Acta Agriculturae Universitatis Jiangxiensis, 2005, 27 (1):119–123.

Wang, X.L., Hu, F., Li, H.X. etc. Characteristics of biological property of erosive degraded red soil under nature restoration. Acta Ecology Sinica,Vol 27(2007), p1404–1405.

Wang, Z.Y., Zuo, C.Q., Cao, W.H. etc. Physical and chemical properties of soils under different vegetation restoration models in red soil hilly region, Acta Pedologica Sinica, Vol 48(2011), p. 715–814.

Zheng, H., Ouang, Z.Y., Wang, X.K. etc. Effects of forest restoration types on soil quality in red soil eroded region, Southern China. ACTA ECOLOGICA SINICA, 2004, 24 (9):1994–2002(in Chinese).

Modelling of electric vehicle penetration in China based on innovation diffusion theory

XunMin Ou, Qian Zhang, Xu Zhang & XiLiang Zhang
Institute of Energy, Environment and Economy, Tsinghua University, Beijing, China
China Automotive Energy Research Center, Tsinghua University, Beijing, China

ABSTRACT: The market penetration of electric passenger vehicles was modelled using innovation diffusion theory and an epidemiological model. The penetration of electric passenger vehicles with various engine displacement levels was calculated under two hypothetical scenarios: a business-as-usual scenario and a scenario of promoted electric vehicle development. The results predict that 1) under the business-as-usual scenario, the penetration rates of electric passenger vehicles according to new sales for all sub-segments are very low, even by 2050; 2) under the scenario of electric vehicle development, the penetration rates vary by sub-segment but are all high in the period from 2030 to 2050; and 3) the major factor controlling the market penetration of electric passenger vehicles is the life-cycle cost of vehicles.

1 INTRODUCTION

Electric vehicles are an important innovation in automobile technologies. As is the case with other innovations, the market acceptance of electric vehicles may be divided into three phases: emergence, subsequent rapid growth, and a final plateau. Accordingly, the market penetration of electric vehicles may follow an S-shaped profile. The marketing scale of electric vehicles may be predicted using innovation diffusion theory. Employing this theory, the present study modelled the market penetration of electric vehicles with various levels of engine displacement under two hypothetical scenarios.

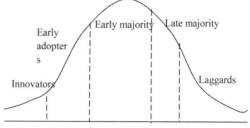

Figure 1. S-curve describing the acceptance of an innovation as a function of time.

2 MODEL

2.1 Innovation diffusion theory

Diffusion of innovations is an influential theory in social media research. The theory suggests that an innovation (e.g., a novel concept, method, or technique) spreads throughout society in a predictable manner. A small number of people accept the innovation immediately they become aware of it, whereas others may require a somewhat longer time to experiment with the innovation before accepting it (Zhang, 2003). Some people may take even longer to accept the innovation. The acceptance of this innovation is thus described by an S-curve as seen in Figure 1. Innovation diffusion theory divides social individuals into five groups according to their ability to accept innovations: innovators, early adopters, early majority, late majority, and laggards. These five groups correspond to different points on the S-curve. Consequently, the S-curve represents the increase in acceptance of an innovation as a function of time.

Similar to the case for other technical innovations, the market acceptance of electric vehicles is divided into three phases, with the market penetration exhibiting an S-curve. Here, we assume that the diffusion of alternative-fuel vehicles (e.g., electric vehicles) follows the logistic growth model. Thus, the diffusion of this innovation is divided into three phases as seen in Figure 2: emergence, subsequent rapid growth, and a final plateau.

On the market penetration curve (Fig. 2), point D is a critical point determining whether the innovation can enter the subsequent period of rapid growth. When the cost of applying this innovation is similar to the cost of the existing technology, the innovation reaches point D (onset). For example, as electric vehicles pass point D, they become economically superior to traditional

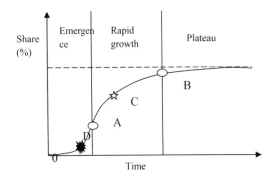

Figure 2. S-curve describing the market penetration of an innovation.

gasoline-fuelled vehicles (according to life cycle cost analysis).

2.2 Innovation diffusion model

The epidemiological model is an important tool in studying the diffusion of an innovation. In this model, an epidemic disease is transmitted via contact between infected and healthy individuals. With the diffusion of the disease, the infected individuals increase in number. Correspondingly, the speed of disease transmission accelerates until only a small number of healthy individuals are available. Applying this diffusion process to the market acceptance of electric vehicles, few customers are initially ready to accept the new technology (or product). However, with time, more customers become aware of (and willing to accept) electric vehicles. Eventually, only a small number of individuals remain unaware of (or unwilling to accept) this new technology. Correspondingly, the rates of new awareness and acceptance decrease until the diffusion ends (Li 1986, Jiang 2013).

Mathematically, the diffusion process can be described as follows. The epidemiological model is based on three assumptions:

1. The population is divided into two groups: "knowers" and "unknowers". At moment t, the proportions of these two groups to the total population (of size N) are denoted $s(t)$ and $i(t)$, respectively.
2. "Agreers" are a subgroup of knowers who agree with the concept. The proportion of agreers in the total population is denoted $g(t)$.
3. In unit time, each knower effectively contacts an average of λ individuals (where λ is known as the unit-time transmission factor). If the knower effectively contacts an unknower, the former converts the latter to a new knower.

According to these assumptions, in unit time, each knower converts $\lambda \times s(t)$ unknowers to knowers. At moment t, there are $N \times i(t)$ knowers. Therefore, in unit time, $\lambda \times s(t) \times N \times i(t)$ unknowers are converted to knowers, meaning the growth rate of the knowers is $\lambda \times N \times s(t) \times i(t)$. We can write:

$$N\frac{di}{dt} = \lambda N s i \quad (1)$$

Obviously, we can also write:

$$s(t) + i(t) = 1 \quad (2)$$

If we denote the initial ($t = 0$) proportion of knowers as i_0, we have:

$$\begin{cases} \frac{di}{dt} = \lambda i (1-i) \\ i(0) = i_0 \end{cases} \quad (3)$$

Equation (3) is known as the logistic model. The solution to Equation (3) is:

$$i(t) = \frac{1}{1 + \left(\frac{1}{i_0} - 1\right) e^{-\lambda t}} \quad (4)$$

Equation (4) can be simply written as $i(t) = f(i_0, \lambda, t)$, meaning the proportion of knowers is a function of i_0 (proportion of knowers at $t = 0$), λ (unit-time transmission factor), and time (t).

Two observations can be made from Equation. (3) and Equation. (4).

1. When $i = 1/2$, (di/dt) reaches a maximum value $(di/dt)_m$. This moment is given by Equation. (5). At this moment, the proportion of knowers is increasing at its fastest rate, indicating the peak of transmission. Moreover, t_m is inversely proportional to λ because a larger λ corresponds to greater transmission ability and thus an earlier transmission peak.

$$t_m = \lambda^{-1} \ln(\frac{1}{i_0} - 1) \quad (5)$$

2. When $t \to \infty$, $i \to 1$. This means that the entire population will eventually be converted to knowers.

3 MARKET PENETRATION OF ELECTRIC PASSENGER VEHICLES

3.1 Division of market

Text Passenger vehicles can be divided according to engine displacement (Tab. 1). (CAERC 2013)

3.2 Scenario set-up

The market penetration of electric vehicles in China was modelled under two hypothetical scenarios: business-as-usual (BAU) and promoted electric-vehicle development (EV). Under the BAU scenario,

Table 1. Division of passenger vehicles according to engine displacement.

Division	Displacement of engine
Mini	<1.0 L
Small	1–1.6 L
Medium	1.6–2.5 L
Large	>2.5 L

Table 2. Penetration of electric passenger vehicles according to new sales under the EV scenario (%).

	2010	2020	2030	2040	2050
Mini	0.1	5.0	40.0	87.0	90.0
Small	0.1	5.0	25.0	50.0	70.0
Medium	0.0	2.0	10.0	25.0	45.0
Large	0.0	2.0	10.0	25.0	40.0
Fleet average	0.1	4.4	26.0	56.1	70.5

Note: the proportions of mini, small, medium and large engine sizes of the fleet are set as 10%, 50%, 30%, and 10%, respectively.

the Chinese government sets no specific goals for the development of vehicular fuels and issues no policies to lead innovations in automobile and fuel technologies. Under this scenario, the current market status and trend of automobile and fuel technologies will persist, meaning there will be no breakthrough in vehicular energy in the middle to long term. In contrast, under the EV scenario, electric vehicles are assumed to contribute maximally to the solution to vehicular energy problems in China (Ou et al., 2010).

3.3 Modelling results

3.3.1 BAU scenario
Our modelling predicted that, under the BAU scenario, the penetration of electric passenger vehicles by new sales is very low for all sub-segments; below 5% even as of 2050.

3.3.2 EV scenario
The penetration of electric passenger vehicles according to new sales by sub-segments, was predicted under the EV scenario is shown in Table 2 and Figure 3. We set a constant mix of different engine sizes in the future, and the prediction for a fleet-average penetration rate of electric passenger vehicles in China was about 70% by 2050.

3.4 Interpretation of results

The major factor controlling the market penetration of electric passenger vehicles is the life cycle cost (Ren et al. 2009, Cai et al., 2012), which is determined by the vehicle price, fuel price, intensity of vehicle use, and rate of fuel consumption in the real world (Wabner et al. 2009, Huo et al., 2011, Huo et al., 2012).

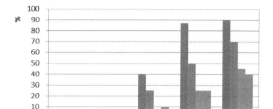

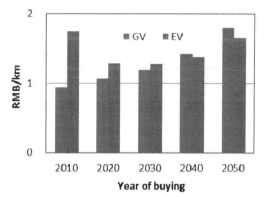

Figure 3. Penetration of electric passenger vehicles according to new sales calculated, assuming the EV scenario.

Figure 4. Comprehensive traffic costs of the mini electric passenger vehicle and its gasoline-fuelled counterpart calculated for the BAU scenario. (GV: gasoline vehicle; EV: electric vehicle).

Under the BAU scenario, the costs of vehicular components (i.e., battery, motor, and control equipment) decrease slowly. Because of the uncompetitive cost (Fig. 4), the mini electric passenger vehicles cannot compete with their gasoline-fuelled counterparts before 2040. This uncompetitiveness also applies to small electric passenger vehicles. Medium and large passenger vehicles cannot be powered purely by batteries because they need to be driven long distances between recharging. Consequently, these vehicles must operate in a plug-in hybrid electric power or extended-range electric power mode. Operating in these modes, medium and large electric passenger vehicles cannot compete with their traditional gasoline-fuelled counterparts in terms of cost (CAERC, 2013).

Under the EV scenario, the development, technical demonstration, and marketing of key techniques related to electric vehicles (i.e., battery, motor, and control equipment) are assumed to undergo major breakthroughs, thereby resulting in a rapid reduction in the cost. Under this scenario (Fig. 5), mini electric vehicles can compete with gasoline-fuelled counterparts from 2025, and will subsequently enter a period of rapid growth. A similar trend applies to small electric passenger vehicles. As for the BAU scenario, medium and large passenger vehicles must operate in a plug-in hybrid electric power or extended-range electric power mode. However, under the EV scenario, they can compete with their gasoline-fuelled counterparts in terms of comprehensive costs from 2025, and they will subsequently enter a period of rapid growth.

4 CONCLUDING REMARKS

The market penetration of electric passenger vehicles can be modelled using innovation diffusion theory and an epidemiological model. Under the BAU scenario, the penetration rates of electric passenger vehicles according to new sales in all sub-segments are very low even by 2050, while under the EV development scenario, the penetration rates vary by sub-segment in the period from 2030 to 2050, and the fleet-average penetration rate of electric passenger vehicles in China will be about 70% by 2050. The major factor controlling the market penetration of electric passenger vehicles is the life cycle cost of vehicles.

REFERENCES

CAERC (China Automotive Energy Research Centre, Tsinghua University). 2013. Sustainable Automotive Energy System in China. Springer Press.

Cai, Z., Ou, X., Zhang, Q., Zhang, X. 2012. Full lifetime cost analysis of battery, plug-in hybrid and FCEVs in China in the near future. Frontier in Energy 6 (2): 107–111.

Huo, H., Yao, Z., He, K., Yu, X. 2011. Fuel consumption rates of passenger cars in China: Labels versus real-world. Energy Policy 39:7130–7135.

Huo, H., Zhang, Q., He, K., Yao, Z. Michael Wang. 2012. Vehicle-use intensity in China: Current status and future trend. Energy Policy 43: 6–16.

Jiang, Q. 1993. Mathematical model. Beijing: Higher Education Press.

Li, Q. 1986. Super numerical analysis. Hubei: Huazhong University of Science and Technology Press.

Ou, X., Zhang, X., Chang, S. 2010. Scenario analysis on alternative fuel/vehicle for China's future road transport: Life-cycle energy demand and GHG emissions. Energy Policy 38(8): 3943–3956.

Ren, Y., Li, H., Sun, R., Guan, L. 2009. Analysis on model of life cycle cost of electric vehicle based on consumer perspective. Technology Economics 28(11): 54–58 (in Chinese).

Wagner, D.V., An, F., Wang, C. 2009. Structure and impacts of fuel economy standards for passenger cars in China. Energy Policy 37(10): 3803–3811.

Zhang, Y. 2003. Technology Transfer: Theory, tools and strategy. Beijing: Enterprise management press.

A study on the readjustment evaluation of a marine nature reserve based on the analytic hierarchy process

WenHai Lu
Ocean University of China, Qingdao, China
National Marine Data and Information Service, Tianjin, China

Yue Wu
National Center of Ocean Standard and Metrology, Tianjin, China

XianQuan Xiang
National Marine Data and Information Service, Tianjin, China

ABSTRACT: The establishment of marine nature reserves in China lags behind international progress. There exist some unreasonable settings for the scope of nature reserves due to the demarcation limited by the levels of investigation, the change of natural environment, and the rapid development of marine economy. It is a serious issue to readjust the scope of nature reserves at the present stage. Taking the Tianjin ancient coast and wetland national marine nature reserve as the study case, this paper mainly carries out the readjustment evaluation of a marine nature reserve based on the analytic hierarchy process, a key multi-criteria decision-making method. Firstly, evaluation indexes are established on the basis of abundant profile collection and data processing, and then weights are determined by the analytic hierarchy process. The calculation score of the evaluation is 76.82, which is higher than the reference value, meaning that it is necessary to readjust the proposed reserve. Through this study, it can be observed that the analytic hierarchy process approach can be preferably tool used to achieve the readjustment evaluation of the marine nature reserve and is, therefore, helpful to marine nature reserve management.

Keywords: marine nature reserve; readjustment; evaluation; analytic hierarchy process

1 INTRODUCTION

Marine nature reserves play a vital role in the protection of marine environments and resources maintenance of the marine ecological balance. The establishment of a marine nature reserve in China lags behind international progress. There exist some unreasonable settings for the scope of nature reserves due to the demarcation limited by the levels of investigation, the change of natural environment, and the rapid development of marine economy (Zhao H. et al. 2013). Furthermore, a series of development strategies and plans released by governments in coastal areas posterior to the establishment of a marine nature reserve in the wake of rapid development of the marine economy, have resulted in contradictions between marine economic development and marine environmental protection (Xiang X. et al. 2013). Therefore, the readjustment of the Marine Nature Reserve has become an inevitable issue. On January 29 2002, the State Council issued an official reply to the Administrative Regulations on the Readjustment of the Scope of National-level Nature Reserve and Readjustment and Rename of Functional Area (No.2 [2002] State Council), proposing specific requirements on the management of readjustment of the scope of national-level nature reserves, with the remaining related scientific verification issues needing to be resolved.

The concept of a marine nature reserve was first explicitly put forward in the *National Marine Functional Zoning* issued by the Chinese government in 2002 as, "maritime space for the protection of rare, endangered marine biological species, commercial biological species as well as related habitats and preservation of marine natural landscapes, natural ecosystems as well as historical sites with significant scientific, cultural and landscape values".

As a relatively specialist and sensitive field, the comprehensive research on marine nature reserves involves multiple disciplines such as ecology and economics as well as social science. The analytic hierarchy process (AHP) is a key multi-criteria decision-making method that is successful in both academic research and engineering applications. The AHP is a structured technique for organizing and analysing complex decisions. Based on mathematics and psychology, it was developed by Thomas L. Saaty in the 1970s and has been extensively studied and refined since then (Saaty, T. L. & Peniwati, K. 2008).

The AHP has been widely applied to numerous real-life problems in the past years. Several literature reviews on the AHP and its applications refer to the surveys of Zahedi (1996), Forman & Gass (2001), Golden & Wasil (2003), and Vaidya and Kumar (2006). It has a particular application in group decision making and is used around the world in a wide variety of decision situations, in fields such as government, business, industry, healthcare, and education (Millet, I. & Wedley, W.C. 2002). Rather than prescribing a "correct" decision, the AHP helps decision-makers find one that best suits their goal and their understanding of the problem. It provides a comprehensive and rational framework for structuring a decision problem, for representing and quantifying its elements, for relating those elements to overall goals, and for evaluating alternative solutions.

In this paper, the AHP method is used to weigh assessment criteria. Three levels, such as the status of the proposed reserve, the socioeconomic status of surrounding areas, and the management status of the reserve, are selected for establishing a necessity evaluation model of the readjustment of the Marine Nature Reserve. The purpose of this study is to provide an objective tool for the management of a marine nature reserve.

2 STUDY AREA

In October 1992, the State Council approved the establishment of the Tianjin ancient coast and wetland national marine nature reserve with a total area of 975.88 square kilometres, including 45.15 km^2 of core area and 43.34 square kilometres of buffer area, as well as 887.39 square kilometres of experimental area. Located in the west of the Bohai Sea and east of Tianjin, the reserve serves as the only national-level marine nature reserve for the protection of rare relics of ancient coast and wetland natural environments, as well as relevant ecosystems formed by shell dikes and oyster reefs. The location of the Tianjin ancient coast and wetland national marine nature reserve is shown in Figure 1.

The establishment of the Tianjin ancient coast and wetland national marine nature reserve has effectively protected the relics of ancient coast and wetland ecosystems of the ancient lagoon, playing a positive role in the promotion of the ecological and environmental quality of Tianjin. However, excessive completed areas and large areas with few protected objects in the reserve, and the related experimental areas, have severely hindered the management of the reserve, and limited the economic construction of the Tianjin Binhai New Area and the benign development of the reserve. In particular, after the issuance of a national general development strategy of the Tianjin Binhai New Area, unreasonable settings for the scope

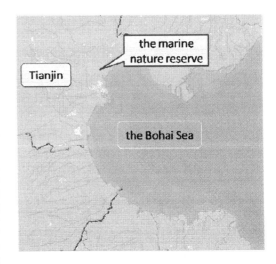

Figure 1. Location of the Tianjin ancient coast and wetland national marine nature reserve.

of the reserve have become increasingly prominent, forcing us to take related readjustment measures.

3 METHODOLOGY

3.1 *The AHP method*

The AHP method uses a typical pairwise comparison method to extract relative weights of criteria based on a hierarchical structure. In a hierarchical problem, each element at a given level is associated with some or all elements at the level immediately below. Elements at a single level are compared in terms of relative importance with respect to an element at the immediate higher level. Such pairwise comparisons are then analysed using an eigenvector method. The AHP method described earlier is a structured, systematic, and effective approach for determining the relative importance of weights. The procedure of AHP can be expressed in a series of steps:

(1) Construct a paired comparison matrix.

A pairwise comparison matrix of criteria is constructed using a scale of relative importance. The judgments are entered using the fundamental scale of the AHP, which is shown in Table 1. In total, $n(n-1)/2$ pairwise comparisons are evaluated for n criteria. Let A represent an $n \times n$ pairwise comparison matrix:

$$A = \begin{bmatrix} 1 & a_{12} & \cdots & a_{1n} \\ a_{21} & 1 & \cdots & a_{2n} \\ \vdots & \vdots & \ddots & \vdots \\ a_{n1} & a_{n2} & \cdots & 1 \end{bmatrix} \quad (1)$$

The diagonal elements in matrix A are self-compared; thus, $a_{ij} = 1$. The values on the left and right sides of the matrix diagonal represent

Table 1. The relational scale proposed by Saaty for pairwise comparisons.

Intensity of Importance	Definition	Explanation
1	Equal importance	Two activities contribute equally to the objective
3	Weak importance of one over another	Experience and judgment slightly favour one activity over another
5	Essential or strong importance	Experience and judgment strongly favour one activity over another
7	Very strong or demonstrated importance	One activity is favoured very strongly over another; its dominance is demonstrated in practice
9	Absolute importance	The evidence favouring one activity over another is of the highest possible order of affirmation
2, 4, 6, 8	Intermediate values between adjacent scale values	Used when compromise is needed
Reciprocals		If activity i has one of the preceding numbers assigned to it when compared with activity j, then j has the reciprocal value when compared with i
Rationals	Ratios arising from the scale	Used if consistency were forced by obtaining n numerical values to span the matrix

the strength of the relative importance degree of the ith element compared to the jth element. Let $a_{ji} = 1/a_{ij}$, where $a_{ij} > 0, i \neq j$.

(2) Calculate the importance degrees.

The average of normalized columns in a reciprocal matrix provides a good estimate of the principal right eigenvector in the deterministic case (Vargas, 1982). Let W_i denote the importance degree for the ith criteria. Then,

$$W_i = \frac{1}{n}\sum_{j=1}^{n}(a_{ij}/\sum_{i=1}^{n}a_{ij}), i,j = 1,2,...,n \quad (2)$$

(3) Test the consistency of the importance degrees.

Due to the limitation of Saaty's discrete nine-value scale and the inconsistency of human judgments when assessing weights during the pairwise comparison process, the aggregation weight vector might be invalid. Examination of consistency of the importance degrees should be made to avoid inconsistencies occurring when using different measurement scales in the evaluation process (Karapetrovic and Rosenbloom, 1999; Kwiesielewicz and van Udem, 2004). Saaty (1980) suggested the maximal eigenvalue λ_{max} be used to evaluate the effectiveness of measurements. To check the consistency between pairwise comparison judgments, the consistency index (CI) and consistency ratio (CR) are calculated using the equations,

$$CI = (\lambda_{max} - n)/(n-1) \text{ and } CR = CI/RI \quad (3)$$

where RI is a random index with a value obtained from Table 2 by different orders of pairwise comparison matrices. If the value of the CR is below 0.1, the evaluation of the importance degrees is considered to be reasonable. In general, the AHP is developed to select the best of a number of alternatives with respect to several criteria (Saaty, T. L. 2008).

3.2 Construction of the evaluation indexes system

The construction of the evaluation indexes system is mainly for the guidance of readjustment evaluation of the Marine Nature Reserve. Fourteen indexes have been selected on three levels (status evaluation, the socioeconomic status of surrounding areas, and the management status of the reserve) for necessity evaluation of the readjustment of the Marine Nature Reserve, based on the analysis of management status and necessity of readjustment of the scope marine nature reserve, as shown in Figure 2.

3.2.1 Status evaluation

Determination of readjustment of the Marine Nature Reserve is based on an objective evaluation of the marine nature reserve.

(1) The proportion of the non-existence of protected objects in a planned readjustment area: more data and information of current conditions of the marine nature reserve can be obtained with the enhanced investigation capability and the change of regional environment, as well as the implementation of special investigation. The higher the proportion, the higher the score will be.

(2) Importance to the country: rarity, preciousness, and the degree of international attention given to protected objects in the marine nature reserve.

(3) Scientific value: the assistance and value in scientific research of protected objects in the marine nature reserve, including marine and coastal natural ecosystems, marine biological species, marine nature relics, and non-biological resources.

(4) Tourism value: the value of protected objects in the marine nature reserve (marine and coastal natural ecosystems, marine biological species, and marine nature relics, as well as non-biological resources) as objects for tourism.

Table 2. Random index (RI) values.

Matrix order	1	2	3	4	5	6	7	8	9
RI	0	0	0.58	0.90	1.12	1.24	1.32	1.41	1.45

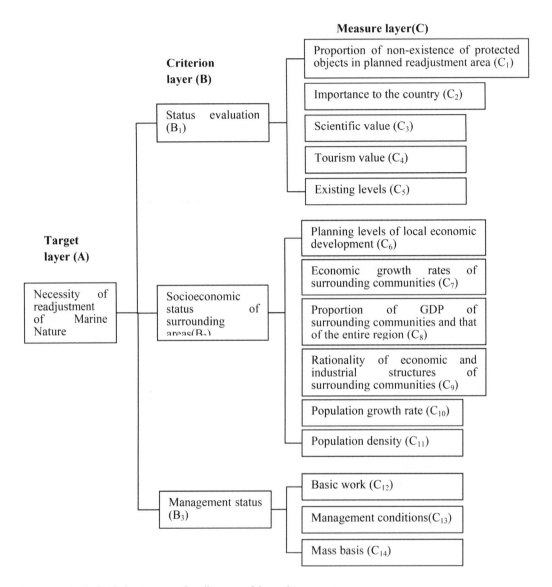

Figure 2. Evaluation indexes system of readjustment of the marine nature reserve.

(5) Existing levels: the levels of the evaluated marine nature reserve, including the national level and the local level.

3.2.2 Socioeconomic status of surrounding areas

The evaluation index of the socioeconomic status includes quality and potential of economic development and demographic situation of communities surrounding the marine nature reserve. The better the socioeconomic status, the higher the necessity of readjustment will be.

(1) Planning levels of local economic development: the regional planning serves as the detailed scheme of regional and industrial development under the guidance of the regional economic development strategy, including regional development, industrial development, and land utilization, as well as urban systems. Related levels of

Table 3. Indexes assignment and quantification.

Evaluation indexes	Quantification methods	Area data	Quantitative data
Proportion of non-existence of protected objects in the planned readjustment area (C_1)	Positive extremum method	84.1%	84
Importance to the country (C_2)	Linguistic quantification method	Relatively great	60
Scientific value (C_3)	Linguistic quantification method	Relatively high	60
Tourism value (C_4)	Linguistic quantification method	Common	100
Existing levels (C_5)	Linguistic quantification method	National-level	30
Planning levels of local economic development (C_6)	Linguistic quantification method	National-level	100
Economic growth rates of surrounding communities (C_7)	Positive extremum method	15%	63
Proportion of GDP of surrounding communities and that of the entire region (C_8)	Threshold-type quantification method	19.30%	60
Rationality of economic and industrial structures of surrounding communities (C_9)	Positive extremum method	99%	98
Population growth rate (C_{10})	Threshold-type quantification method	1.12%	100
Population density (C_{11})	Positive extremum method	492	1
Basic work (C_{12})	Linguistic quantification method	Completion of multi-disciplinary scientific research and collection of most samples	60
Management conditions (C_{13})	Linguistic quantification method	Basic full of necessary infrastructures and deficient administrative institutions	60
Mass basis (C_{14})	Linguistic quantification method	Incomprehension and misunderstanding of local people on the work of the reserve	100

planning include plans at county/district/municipal levels and national-level planning, as well as international-level planning.

(2) Economic growth rates of surrounding communities: measurement of economic growth rates of surrounding communities. Let ΔY_t be increment of the total economic output of this year, Y_{t-1} be the total economic output of last year, and the economic growth rate (G) shall be: $G = \Delta Y_t/Y_{t-1} - 1$.

(3) Proportion of GDP of surrounding communities and that of the entire region: the proportion of GDP of surrounding communities and that of the entire region (usually for one year); an index for the measurement of the importance of surrounding communities in the regional economy.

(4) Rationality of economic and industrial structures of surrounding communities: the measurement of the rationality of economic and industrial structures through calculation of the output of secondary and tertiary industries and their respective proportion, in the total economic output of communities surrounding the marine nature reserve.

(5) Population growth rate: natural increase of population (the number of births minus the number of deaths) in a certain period (usually for one year) of communities surrounding the marine nature reserve.

(6) Population density: population number per unit of land area of communities surrounding the marine nature reserve reflecting the population density of communities surrounding the marine nature reserve.

3.2.3 *Management status*

The stronger the management capability, the smaller the necessity of readjustment will be.

(1) Basic work: completion of scientific research and collection of samples.
(2) Management conditions: management of necessary infrastructures, administrative organizations, and staffing, as well as financial support plays a role in influencing the readjustment of the Marine Nature Reserve.
(3) Mass basis: understanding and support from local people on the work of the marine nature reserve plays a role in influencing the management of the related reserve.

3.3 *Index quantification and weight calculation*

This paper grades qualitative and quantitative indexes, or indexes of different dimensions in accordance with attributes or quantitative characteristics, and assigns them with proper values for standardized processes due to disunity of dimensions and lack of comparability. Furthermore, quantification processing is applied to numeric, threshold-types, as well as linguistic evaluation indexes.

Table 4. Weight of C-layer evaluation indexes on necessity of readjustment of the marine nature reserve

A	C_1	C_2	C_3	C_4	C_5	C_6	C_7	C_8	C_9	C_{10}	C_{11}	C_{12}	C_{13}	C_{14}
W	0.248	0.071	0.022	0.030	0.083	0.262	0.035	0.036	0.036	0.049	0.036	0.059	0.020	0.012

Index weight is determined by the analytic hierarchy process. Related experts and decision-makers grade all layers, layer by layer, by evaluating the significance of various indexes through pairwise comparison, and then determining the weights of the indexes according to the eigenvector, providing advice and basis for relevant decision-making analysis (Meng L. 1998).

4 RESULTS AND DISCUSSION

4.1 Assignment and quantification of evaluation indexes

Fourteen evaluation indexes have been assigned and processed with non-dimensional methods such as the positive extremum method and the threshold-type quantification method, as well as the linguistic quantification method, in accordance with related data of socioeconomic statistics and geological exploration, as well as statistical analysis of significant indexes through questionnaire surveys. Relevant quantitative data are listed in Table 3.

4.2 Weight calculation

Weights are determined by the analytic hierarchy process: first of all, grade all layers, layer by layer by evaluating the significance of various indexes through pairwise comparison and then determine the comprehensive weights of the indexes according to the eigenvector (Wu Y. et al. 2008).

4.2.1 Single taxis of layers
(1) Determination of weight of Matrix A-B (W)

A	B_1	B_2	B_3	W
B_1	1	1	5	0.455
B_2	1	1	5	0.455
B_3	1/5	1/5	1	0.091

$\lambda_{max} = 3.000$, CI $= 0$, RI $= 0.580$, CR $= 0 < 0.1$, a satisfactory result of consistency.
(2) Determination of weight of Matrix $B_1 - C$ (W)

B_1	C_1	C_2	C_3	C_4	C_5	W
C_1	1	5	9	7	3	0.546
C_2	1/5	1	5	3	3/5	0.156
C_3	1/9	1/5	1	7/9	1/3	0.049
C_4	1/7	1/3	9/7	1	3/7	0.067
C_5	1/3	5/3	3	7/3	1	0.182

$\lambda_{max} = 5.134$, CI $= 0.034$, RI $= 1.120$, CR $= 0.030 < 0.1$, a satisfactory result of consistency.
(3) Determination of weight of Matrix $B_2 - C$ (W)

B_2	C_6	C_7	C_8	C_9	C_{10}	C_{11}	W
C_6	1	6	8	7	5	9	0.576
C_7	1/6	1	3/4	6/7	6/5	2/3	0.077
C_8	1/8	4/3	1	7/8	5/8	9/8	0.079
C_9	1/3	7/6	1/2	1	5/7	7/9	0.080
C_{10}	1/5	5/6	8/5	7/5	1	9/5	0.108
C_{11}	1/9	3/2	8/9	9/7	5/9	1	0.080

$\lambda_{max} = 6.136$, CI $= 0.027$, RI $= 1.240$, CR $= 0.022 < 0.1$, a satisfactory result of consistency.
(4) Determination of weight of Matrix $B_3 - C$ (W)

B_3	C_{12}	C_{13}	C_{14}	W
C_{12}	1	3	5	0.652
C_{13}	1/3	1	5/3	0.217
C_{14}	1/5	3/5	1	0.130

$\lambda_{max} = 3.000$, CI $= 0$, RI $= 0.580$, CR $= 0 < 0.1$, a satisfactory result of consistency.

4.2.2 Total taxis of layers
The comprehensive weights of $C_1 - C_{14}$ calculated through the total taxis of layers are shown in Table 4.

4.2.3 Comprehensive score worked out through weighted summation

$$H = \sum_{i=1}^{n} w_i f_i \qquad (4)$$

Where H stands for index of necessity of readjustment of the scope;
w_i stands for the weight of the i evaluation index,
f_i stands for the score of the i evaluation index.
Upon calculation, the index of necessity of readjustment of the scope of the Tianjin ancient coast and wetland national marine nature reserve is: $H = 76.82$.

4.2.4 Analysis of the result
According to the assignment and quantitative criteria for the indexes, the reference value for readjustment of the marine nature reserve is 60. As the comprehensive score of the evaluation is 76.82, which is higher than the reference value, so it is necessary for the Tianjin ancient coast and wetland national marine nature reserve to carry out readjustment of the scope.

5 CONCLUSIONS

In this paper, exploratory research has been conducted on technical methods related to readjustment of a marine nature reserve. Taking the Tianjin ancient coast and wetland national marine nature reserve as the study case, firstly evaluation indexes are established on the basis of abundant profile collection and data processing, and then weights are determined by the analytic hierarchy process. The results of the evaluation show that it is necessary for the proposed reserve to readjust the scope. Through this study, it can be observed that the analytic hierarchy process approach can be the tool of preference to achieve the readjustment evaluation of the marine nature reserve.

The evaluation indexes system, the standardized processing method for the indexes, and the proposed evaluation method have provided references for the readjustment evaluation of the marine nature reserve, and have both theoretical and practical significance to the improvement of construction and management of marine nature reserves in China.

REFERENCES

Forman, E. & Gass, S. I. 2001. The analytic hierarchy process: an exposition. *Operations Research*, 49: 469–486.

Golden, B.L. & Wasil, E.A. 2003. Celebrating 25 years of AHP based decision making. *Computers and Operations Research*, 30: 1419–1420.

Meng L. 1998. Application of Analytic Hierarchy Process in Assessment of Grassland Resource. *Pratacultural Science* 15(6):1–4.

Millet, I. & Wedley, W.C. 2002. Modelling Risk and Uncertainty with the Analytic Hierarchy Process. *Journal of Multi-Criteria Decision Analysis*, 11: 97–107.

Saaty, T. L. 2008. Relative Measurement and its Generalization in Decision Making: Why Pairwise Comparisons are Central in Mathematics for the Measurement of Intangible Factors-The Analytic Hierarchy/Network Process. *Review of the Royal Spanish Academy of Sciences, Series A, Mathematics*, 102 (2): 251–318.

Saaty, T. L. & Peniwati, K. 2008. *Group Decision Making: Drawing out and Reconciling Differences*. Pittsburgh, Pennsylvania: RWS Publications.

Vaidya, O.S. & Kumar, S. 2006. Analytic hierarchy process: an overview of applications. *European Journal of Operational Research*, 169: 1–29.

Wu Y, Liu X, & Guo J 2008. *Quantitative Methods of Economic Management*. Beijing: Economic Science Press.

Xiang X., Xu X. & Tao J. 2013. Modelling chlorophyll-in Bohai Bay based on hybrid soft computing approach. *Journal of Hydroinformatics*, 15 (4): 1099–1108.

Zahedi, F. 1996. The analytical hierarchy process: a survey of the method and its applications. *Interfaces*, 16: 96–108.

Zhao H., Tao J., Li Q., et al. 2013. Microbial ecological characteristics in the Red Tide-Monitoring area of Bohai Bay. *Journal of Hydro-environment Research*, 7: 141–151.

A study on regional comparison of farmland resources' value and innovative strategies for farmland protection compensation in China

XiaoPing Zhou, Qing Wang & Duo Chai
School of Government, Beijing Normal University, P. R. China

ABSTRACT: The farmland protection system has always been a basic state policy for China. Due to the difference among regions of the aspects of nature, economics and society, it is of high significance to realize the farmland protection target and specify the comprehensive value of different farmland resources and their regional distribution characteristics. Based on the economic value, social value, and ecological value of the farmland resources, this paper calculates their comprehensive value systematically and further probes into the regional differences in the quantitative feature and structural feature of farmland resources. According to the research results, at the national level, its farmland resources' comprehensive value is considerably large, and its noneconomic value is bigger than the economic value; at the regional level, the economic value, social value and ecological value of farmland resources mainly come from the contribution of production areas of agricultural products, developed areas, and key ecological preservation areas, respectively. Therefore, two suggestions are proposed: on the one hand, we should strengthen its understanding of the farmland resources' noneconomic value, and make compensation for the farmland resources' positive externality; on the other hand, we should formulate differentiated farmland protection policies and increase the compensation efficiency of farmland protection.

1 INTRODUCTION

The report of the Chinese 18th National Congress of the Communist Party points out: "we should improve the system for providing the strictest possible protection for farmland…a system that responds to market supply and demand, and resource scarcity, recognizes ecological values and requires compensation in the interests of later generations"; the No.1 Central Document in 2013 states: "we should strengthen the agricultural compensation policy, improve measures in the compensation for the interest of major grain-producing areas, farmland protection, and ecology; we should speed up efforts to make agriculture obtain reasonable profits, and the financial resources of major grain-producing areas reach the average level of the nation or the provinces"; and according to the Decision from the third session of the 18th National Congress of Communist Party of China, "we should improve the agriculture support and protection system, reform the agriculture compensation system, and perfect the compensation mechanism for the interest of major grain-producing areas". All these documents demonstrate China's resolution to promote the compensation mechanism for farmland protection.

However, in the specific practice of the current farmland protection compensation, motivation and legal basis are lacking in the protection of farmland quality and ecology, because there is not enough emphasis on the noneconomic value of farmland resources, and since China does not have a comprehensive understanding on the regional differences of farmland resources' value, its unified policy for protecting farmland may not suitable, for example, general areas or those who have low efficacy, which results in the limited effectiveness of farmland protection and difficulty in achieving the farmland protection target. In order to have a correct understanding of the value system of farmland resources, theoretical support is required for promoting the equalization of returns of the agriculture value, and increasing farmers' benefits from the land property rights. It is the basic premise to specify the focus of farmland resources protection, but also a necessary guarantee to advance the development of the compensation mechanism for farmland resources.

Currently, most overseas scholars advocate the farmland protection compensation modes based on land development rights[1–5]. But, some Chinese scholars have put forward the new farmland protection compensation ideas based on the establishment of a farmland resources' value system[6–8]. However, as the connotation of the social value and ecological value of farmland resources have not been understood comprehensively, and the regional differences have not been taken into account fully on the existing research, there is still the need for continuous improvement.

Therefore, based on the statistics from the Chinese government's official website and the research data in local provinces, such as Fujian, Shandong and Hunan, this paper makes a systematic calculation of the farmland resources' value of Chinese provinces, and draws a regional comparison on their value from the quantitative feature and structural feature using SPSS and GIS, so as to provide an empirical reference for the development of the Chinese compensation mechanism for farmland resources.

2 CALCULATION OF THE FARMLAND RESOURCES' VALUE OF CHINESE PROVINCES

2.1 Establishment of a farmland resources' value system

From the perspective of the meaning of "farmland resources' value", the farmland resources' diversified value, as a compound system of production, life and ecology, contains the "economic" value of strategic security which produces agricultural products, ensures the necessary self-sufficiency ratio of national grain, and provides basic raw materials; the "social" value of the public service which absorbs the rural labour force, reduces the necessary social costs of living security, and maintains social stability; the "ecological" value of environmental preserving which keeps the material and energy circulation in the farmland ecosystem, protects and purifies the environment, and sustains the diversification of the farmland ecosystem. Therefore, consulting the classification standard[9] which has been widely recognized and employed in practice, the paper divides the comprehensive value of Chinese farmland resources into the economic value which produces and develops agricultural products, the social value which safeguards survival security and stability, and the ecological value which maintains a hospitable environment.

2.2 Calculation methods of farmland resources' value

According to the above meaning of the comprehensive value of farmland resources, there is Formula (1) of farmland resources' comprehensive value, where V_e, V_s, and V_z represent the economic value, social value and ecological value of farmland resources, respectively:

$$V = V_e + V_s + V_z \tag{1}$$

2.2.1 Calculation method of economic value

From the practices of Chinese development, farmland resources' economic value refers to the value from farmland application in agricultural production, that is to say, farmland is used in agricultural production as a basic element and could earn normal economic benefits, also called the production yield value. Therefore, the economic value could be calculated based on the market value of agricultural products: farmland economic value = agricultural products output × products unit price. According to the conversion factor of standard grain[1], the annual yield of agricultural products of each hectare of farmland in Chinese provinces could be calculated to be Q, the yield of standard grain[2]. And, based on the structure of standard grain, the market price of standard grain of each unit, represented by p, could be calculated according to the average purchase price of all kinds of grains in some years, so the formula below can be developed:

$$V_e = Q \times P \tag{2}$$

2.2.2 Calculation method of social value

From the basic national conditions of China, farmland itself possesses the functions of nurturing, bearing, value appreciation and maintenance, and credit guaranty, which are still the reliable material bases and safeguard measures for Chinese farmers in basic living, pension, employment, medical insurance, and credit guaranty. Currently, there are 2 billion mu[3] of farmland resources bearing a large amount of the rural labour force, which is also a measure to ensure social security. Therefore, the public service function of Chinese farmland could be summarized into the social stability value, V_α, which absorbs the rural labour force and ensures social stability, and the social security value, V_β, which maintains the necessary living expenses of the agricultural population and reduces the necessary social costs of living security. In order to ensure the availability and reliability, the paper, learning from the value substitution method[10-12], amends the calculation of farmland resources' social value, V_s, with the urban minimum wage standard, m, and the social pension insurance payment, n, multiplying respectively, the disposable income ratio of urban and rural residents (I_1/I_2):

$$V_s = V_\alpha + V_\beta = [m \times (I_1/I_2) + n \times (I_1/I_2)] \times R^4 \tag{3}$$

2.2.3 Calculation method of ecological value

From the course of Chinese industrialization and urbanization, human's activities have seriously endangered the ecological environment closely connected with human survival and development, and the ecological function of farmland has also played an increasingly important role. The ecological value of farmland resources could be interpreted as their ecological services, including atmosphere regulation, V_1, environmental purification, V_2, water and soil conservation,

---

[1] The conversion factor of standard grain, based on Chinese designated base crop, is the ratio of the base crop real output at a unit area to the real output of other designated crops at a unit area.

[2] The yield of standard grain, based on the output of Chinese designated base crop, could be calculated after the output of other designated crops is converted with the conversion factor of standard grain.

[3] 1 hm^2 = 15 mu.

[4] R represents the amount of rural labour force borne in each unit area.

V_3, nutrient substance circulation, V_4, and biodiversity maintenance, $V_5^{[13]}$. V_1 could be calculated by the carbon tax price method well recognized internationally; V_2 and V_4 by the "shadow price method"; V_2 and V_4, based on the researches of scholars like Ouyang Zhixun[14] and Xie Gaodi[15], could be figured out after converting the ecological green equivalent per unit into artificial grassland and forest land with the "shadow engineering approach".

$$V_z = V_1 + V_2 + V_3 + V_4 + V_5 \quad (4)$$

2.3 Calculation result of farmland resources' value

Based on the above calculation methods, the farmland resources' economic value $V_{e(it)}$, social value $V_{s(it)}$, and ecological value $V_{z(it)}$ per hectare each year in Chinese provinces, could be calculated by consulting relevant statistics of the thirty-one Chinese provinces, municipalities, autonomous regions (excluding Taiwan, Hong Kong and Macao), and large amount of survey data in Fujian, Shandong and Hunan. Thus, from the above values of farmland resources, the comprehensive value $V_{(it)}$ could be concluded: i represents the code of each province, 1-31; t represents the year from 2009 to 2012.

Table 1. The average farmland resources' value per unit area in China from 2009 to 2012
Unit: Yuan / hm² · Year

Area	V	V_e	V_s	V_z
Beijing	181525	24978	141825	14722
Shanghai	165534	27528	125342	12663
Zhejiang	165080	30325	118478	16277
Guangdong	147411	26681	106401	14328
Fujian	140912	32945	92734	15233
Hunan	114644	26230	67250	21163
Jiangsu	113937	29441	81248	3247
Tianjin	105063	17293	77558	10212
Jiangxi	97562	24498	53728	19337
Shandong	95129	23354	61510	10264
Henan	94798	24354	52224	18220
Sichuan	91061	19460	54363	17238
Hebei	85484	26603	50325	8557
Hubei	80069	18496	48401	13171
Anhui	72677	18991	43141	10545
Hainan	71806	12171	41712	17923
Guangxi	68549	13730	37702	17117
Chongqing	67447	13941	42653	10852
Liaoning	61618	15312	37617	8688
Xizang	59910	10862	33106	15942
Shanxi	52803	9663	23117	20023
Qinghai	52360	9582	27583	15195
Shanxi	49730	10597	30717	8415
Guizhou	46108	9309	20057	16742
Jilin	45795	10711	23030	12053
Xinjiang	44671	8505	25328	10837
Yunnan	44649	10296	24061	10292
Ningxia	40247	7687	22727	9832
Gansu	36981	7041	14648	15292
Heilongjiang	29181	7931	17208	4042
Neimongol	27696	7788	16668	3240
National	70047	16392	41909	11746

3 REGIONAL COMPARISON OF CHINESE FARMLAND RESOURCES' VALUE

3.1 Comparison of farmland resources' value by the quantitative feature

Chinese average comprehensive value of farmland resources reaches 70,047 Yuan/hm²·Year, including the noneconomic value of 53,655 Yuan/hm²·Year, about 3.27 times the economic value. Beijing has the highest value, 181,525 Yuan/hm²·Year, 6.5 times of that of Nei Mongol, which has the lowest value of 27,696 Yuan/hm²·Year. In advanced provinces and municipalities, such as Shanghai, Zhejiang, Guangdong, Fujian, Hunan, Jiangsu, and Tianjin, their values have reached over 100,000 Yuan/hm²·Year, while Ningxia, Gansu, Heilongjiang, and Nei Mongol have lower values, less than half of the national average value (Table 1 and Figure 1).

After a further study on the composition of farmland resources' comprehensive value, it could be found that Fujian, Zhejiang, Jiangsu, Shanghai, and Guangdong are the top five according to the economic value of farmland resources per unit area; Beijing, Shanghai, Zhejiang, Guangdong and Fujian rank the top five according to the social value of per unit area of farmland resources; Hunan, Shaanxi, Jiangxi, Henan and Hainan are in the top five according to the ecological value of each unit area of farmland resources (Table 1). It is thus evident that the economic value and social value account more for the comprehensive value of

Table 2. The farmland resources' comprehensive value per unit area in China from 2009 to 2012
Unit: Yuan / hm² · Year

Area	2009	2010	2011	2012
Beijing	152618	180250	202115	191115
Shanghai	135348	191835	199079	135872
Zhejiang	147200	161824	171165	180131
Guangdong	128689	136951	151950	172055
Fujian	139640	138467	145631	139908
Hunan	96499	109202	116337	136537
Jiangsu	109368	109903	114429	122047
Tianjin	104575	115683	99497	100497
Jiangxi	82297	100239	104507	103206
Shandong	91927	97657	93660	97270
Henan	85325	96070	98609	99189
Sichuan	80535	92597	94571	96540
Hebei	75256	86217	92726	87738
Hubei	69993	79779	83606	86896
Anhui	61090	74585	76415	78619
Hainan	60993	69750	77199	79280
Guangxi	58845	67268	73897	74185
Chongqing	61350	68827	69444	70167
Liaoning	60156	61685	64447	60182
Xizang	44830	53288	69814	71707
Shanxi	46824	52572	55034	56781
Qinghai	45366	49724	55880	58470
Shanxi	46239	50227	52472	49982
Guizhou	41388	46904	48923	47218
Jilin	42105	47293	44352	49430
Xinjiang	39693	43588	48662	46741
Yunnan	38923	44047	46491	49136
Ningxia	35121	37794	41539	46532
Gansu	32342	36148	41093	38340
Heilongjiang	28956	25954	34232	27581
Neimongol	25367	28489	28650	28279
National	63201	69644	73159	74177

Figure 1. The distribution map of farmland resources' comprehensive value per unit area in China from 2009 to 2012.

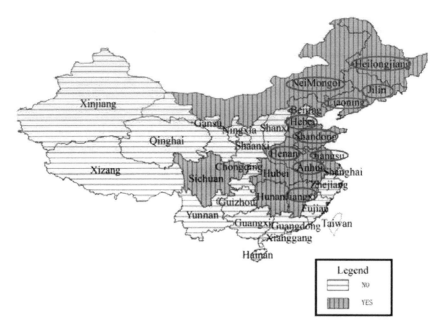

Figure 2. The distribution map of the main producing areas of agricultural product in China.

farmland resources in developed areas, so the developed provinces along the south-eastern coast have higher a comprehensive value of farmland resources.

From the changing trend of farmland resources' value, sixteen Chinese provinces have an ever-increasing value of farmland resources per unit area with each passing year from 2009 to 2012, namely Henan, Guangxi, Shaanxi, Anhui, Tibet, Hainan, Sichuan, Chongqing, Jiangsu, Yunnan, Zhejiang, Qinghai, Hubei, Hunan, Guangdong, and Ningxia; from 2009 to 2010, the farmland resources value per unit area increased in twenty-nine of the all thirty-one

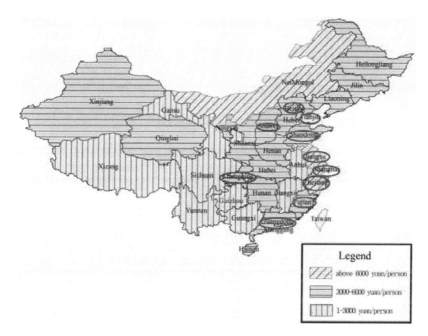

Figure 3. The distribution map of Chinese real GDP per capita in 2012.

provinces, except Fujian and Heilongjiang; from 2010 to 2011, except Shandong, Jilin and Tianjin, the other twenty-eight provinces have a higher value than that of the former year; from 2011-2012, the farmland resources' value decrease in a total of twelve provinces. Therefore, it can be seen that with the advance of urbanization, the enhancement of land development and the common phenomena of cultivating fertile land while supplementing land of low quality, the overall farmland quality deteriorates gradually in more and more provinces, which results in the reduction of the farmland resources' economic value and ecological value. And, since the social value of farmland resources, driven by the national economic and social development, have few changes, the comprehensive value drops down as a consequence.

3.2 Comparison of farmland resources' value by the structural feature

According to the ratio of economic value in the comprehensive value, the top ten provinces are Hebei, Nei Mongol, Heilongjiang, Anhui, Jiangsu, Henan, Jiangxi, Liaoning, Shandong, and Jilin, all of which have exceeded the national average of 21.84%. It could be seen from Figure 2 that all those provinces are in Chinese major grain-producing areas with farmland resources of high quality, so their economic output efficiency is also higher than other areas.

According to the ratio of social value in the farmland resources' comprehensive, Beijing, Shanghai, Tianjin, Guangdong, Zhejiang, Jiangsu, Fujian, Shandong, Chongqing, and Shanxi are the top ten, with all of them accounting for 60%, and the top six reaching over 70%. As is shown in Figure 3, there is a positive correlation between the farmland resources' social value and the local economic development level. For the developed areas, since their opportunity costs increase, they have higher requirements for the farmland resources' social value, which ensures farmers' survival and mitigates the employment pressure of farmers.

Also, based on the ratio of ecological value in the comprehensive value, the top ten are Gansu, Shanxi, Guizhou, Qinghai, Tibet, Jilin, Guangxi, Hainan, Ningxia, and Xinjiang, all of which have accounted for over 20%, far more than the national average of 13.24%, and the top three have reached more than 35%. According to the Chinese Outline of National Main Functional Regions published in June 2011, the above provinces belong to the key ecological function areas, where the farmland resources have played an important role in their ecosystems. They bear the multiply ecological functions, such as atmosphere regulation, environmental purification, water and soil conservation, nutrient substance circulation, and biodiversity maintenance, thus providing huge positive externality value for all members of society.

4 CONCLUSION AND SUGGESTION

From the above analysis, three characteristics have been found in Chinese farmland resources' value: firstly, the overall farmland resources' comprehensive value is considerably large, and its noneconomic value is bigger than the economic value; secondly, farmland resources in developed areas have higher comprehensive value in general, and the major part of the comprehensive value comes from the economic value and social value; thirdly, there exists regional differences in the value composition of farmland resources, and such difference presents good consistency with Chinese agricultural function zones and ecological function zones. Based on those characteristics, the paper proposes the following innovative suggestions on Chinese compensation mechanism for farmland protection.

(1) To strengthen the understanding on farmland resources' noneconomic value and make compensation for the farmland resources' positive externality

China has a large amount of comprehensive value of farmland resources, far more than that in the traditional understanding of farmland resources. Besides, the underappreciated noneconomic value by the state and all provinces for a long time is more than the economic value, and the social value accounts for most of the comprehensive value, reaching over 50% in most provinces.

Under the current value system, most of farmland resources' value exists outside of the market, so it is actually not embodied in the final agricultural products. In reality, farmland resources provide society with basic survival materials, but also the indispensable functions of maintaining social stability and serving the ecosystem with their strong positive externality. Therefore, the necessity and reasonability of the farmland protection system could by no means be measured only by the economic benefits of farmland utilization, which is exactly the limitation of former studies on the farmland protection system.

Therefore, when designing the future farmland protection system, China should strengthen its understanding of the noneconomic value of farmland resources and gradually implement the measures of compensating for farmland resources' positive externality, so as to achieve the target of the main bodies of farmland protection, preserving the farmland and promoting the equity among different regions and social subjects. Meanwhile, during the process of implanting farmland protection policy, all regions should keep exploring and selecting appropriate financial security mechanisms, and collect large amounts of money for compensation, in order to ensure the practical feasibility of compensation policies of farmland protection based on the comprehensive value.

(2) To formulate differentiated farmland protection policies and increase the compensation efficiency of farmland protection

The difference concerning the value composition of farmland resources among Chinese provinces is related to the farmland qualities, economic development, and ecological productivity of those areas. Thus, local governments should formulate differentiated farmland protection policies and employ diversified compensation measures in line with their own conditions, to increase the efficiency of their policies in preserving farmland.

The regions with remarkable economic function of farmland resources are mostly Chinese major grain-producing areas, which result from the Chinese government's preferential policies of farmland protection and agricultural support on those grain-producing areas. Accordingly, those areas could increase compensation for quality enhancement in farmland, step up investment in land reclamation and agricultural development projects, and promote the development of basic farmland of high quality to steadily improve the economic productivity of farmland.

The regions with strong social functions of farmland resources are largely located in developed areas. With the structural adjustment, transformation, and upgrading of local industries in those areas, their economy will definitely grow gradually, so will their farmland resources' social value, which means that higher requirements will be needed on farmland resources' functions of social stability and social security. Therefore, in the long term, the increase of economic value in construction land resulting from economic growth should not be an excuse for the loosening of the farmland protection system. As long as there are farmland resources and farmers depending on those farmland resources in these areas, the country should provide guidance on implementing compensation policies for farmland protection in these areas, and adjust the compensation level by reference to the increase of urban social security. The Chinese government could try to establish special funds for social security to ensure the social value of farmland resources.

In the regions where the ecological function of farmland resources is most evident, the national key ecological function zones are mainly distributed. As shown by the result of value calculation, on the whole, the ability of Chinese farmland resources to provide ecological products is weakening, but people are demanding more on these products. Therefore, more efforts should be put into ensuring the ecological functions of the farmland resources in these areas. The central government could increase transfer payments on the compensation for farmland's ecological value in these areas, and at the same time, local governments should intensify their support for the development of ecological agriculture and increase the ecological productivity of farmland as an ecological product, so as to keep the green ecological space on a reasonable scale and the ecosystem more stable, thereby further ensuring the ecological security in a relatively large area or, indeed, across the nation itself.

REFERENCES

Barrows R L, Prenguber B A. 1975. Transfer of development rights: an analysis of a new land use policy tool. *American Journal of Agricultural Economics* 57(4): 549–557.

Kopits E, McConnell V, Walls M. 2003. A Market Approach to Land Preservation. *Resources for the Future* 1: 15–18.

Wichelns D, Kline J D. 1993. The impact of parcel characteristics on the cost of development rights to farmland. *Agricultural and Resource Economics Review* 22(2): 150–158.

Pizor P J. 1978. A review of transfer of development rights. *The Appraisal Journal* 46(3): 386–396.

Machemer P L, Kaplowitz M D. 2002. A framework for evaluating transferable development rights programmers. *Journal of Environmental Planning and Management* 45(6): 773–795.

Qu Futian, Feng Shuyi, Yu Hong. 2001. Research on non-agricultural land prices and economic mechanisms for the relationship between agricultural land allocations: an example of the economic developed regions. *Chinese Rural Economy* 12: 54–60.

Yu Fengqing, Cai Yunlong. 2003. A new insight of cultivated land resource value. *China Land Science* 3: 3–9.

Yu Fengqing, Cai Yunlong. 2004. Re-establishing the value of cultivated land resource for subsidizing agriculture. *China Land Science* 1: 18–23.

Li Jia, Nan Ling. 2010. Connotation and calculation methods for the cultivated land resources values: A case of Shanxi Provinc. *Journal of Arid Land Resources and Environment* (9): 10–15.

Wu Zhaojuan, Wei Chaofu. 2012. Finite Element Analysis of Strength of Curved Sub: soiling Shovel Based on Pro / E 5.0 and ANSYS Workbench. *Journal of Agricultural Mechanization Research* 1: 29–32, 40.

Jin Shulan, Jin Wei, Xu Lei, Hou Lichun. 2011. Compensation Standard of Land Expropriation in Jiangxi according to Value of Cultivated Land. *Hubei Agricultural Sciences* 15: 3054–3057.

Ye Shan, Li Shiping. 2013. Assessment On the Social Value of Cultivated Land Resources: Taking Xi'An as an example. *Chinese Journal of Agricultural Resources and Regional Planning* 2: 27–32.

Wang Shiju, Huang XianJin, Chen Zhigang, Tan Dan, Wang Guanghong. 2008. Study on Compensation Standard of Land Expropriation Based on Value of Cultivated Land. *China Land Science* 22(11): 44–50.

Ouyang Zhiyun, Wang Xiaoke, Miao Hong. 1999. A primary study on Chinese terrestrial ecosystem services and their ecological-economic values. *ACTA ECOLOGICA SINICA* 19(5): 607–613.

Xie Gaodi, Xiao Yu, Zhen Lin, Lu Chunxia. 2005. Study on ecosystem services value of food production in China. *Chinese Journal of Eco-Agriculture* 13 (3): 10–13.

Energy and Environmental Engineering – Wu (Ed.)
© 2015 Taylor & Francis Group, London, ISBN 978-1-138-02665-0

The application of the ETKF data assimilation method in the Lorenz-96 system

JiChao Wang, Jian Hou, QingJun Du & RongXia Xu
China University of Petroleum, Qingdao, China

ABSTRACT: The basic principle of the ETKF (Ensemble Transform Kalman Filter) method is reviewed, and the application of the ETKF based on the nonlinear dynamic system Lorenz-96 is investigated. Comparative analysis of the ETKF and the EnKF method are performed with different observation intervals, ensemble numbers and observation numbers. Not only Gaussian perturbation but also non-Gaussian perturbation is carried out in order to analyse the results of assimilation. Generally, efficiency of computation and the stability of the ETKF method are better than the EnKF method. The evolution of the ensemble members is different between the two methods. When ensemble members are relatively lacking, the assimilation effect of the ETKF method is poor; the "localisation" strategy should be introduced to solve the filter divergence problem caused by a small ensemble.

1 INTRODUCTION

The Variational and Ensemble Kalman Filter methods are the main data assimilation methods that have been successfully applied in atmospheric, reservoir and oceanic models at present (Bouttier and Courtier, 2002). Since the 1990s, the operational running of the Variational method was achieved in only a few countries and was gradually becoming the mainstay for the development of the data assimilation methods. The EnKF method was proposed by Evensen (1994), and comprehensively studied both experimentally and theoretically (Evensen, 2003). The EnKF method not only solves the problems of strong nonlinearity and parallelize easily applied the thought of mathematical set, but also does not need the quite complicated adjoint code. So this method demonstrates the strong vitality in the past decade.

The ETKF (Ensemble Transform Kalman Filter) method which is the improved method of EnKF, was proposed by Bishop et al. (2001), then the "localisation" strategy was introduced by Ott et al. (2004), and the LETKF (Local Ensemble Transform Kalman Filter) method was developed. Since then, many scholars have worked very much on this study. Because of the simple method and high computational efficiency, and also considering the background error covariance based on flow-dependent, the research and applications of the ETKF method are becoming a hot topic in data assimilation methods (Hunt et al., 2007, Bishop and Hodyss, 2009, Harlim and Hunt, 2007, Jiang et al., 2011).

Let n be the state dimension, m is the ensemble size, and p is the number of observations. Let an n × m matrix of an ensemble of model states be denoted by E, the ensemble average by X, so that:

$$E = (x^1, x^2, \cdots, x^m) \in \mathfrak{R}^{n \times m} \quad (1)$$

The EnKF method updates the ensemble with the equation:

$$E^f = ME^a + \varepsilon \quad (2)$$

The ensemble anomalies denoted by A is:

$$A = E - E1_m \quad (3)$$

where **1** is a vector with all elements equal to one. Then background error covariance $P_e \in \mathfrak{R}^{n \times n}$ could be expressed as:

$$P_e = \frac{AA^T}{m-1} \quad (4)$$

where superscript T denotes matrix transposition.

The Kalman gain matrix is:

$$K_e = AA^T H^T (HAA^T H^T + R) \quad (5)$$

where $R \in \mathfrak{R}^{p \times p}$ is the observation error covariance matrix and $H \in \mathfrak{R}^{p \times n}$ is a p × n matrix of interpolation coefficients from states to observations.

The EnKF method uses the idea of ensemble to solve the problem where the background error covariance matrix is difficult to estimate and predict. The analysis value and the ensemble average can be obtained by the follow equations:

$$E^a = E^f + K_e(D - HE) \qquad (6)$$

$$X^a = X^f + K_e(y^o - HX^f) \qquad (7)$$

where $D = (y_1^o, y_2^o, \ldots, y_m^o) \in \Re^{p \times m}$ and $y^o \in \Re^p$ is observation vector, $y_j^o (j = 1, 2, \ldots, m)$ are vectors which are perturbed by the y^o.

The ETKF method was proposed by Bishop (2001), the unbiased form of this method was proposed by Wang et al. (2004). In the EnKF method, the update of the ensemble members and the ensemble average are defined by equations (6) and (7). It is important to note that D in (6) is a matrix obtained by the disturbance of observation vector yo, this artificial disturbance could increase unnecessary experiment error. The EnKF method which needs the disturbance of observation is stochastic EnKF, and the EnKF method which no need the disturbance of observation is deterministic EnKF. The ETKF is a deterministic EnKF, essentially. The advantage of the ETKF method is that it could analyse problems in the m dimension based on ensemble transform so that it has high computational efficiency (Hunt, 2005). Compared to (6) and (7), the ETKF calculates the analytical value based on the update of the gains and ensemble average. The analysis formula of the ensemble average is:

$$X^a - X^f = A^f G R^{-1/2}(y^o - HX^f) \qquad (8)$$

Here, $G = (I + S^T S)^{-1} S^T$

The analysis formula of the ensemble gain is:

$$A^a - A^f = A^f T \qquad (9)$$

where $T = (I + S^T S)^{-1/2} - I$ is the Transform Matrix and $S = R^{-1/2} H A^f / \sqrt{m-1}$.

2 LORENZ-96 SYSTEM

The system Lorenz-96 is a highly nonlinear system, which has high sensitivity on initial value. It is a simplified model deduced by dynamic model by Lorenz (1996), which is widely used to verify all kinds of data assimilation methods. The dynamic equation is:

$$\frac{dx_j}{dt} = (x_{j+1} - x_{j-2})x_{j-1} - x_j + F, \quad (j = 1, 2, \cdots, n) \quad (10)$$

where $x_{-1} = x_{n-1}$, $x_0 = x_n$, $x_{n+1} = x_1$. The system Lorenz-96 has three properties: firstly, it has a nonlinear dissipation that could reduce the total energy; secondly, external forcing terms F could increase or reduce the total energy; third, the quadratic advection term could 'reserve' the total energy, that is, the quadratic advection term does not contribute to dV/dt.

In this study, the external forcing terms F = 8, the number of variables $n = 40$ and the time step $\Delta t = 0.1$,

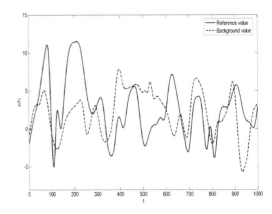

Figure 1. Sensitivity study on the initial value of Lorenz-96 system.

use 4-order Runge-Kutta to solve it. All of the simulation results are only aimed at the first variables (x_1) in this study, other results are not listed.

The numerical simulation is based on this system that is not considered the model errors. That is, all of the initial conditions calculated by the model could get 'Truth values', The system Lorenz-96 has high sensitivity on initial value: Figure 1 shows that after $x_1 = -1.90$, increases turbulence 0.12, time to run 1000 step, the time series of the two solutions are compared. The next graphics in this study do not add the turbulence-reference solutions of state variables ('background values' or 'forecasted values') in order to make the graphics not 'messy'.

With adding different turbulence value on the observation value, all kinds of observation errors are tested to the assimilating effects. This study uses Gaussian distribution $\phi(x)$ and non-Gaussian distribution $\varphi(x)$, the function is:

$$\phi(x) = \frac{1}{\sqrt{2\pi}\sigma} e^{-\frac{(x-\mu)^2}{2\sigma^2}} \qquad (11)$$

$$\varphi(x) = \frac{1}{2\sqrt{2\pi}}(e^{-\frac{1}{2}(x-v)^2} + e^{-\frac{1}{2}(x+v)^2}) \qquad (12)$$

where $\varphi(x)$ was defined by Lawson and Hansen (2004). The two functions are shown in Figure 2.

Both the EnKF and the ETKF method are based on the ensemble forecasting. Figure 3 shows that initial value, mean and true value of the ensemble where variable numbers of Lorenz-96 are 40, and the ensemble numbers are 50. It is found that initial mean value and true value vary widely.

Using the mean square root error to analyse the assimilating effects, RMS is defined as:

$$RMS = \sqrt{\frac{1}{N}\sum_{i=1}^{N}(x_i^a - x_i^t)^2}, \qquad (13)$$

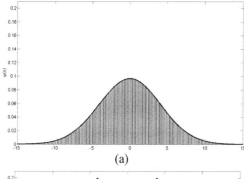

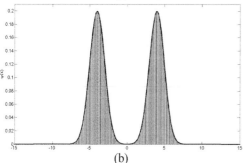

Figure 2. Graphs of $\phi(x)$ (a) and $\varphi(x)$ (b), respectively.

Table 1. The RMS of EnKF and ETKF with different conditions

No.	r	m	p	Per	RMS_EnKF	RMS_ETKF
1	5	50	40	Yes	0.4579	0.3873
2	5	50	40	Yes	0.4958	0.7287
3	5	50	40	No	0.4174	0.3040
4	20	50	40	No	0.9247	0.5699
5	5	30	40	No	0.7146	2.9570
6	5	30	40	No	——	0.2359
7	5	50	20	No	1.3858	0.5604

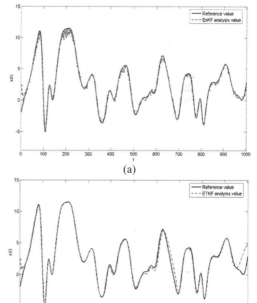

Figure 4. Assimilation results of EnKF (a) and ETKF (b) from experiment 2 with r of 5, m of 50, p of 40 and with non-Gaussian perturbation.

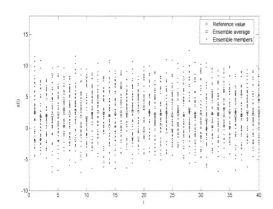

Figure 3. Initial value, mean and true value of the ensemble where variable numbers of Lorenz-96 are 40.

where x_i^a is the analysis value and x_i^t is the reference value. The RMS of the two methods is expressed by RMS_EnKF and RMS_ETKF, respectively.

3 NUMERICAL EXPERIMENTS

Using different frequency of observations, ensemble numbers, observation numbers and added (or not) turbulence values, seven experiments have been conducted in this paper. The results can be seen in Table 1 and the assimilation results can be found in Figures 4–6. Here, r is the observation interval.

In this study, experiment 1 is perturbed by Guassian noise and experiment 2 is perturbed by non-Guassian noise. Experiment 6 makes use of the strategy of localisation and the method named LETKF. The assimilation effects of only experiments 2, 4, 5 and 6 are listed as follows (Fig. 4, Fig. 5 and Fig. 6).

The two methods change the ensemble members in different ways when running the mode. Figure 7 shows the development results of ensemble members and the mean of EnKF and ETKF after running 1000 steps.

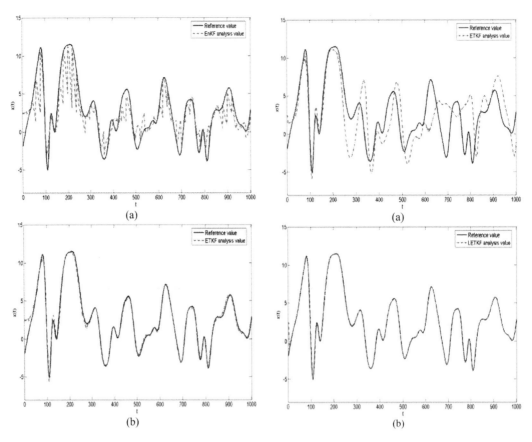

Figure 5. Assimilation results of EnKF (a) and ETKF (b) from experiment 4 with r of 20, m of 50 and p of 40.

Figure 6. Assimilation results of ETKF (a) and LETKF (b) from experiment 5 and 6 with r of 20, m of 30 and p of 40.

4 RESULTS AND DISCUSSION

In this study, the basic formula of the EnKF and ETKF method is introduced. Then, based on system Lorenz-96, the EnKF and ETKF filter methods are simulated under different conditions. Using seven experiments, the main conclusions are drawn as follows:

(1) When there are observation errors of non-Gaussian noise, the assimilation results of the ETKF method significantly reduces. It could cause filter divergence when the model integration time is increased (t > 950) (Fig. 4b). The ETKF method could be improved when there are observation errors of non-Gaussian noise.
(2) If the observation interval increases, the ETKF method exhibits good stability, and the assimilation results of the EnKF method significantly reduce. Obviously, if the observation interval is too large, both the assimilation results of those two methods do not perform satisfactorily.
(3) If the ensemble numbers are relatively small, the assimilation results of the ETKF method significantly reduce in experiment 5. The "localisation" strategy is an effective process to achieve this problem and the method is named the LETKF method. But the computing time of the LETKF method is 26.9 s, which is much slower than the ETKF method (3.3 s).
(4) Compared to the EnKF method, the assimilation results of the ETKF method are less affected by the poor observation numbers, it is also indicated that ETKF has more stability.
(5) The computing time of the ETKF method is about 3 s, which is 40% less than the EnKF (5 s). There is no doubt that the ETKF is highly efficient.

All of the conclusions about the ETKF method are only aimed at the ideal system, Lorenz-96, in this paper. When aimed at the real atmospheric and oceanic models, the conclusions above only have reference value. It was found that the ETKF method needs to reverse the matrix in the calculation process. When the matrix is large, it needs regularization. There are still some problems worth studying in the aspects of regularization aimed at the ETKF method.

When the ensemble numbers are relatively small, the assimilation results of the ETKF method perform satisfactorily. Using the "localisation" strategy is an effective process to achieve this problem, but it is at

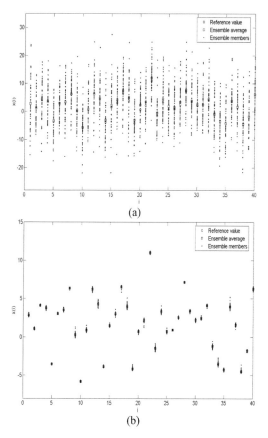

Figure 7. Development results of ensemble members and mean of EnKF (a) and ETKF (b).

the expense of the calculation time. It is worth studying the effective localisation strategy based on the ETKF.

ACKNOWLEDGEMENTS

The Fundamental Research Funds for the Central Universities (No.13CX05016A, 14CX02035A), the writers greatly appreciate the financial support of the National Natural Science Foundation of China (No.11102236), Open fund of State Oceanic Administration Laboratory of Data Analysis and Application (No. LDAA-2013-03).

REFERENCES

Bishop C., Etherton, B. and Majumdar S., 2001, Adaptive sampling with the ensemble transform kalman filter part i: the theoretical aspects. Monthly Weather Review, 129, 420–436.

Bishop C., Hodyss D., 2009, Ensemble covariances adaptively localized with ECO-RAP. Part 2: a strategy for the atmosphere. Tellus, 61A: 97–111.

Bouttier F. and Courtier P., 2002, Data assimilation concepts and methods. Meteorological Training Course Lecture Series, ECMWF.

Harlim J. and Hunt B., 2007, A non-Gaussian ensemble filter for assimilating infrequent noisy observations. Tellus, 59A, 59A: 225–237.

Hunt B., 2005, Efficient data assimilation for spatiotemporal chaos: a local ensemble transform Kalman filter. http://arxiv.org/abs/physics/0511236.

Hunt B., Kostelich E., and Szunyogh I., 2007, Effcient data assimilation for spatiotemporal chaos: A local ensemble transform Kalman Filter. Physica D, 230: 112–126.

Evensen G., 1994, Sequential data assimilation with a nonlinear quasi-geostrophic model using monte carlo methods to forecast error statistics. Journal of Geophysical Research, 99, 10143–10162.

Evensen G., 2003, The ensemble Kalman filter: theoretical formulation and practical implementation. Ocean Dyn. 53, 343–367.

Jiang Z., Fei Z. and Xichen L., 2011, A new localization implementation scheme for ensemble data assimilation of non-local observation. Tellus, 63A, 244–255.

Lawson W.G. and Hansen J.A., 2004, Implications of Stochastic and Deterministic Filters as Ensemble-Based Data Assimilation Methods in Varying Regimes of Error Growth. Mon. Wea. Rev., 132, 1966–1981.

Lorenz E., 1996, Predictability: a problem partly solved. Proceedings on predictability, held at ECMWF on 4-8 September 1995, 1–18.

Ott E., Hunt B., Szunyogh I. et al., 2004, A local ensemble kalman filter for atmospheric data assimilation. Tellus, 56A, 415–428.

Wang X., Bishop C. and Julier S.J., 2004, Which is better, an ensemble of positive-negative pairs or a centered spherical simplex ensemble? Mon. Wea. Rev., 132, 1590–1605.

Energy and Environmental Engineering – Wu (Ed.)
© 2015 Taylor & Francis Group, London, ISBN 978-1-138-02665-0

Robust suboptimal control of a nonlinear enzyme-catalytic dynamical system in microbial continuous culture of glycerol to 1,3-propanediol

Juan Wang
College of Science, China University of Petroleum, Qingdao, China

ZhiLong Xiu
School of Environmental and Biological Science and Technology, Dalian University of Technology, Dalian, Liaoning

ABSTRACT: A nonlinear enzyme-catalytic dynamical system is investigated, which can describe the continuous fermentation of glycerol to 1, 3-propanediol by Klebsiella pneumoniae. With the dilution rate and the initial glycerol concentration as the control parameters, we present a robust suboptimal control problem for the enzyme-catalytic system aiming at maximizing the productivity of 1, 3-propanediol, together with minimizing the system sensitivity which can be generated by solving an auxiliary dynamical system. Finally, a discrete computational procedure is developed based on the discretizations of the kinetic system and the auxiliary system.

1 GENERAL INSTRUCTIONS

The bioconversion of glycerol to 1, 3-propanediol (1, 3-PD) by Klebsiella pneumoniae is particularly attractive to industry because of renewable feedbacks and potential uses of 1, 3-PD (Biebl et al., 1999). The latter is discussed as a bifunctional chemical reagent on a large commercial scale, especially as a monomer for polyesters, polyethers and polyurethanes (McCoy, 1998).

The fermentation of glycerol is a complex bioprocess, since microbial growth is subjected to multiple inhibitions of substrate and products, e.g. glycerol, 1, 3-PD, ethanol and acetate (Liu et al., 2001). In continuous culture, the dilution rate and the initial glycerol concentration are usually regarded as the control variables to maximize the productivity of 1, 3-PD (Li et al., 2006). So how to find the optimal control is important and interesting. On the other hand, since the mathematical model is only an approximation of the real fermentation system, it is imperative that the optimal control be robust with respect to modelling errors, which mainly refer to the uncertainty of model parameters as to their exact values (Loxton et al., 2011). More precisely, the performance of the actual metabolic system when the optimal control is applied should be similar to the theoretical performance predicted by the mathematical model. Therefore, when we compute the control variables, we wish to balance two objectives. The first is system cost, while the second is system sensitivity (i.e. the variation of system cost with respect to model parameters). The optimal control chosen on this method is also called the robust suboptimal control, which has already been extensively studied and applied since 1992 (Rehbock et al., 1992, Defeng et al., 2007).

In our preceding work (Wang et al., 2012), we proposed a nonlinear enzyme-catalytic dynamical system of glycerol continuous fermentation and carried out its system identification. In the present work, with the dilution rate and the initial glycerol concentration as the control parameters, a robust suboptimal control problem for the nonlinear enzyme-catalytic dynamical system is proposed, aiming to maximize the productivity of 1,3-PD at the terminal time and to minimize the system sensitivity. To calculate system sensitivity, we then present an auxiliary dynamical system which is coupled with the enzyme-catalytic system. Besides, by discretizing the dynamics of both the kinetic system and the auxiliary system, we obtain the discrete-time approximations of those two system, where the approximation errors are bounded. Finally, a discrete computational method is developed.

2 NONLINEAR ENZYME-CATALYTIC DYNAMICAL SYSTEM

According to the preceding work (juan1 et al., 1989), the continuous fermentation process of glycerol can be formulated as the following nonlinear enzyme-catalytic dynamical system:

$$\begin{cases} \dot{x}(t) = F(x,u,p), & t \in [0,T] \\ x(t_0) = x^0. \end{cases} \quad (1)$$

where the elements of the state vector $x(t) := (x_1(t), x_2(t), \ldots, x_8(t))^T$, denote the concentrations of biomass, glycerol, 1,3-PD, acetate, ethanol, intracellular glycerol, 3-hydroxypropionaldehyde (3-HPA) and intracellular 1,3-PD in the reactor at time $t \in [0,T]$, respectively; T is the given terminal time and x^0 is a given initial state. $u := (D, C_0)^T$ denotes the control parameter vector with the components representing the dilution rate D and the initial glycerol concentration C_0. p is the known kinetic parameter vector with the value given in (juan1 et al. 2012}. The right hand side of (1) is of the form $F := (f_1, \ldots, f_8)^T$ with the components defined as (Wang et al. 2012, Sun et al. 2008):

$$\begin{cases} f_1 = (d-D)x_1, \\ f_2 = D(C_0 - x_2) - q_2 x_1, \\ f_i = q_i x_1(t) - D x_i, \quad i = 3, 4, 5, \\ f_6 = \dfrac{1}{p_7}(p_8 \dfrac{x_2}{x_2 + p_9} + p_{10}(x_2 - x_6) - q_0) - d x_6, \\ f_7 = \dfrac{p_{11} x_6}{K_1(1 + \frac{x_7}{p_{12}}) + x_6} - \dfrac{p_{13} x_7}{K_2 + x_7(1 + \frac{x_7}{p_{14}})} - d x_7, \\ f_8 = \dfrac{p_{11} x_7}{K_2 + x_7(1 + \frac{x_7}{p_{14}})} - \dfrac{p_{15} x_8}{x_8 + p_{16}} - p_{17}(x_8 - x_3) - d x_8. \end{cases}$$
(2)

where d and q_i, $i = 0, 1, \ldots, 5$ are known specific rate with the formulation of

$$\begin{cases} d = d_m \dfrac{x_2}{x_2 + K_3} \prod_{i=2}^{5}\left(1 - \dfrac{x_i}{x_i^*}\right), \\ q_1 = m_1 + \dfrac{d}{Y_1} + \Delta q_1 \dfrac{x_2}{x_2 + k_1^*}, \\ q_2 = p_1 \dfrac{x_2}{x_2 + p_2} + p_3(x_2 - x_6), \\ q_3 = p_4 \dfrac{x_8}{x_8 + p_5} + p_6(x_8 - x_3), \\ q_4 = m_2 + Y_2 d + \Delta q_2 \dfrac{x_2}{x_2 + k_2^*}, \\ q_5 = m_3 + Y_3 d. \end{cases}$$
(3)

where d_m, K_2^*, K_4^*, m_2, m_4, m_5, Y_2, Y_4, Y_5, Δq_2, Δq_4 are all constants whose concrete biological meanings and values can be referred to the previous literature (Xiu et al, 2000).

Denote the allowable sets of x, u and p by X_a, U_a and P_a, respectively, and they are all assumed compact. Then, we can prove the following properties of the dynamical system (1).

Property 1. The function F defined in (2)–(3) satisfies:

1) F is twice continuously differentiable in $u \times p$ on $U_a \times P_a$;
2) for fixed $u \times p \in U_a \times P_a$, there exist positive constants a, b, such that the linear growth condition holds, i.e. $\|F\| \leq a \|x\| + b$, where $\|\cdot\|$ is the Euclidean norm.

Property 2. For fixed $u \times p \in U_a \times P_a$, there exists a unique solution to system (1), denoted by $x(\cdot;$ u, p). Furthermore, $x(\cdot;$ u, p) is twice continuously differentiable in $u \times p$ on $U_a \times P_a$.

Define $S_1(p)$ as the set of solutions of the dynamical system (1) with fixed $p \in P_a$. According to the compactness of U_a and the continuity of the mapping from $u \in U_a$ to $x(\cdot;$ u, p) $\in S_1(p)$. By Property 1, we can prove the following consequence:

Theorem 1. For fixed $p \in P_a$, the set $S_1(p)$ is compact in C^1 ([0,T];R^8).

3 SYSTEM SENSITIVITY

Sensitivity analysis deals with the influence that small changes in nominal values of model parameters exert on model performance (Zangemeister et al., 1981). For the nonlinear enzyme-catalytic dynamical system (1), we define system cost of the dynamical system (1) as:

$$J_1(u; p) := -u_1 x_3(t_f; u; p) \tag{4}$$

According to the literature (Loxton et al., 2011, Rehbock et al., 1992), system sensitivity corresponding to a control $u \in U_a$ can be defined as:

$$sen(u; p) := \left(\dfrac{\partial J_1(u; p)}{\partial p}\right)^T \dfrac{\partial J_1(u; p)}{\partial p} \tag{5}$$

Substituting (4) into (5), we can rewrite (5) as:

$$sen(u; p) := u_1^2 \left(\dfrac{\partial x_3(t_f; u, p)}{\partial p}\right)^T \dfrac{\partial x_3(t_f; u, p)}{\partial p} \tag{6}$$

Aiming to compute the system sensitivity in (6), we consider the following auxiliary system:

$$\begin{cases} \dot{y}(t) = \dfrac{\partial F(x,u,p)}{\partial x} y(t) + \dfrac{\partial F(x,u,p)}{\partial p}, \quad t \in [0,T] \\ y(t_0) = I. \end{cases} \tag{7}$$

where I is a 8×17 matrix with all elements having the value 0.

According to Property 2, we can prove that, for given $u \times p \in U_a \times P_a$, the dynamic system (7) has a unique solution, denoted by $y(\cdot;$ u, p). Furthermore, it is similar to Theorem 1 in (Loxton et al. 2011) that, for each $u \in U_a$,

$$\dfrac{\partial x_i(t; u, p)}{\partial p} = y_i(t; u, p), t \in [0,T], i = 1, 2, \cdots, 8 \tag{8}$$

Then, for fixed $p \in P_a$, let $S_2(p)$ be the set of solutions of the auxiliary system (7). In view of Property 2, we can have the following consequence:

Theorem 2. For fixed $p \in P_a$, the set $S_2(p)$ is compact in C^1 ([0,T]; $R^8 \times R^{17}$).

Proof: it follows from Property 2 that $x(\cdot; u, p)$ is twice continuously differentiable in $u \in U_a$. Therefore, $y(\cdot; u, p)$ is also continuous in $u \in U_a$. According to the compactness of U_a, one can easily verify that the set $S_2(p)$ is compact in $C^1([0,T]; R^8 \times R^{17})$.

4 ROBUST SUBOPTIMAL CONTROL PROBLEM

As illustrated in the literature (Loxton et al., 2011, Rehbock et al., 1992), the ideal control is one that minimizes both system cost and system sensitivity. Such a control, however, is unlikely to exist. Thus, we need a compromise and try to find a control $u \in U_a$ to minimize the function:

$$J(u; p) := J_1(u; p) + w \operatorname{sen}(u; p)$$
$$= -u_1 x_3(T; u, p) + w u_1^2 y_3(T; u, p)^T y_3(T; u, p) \quad (9)$$

where the weight w can adjust the relative importance of each term in $J(u; p)$. Thus, the robust suboptimal control problem for the nonlinear enzyme-catalytic dynamical system (1) can be expressed by:

s.t. $x(\cdot; u, p) \in S_1(p)$, $y(\cdot; u, p) \in S_2(p)$
$x(t_0) = x^0$, $y(t_0) = I$, $u \in U_a$, $t \in [0, T]$

Note that the problem (P) is subjected to both the enzyme-catalytic system (1) and the auxiliary system (7), and those two systems are coupled according to the formula (8).

Theorem 3. There exists an optimal solution u^* to the problem (P), that is, there exists $u^* \in U_a$ such that:

$$J(u^*; p) \leq J(u; p), \quad \forall u \in U_a.$$

Proof: by Property 2, both $x(\cdot; u, p)$ and $y(\cdot; u, p)$ are continuous in $u \in U_a$, so the mapping from $u \in U_a$ to $J(\cdot; p)$ is continuous. By the compactness of U_a, we can conclude that (P) has an optimal solution, denoted by u^*, such that:

$$J(u^*; p) \leq J(u; p), \quad \forall u \in U_a.$$

which completes our proof.

5 A DISCRETE COMPUTATIONAL PROCEDURE

Let Q be a partition of $[0,T]$ given by:

$Q := \{ 0 = t_0 < t_1 < \cdots < t_{N-1} < t_N = T \}$

with $t_i := iT/N$, $i = 0, 1, \ldots, N$. Then, we can obtain the discrete-time approximations to the nonlinear dynamical system (1) and the auxiliary dynamical system (7) using the Euler method of integration.

Given an $u \in U_a$, the Euler integration formula computes a sequence of vector $x_Q := \{x_Q(t_i; u,p), i = 0,1, \ldots, N-1\}$, according to the Euler recursion (which approximates the differentiation in (1) by a finite difference):

$$\begin{cases} x_Q(t_{i+1}; u, p) - x_Q(t_i; u, p) = \dfrac{T}{N} F(x_Q(t_i; u, p), u, p), \\ x_Q(t_0; u, p) = x^0. \quad i = 0, 1, \cdots, N-1 \end{cases} \quad (10)$$

Similarly, we have a sequence of matrix $y_Q := \{y_Q(t_i; u,p), i = 0, 1, \ldots, N-1\}$, in $R^8 \times R^{17}$ according to the Euler recursion

(The differentiation in (1) by a finite difference):

$$\begin{cases} y_Q(t_{i+1}; u, p) - y_Q(t_i; u, p) \\ \quad = \dfrac{T}{N} [\dfrac{\partial f(x_Q(t_i), u, p)}{\partial x} y_Q(t_i; u, p) + \dfrac{\partial f(x_Q(t_i), u, p)}{\partial x}] \\ y_Q(t_0; u, p) = I. \quad i = 0, 1, \cdots, N-1 \end{cases} \quad (11)$$

Clearly, there is no question that solutions of (10) and (11) exist. By Theorem 5.6.16 in (Polak,1997), we have that $x_Q(t_i; u, p)$ and $y_Q(t_i; u, p)$, $i = 0,1, \ldots, N-1$ are, respectively, Lipschitz continuous on bounded subsets of X_a and that of $R^8 \times R^{17}$. With solutions of (10) and (11), we can associate the piecewise constant time functions $x_Q(\cdot; u, p)$ and $y_Q(\cdot; u, p)$ defined by:

$$x_Q(\cdot; u, p) := \sum_{i=0}^{N-1} x_Q(t_i; u, p) \chi_{Q,i}(t) \quad (12)$$

and

$$y_Q(\cdot; u, p) := \sum_{i=0}^{N-1} y_Q(t_i; u, p) \chi_{Q,i}(t) \quad (13)$$

where

$$\chi_{Q,i}(t) := \begin{cases} 1, t \in [t_i, t_{i+1}) \\ 0, otherwise \end{cases}$$

Obviously, $x_Q(\cdot; u, p)$ and $y_Q(\cdot; u, p)$ are approximate solutions of (1) and (7), respectively. By Theorem 5.6.23 in (Polak, 1989), there exist constants L_1, $L_2 < +\infty$, such that:

$$\|x(\cdot; u, p) - x_Q(\cdot; u, p)\| \leq \dfrac{L_1}{N} \quad (14)$$

and

$$\|y(\cdot; u, p) - y_Q(\cdot; u, p)\| \leq \dfrac{L_2}{N}. \quad (15)$$

Let

$$J_Q(u; p) := -u_1 x_{Q,3}(t_N; u, p) + w u_1^2 y_{Q,3}(t_N; u, p) \quad (16)$$

where $x_{Q,3}(\cdot; u, p)$ denotes the third element of $x_Q(\cdot; u, p)$ and $y_{Q,3}(\cdot; u, p)$ is the third column of $y_Q(\cdot; u, p)$. Furthermore, according to the inequalities (14) and (15), it is easy to verify that:

$$\|J_Q(u; p) - J(u; p)\| \leq L/N, L = u_1(L_1 + w u_1 L_2) \quad (16)$$

Next, inspired by the numerical method in the literature (Degond et al. 2004), we construct a discrete computational procedure for the robust suboptimal control problem (P), where N and M are all the sizes determined by the Monte-Carlo method.

Algorithm 1

Step 1. Given the initial values x^0, the terminal time T, the nominal value of parameter p, the weight $w > 0$ and the positive integers N, M, set m: = 0 .t_i: = iT/N, i = 0,1,...N.

Step 2. Randomly generate the initial sample point u^0 from U_a, calculate $x_Q(\cdot; u^0, p), y_Q(\cdot; u^0, p)$ and $J_Q(u^0, p)$. Set $u^* = u^0$, J∗: = $J_Q(u^0, p)$ and m: = m + 1.

Step 3. If m < M, randomly generate a sample point u from U_a, denote $u^m = u$ and calculate $x_Q(\cdot; u^m, p), y_Q(\cdot; u^m, p)$ and $J_Q(u^m, p)$.

Step 4. If $J_Q(u^m, p) > $ J∗, goto Step 5; else set $u^* = u^m$ and J∗: = $J_Q(u^m, p)$, goto Step 5.

Step 5. Set m: = m + 1, goto Step 3.

6 CONCLUSIONS

In this paper, a nonlinear dynamical system of glycerol continuous fermentation was investigated. We discussed its system sensitivity by solving an auxiliary dynamical system. Moreover, a robust suboptimal control problem for the nonlinear dynamical is proposed, in which the control target is minimizing the weight sum of system cost and system sensitivity. Finally, a computational method was constructed via the discretization of the dynamical system and its auxiliary system.

REFERENCES

Biebl, Menzel, Zeng, Deckwer, 1999, Microbial production of 1,3-propanediol, Appl Microbiol Biotechnol.

McCoy, 1998, Chemical makers try biotech paths, Chem. Eng. News.

Liu et al, 2001, Fermentative production of 1, 3-propanediol by Klebsiella pneumoniae in fedbatch culture, Food Ferment Ind.

Li et al, 2006, Stability and optimal control of microorganisms in continuous culture, J. Appl. Math. Computing.

Loxton et al, 2011, Robust suboptimal control of nonlinear systems, Appl. Mathe. Compu. 2

Rehbock et al, 1992, A computational procedure for suboptimal robust controls, Dynamics and Control.

Defeng et al, 2007, On robustness of suboptimal min-max model predictive control, WSEAS Transactions on Systems and Control.

Wang et al, 2011, Complex metabolic network of glycerol fermentation by Klebsiella pneumoniae and its system identification via biological robustness, Nonlinear Anal. Hybrid Syst.

Sun et al, 2008, Mathematical modeling of glycerol fermentation by Klebsiella pneumoniae: Concerning enzyme-catalytic reductive pathway and transport of glycerol and 1, 3-propanediol across cell membrane, Biochem. Eng. J.

Xiu et al, 2000, Mathematical modelling of kinetics and research on multiplicity of glycerol bioconversion to 1, 3-propanediol, J. Dalian Univ. Technol.

Zangemeister et al, 1981, Sensitivity Analysis and Optimization for a Head Movement, Model, Biol. Cybern.

Polak, 1997, Optimization: Algorithms and Consistent Approximations, Springer-Verlag, New York.

Degond et al., 2004, Modeling and Computational Methods for Kinetic Equations, Modeling and Simulation in Science, Engineering and Technology, Brikhauser, Boston.

Energy and Environmental Engineering – Wu (Ed.)
© 2015 Taylor & Francis Group, London, ISBN 978-1-138-02665-0

Development and application of deep foundation pit internal bracing support technology in Nanchang

YiHui Li
College of Science and Technology, Nanchang University, China

ZiYing He
Nanchang Urban Rail Group Co. Ltd., China

XiaoPin Wang
College of Science and Technology, Nanchang University, China

ABSTRACT: The study describes the geological conditions and environmental characteristics of Nanchang, and the development and application characteristics of deep foundation pit support technology in Nanchang. Through the typical engineering examples, the support system and the optimized design scheme for deep, big and rich water layers were introduced, and future development ideas for deep foundation pit support technology in deep, big and rich water layers were discussed.

Keywords: Deep foundation pit, support technology, internal bracing

1 INTRODUCTION

Foundation pit support technique in the Nanchang region is accompanied by economic development and urban construction and development. Due to the limitation of urban construction land, building development models from planar to underground space development have become a major trend. With the increase in tall and super-tall buildings, the excavation pit depth is also deeper. From its early development of about 5 metres, to the current depth of more than 20 metres, the basement area in the past was one layer and it is now up to three layers, and the basements of tall buildings have generally moved up from two to three layers. Meanwhile, the environmental conditions surrounding the pit have become increasingly complex, environmental effects of deep excavation problems have become increasingly prominent, and deep excavation techniques put forward necessitate higher requirements. To do technologically advanced, economical, safety and environmental protection applicable, so deep excavation technology to become one of the hot spots of Nanchang construction projects.

2 GEOLOGICAL CONDITIONS AND ENVIRONMENTAL CHARACTERISTICS

The engineering, geological, and hydrogeological conditions in Nanchang are more complicated, and the causes of many types of soil. Mainly by artificial fill, Q4al, E1-2xn. Lithology and their engineering properties, from top to bottom were plain fill, silty clay, sand, coarse sand, gravel sand, gravel lowermost sand silt or clay bedrock, according to the degree of weathering can be divided: intense weathering, weathered and weak weathering.

Nanchang, near the Poyang Lake and the Gan River, is rich in water resources. The pattern of the urban water system is composed by the Gan River and several tributaries. The Gan River is the seventh largest tributary of the Yangtze; it is also the largest river in Jiangxi Province, and is the backbone of this area, followed by the Yingshang Lake (Diezi Lake), the Fu River, the East Lake, the Yudai River, the Aixi Lake, and the Yao River.

The ground water level is high and abundant in the Nanchang area. Taking the Gan River as the boundary, the burial depth to the east of the Gan River is generally 3.20~10.50 m, elevation is usually 13.06~18.98 m, and the water lever changes with the Gan River water level; the burial depth to the west of the Gan River is generally 7.00~8.00 m, elevation is usually 5.02~15.17 m, and the water lever changes with the Gan River water level; annual water level changes are in the range of 2~4 m. Affected by weather conditions, the annual five-month-long rainy season is a great threat to foundation pits.

3 THE MAIN TYPES OF BUILDING FOUNDATION PIT SUPPORT TECHNOLOGY[1–8]

Due to the different conditions of the surrounding environment, the engineering-geologics, and hydrologic geology, there are a lot of foundation pit supporting forms. Currently, the main building foundation pit supporting methods are piles with anchors, piles with internal bracing, soil nail walls, and double piles. Among these techniques, because the soil nail wall requires simple equipment, the production of soil nails and making holes does not require complex technology and large machinery, it is low-cost, and there is little interference to the surrounding environment, it has become one of the fastest growing technologies in Nanchang. But with the rapid development of highrise buildings, the depth of the foundation pits is also deeper, especially in the construction of the Nanchang Metro subway, and these subway deep foundation pit supporting techniques are increasingly being applied to the deep foundation pits of buildings, especially that of the application of piles with internal supporting systems.

3.1 Piles in row

Piles in rows means that the retaining structure is composed of an interval arrangement of RC hand-dug piles, bored piles, tube-sinking cast-in-situ piles, prestressed pipe piles, etc. Piles in rows can be divided into cantilever supporting structures, anchored retaining piles, and piles with internal bracing. All these types are used in Nanchang, but the most common type is the anchored retaining pile.

3.2 Soil nail wall and composite nailing wall

A soil nail wall (also known as shotcrete and rock bolt support, soil nailing, etc.) is a new form of foundation fit support developed in recent years, and because it has two advantages of being low-cost and of fast construction, it has been more and more widely used. The Nanchang area has since begun to apply this technology for foundation pit supporting, and it has been successfully applied to hundreds of projects. With the development of soil nail wall technology, a lot of composite nail wall technologies have appeared such as soil nail walls combined with waterproof curtains, soil nail walls combined with prestressed anchors, soil nail walls combined with row-in-piles, etc. This new technology is namely in the upper pit slope using soil nails, the lower pit uses slope piles in order to save costs, thus, making the soil nail wall technology occupy the main market of pit supporting structures.

3.3 Other supporting technologies

In addition to the above major support structures, the Nanchang area also has some other supporting types, including diaphragm walls, bored piles, deep mixing piles, gravity retaining walls, steel sheet piles, and prestressed pipe piles, but these technologies are used less.

4 THE APPLICATION OF PILES WITH INTERNAL BRACING SUPPORTING TECHNOLOGY IN THE WATER-RICH SAND LAYER OF NANCHANG

4.1 Project overview

A pit for the three-storey basement in Nanchang, and the basement of the monolayer area is about 10000 m^2, the storey height (5.2 m + 5.2 m + 4.9 m) is higher, the foundation plane size is 116.47 m × 117.3 m, the absolute ground elevation is 23.0 m, the absolute elevation of the ground water level is 15.0 m, the absolute elevation of the basement floor is 5.95 m, located in water-rich sand layer, and the absolute elevation of the bottom plate of core tube is −0.8 m, located in intense weathering glutinite.

The excavation depth is 17.05 m, the core tube local deepen is 6.75 m, the excavation depth is 23.8 m, the deepen of the section plane size is 26.5 m × 23.18 m, and the pit is exposed for a long time, higher underground water level, So the foundation pit waterproof requirements are very strict. It is the longest, widest, and deepest pit in Nanchang, and such a pit has never been encountered before, so no design and construction experience can be followed. The original design of this pit is a pile with anchor supporting structure.

4.2 Supporting structure design

(1) The original design.
Considering the project foundation pit's surrounding environment and its size, the excavation depth, the engineering geology and hydrogeology, the region construction equipment, the construction technology, and construction conditions such as the season, a "soil nail wall", piles in rows, and jet grouting piles as waterproof curtains and anchors were used (Fig. 1) for the foundation pit supporting structure. That is to say, the plain fill and silty clay at the top with the 1:1.25-slope excavation, slope protection with wire-netting shotcrete, in the following using piles in row and jet grouting pile as waterproof curtain and anchor.

Pile in rows: Using an adaptable, good quality mechanical drilling pile embedded in rock-socketed pile, Φ1.20 m, pile spacing is 1.50 m and the concrete strength is C30. Top beam: In the top of the piles using a C25 reinforced concrete top beam to

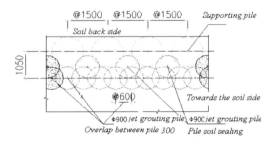

Figure 1. Original waterproof curtain programme.

improve the integrity of the piles. Anchor: in the midpoint between bored piles three rows of anchor are arranged. Breast beam: set breast beam at the anchor position, the breast beam using two root [20a -type channels. Using jet grouting piles as a waterproof curtain: between the bored piles, two rows of $\phi 900$ triple tube jet-grouting pile were used, together with bored piles, to form a closed waterproof curtain. Triple tube jet-grouting pile. Soil between pile protection: slope protection with C20 shotcrete, thickness 40~60 mm.

Drainage system of foundation pit: On top of the foundation pit position drains were built with cement blocks, used to intercept surface water. The drains are connected with the municipal drainage system by a sedimentation tank; along the bottom of the pit drains were also built with cement blocks, and meanwhile each corner point set inlet well. When there is a lot of water, every 30 m or encryption settings sump, to exclude the pit water.

(2) The optimal design.
Based on safety aspects, one design is based on the results of test triple tube jet-grouting piles; it shows that the interface between gravel and rock core samples is not ideal, therefore the sealing effect is difficult to guarantee, and in the early pile construction it was found that the east and south sides of the pit had abandoned RC structures such as sluices. The southwest corner had concrete drainage pipes, etc. These underground obstructions had serious influences on the hole-forming of the triple tube jet-grouting piles; Secondly, that with the anchoring section of prestressed anchors located in coarse sand, gravel sand, and the sand affected by the dynamic water pressure, the bond stress of the anchor is difficult to guarantee;, if the prestressed anchor fails , the consequences would be disastrous; additionally, the bond stress in the sand required to verify this experiment, will adversely affect the construction period. While the surrounding land development, one is due to the design of prestressed anchor cable lengths up to 22m, while the pit slope of the land has been close to the red line, it will invade surrounding land nearly 20m, it will have adverse effect on the surrounding land to be developed and nearby building project construction.

Based on the above reasons, the design was optimized by using a combined support system such as the soil nail wall, pile in rows, waterproof curtains (outside a row of triaxial cement deep mixing piles and inside triple tube jet-grouting piles), and internal bracing(gussets). These are shown in Figures 2 and 3. Due to changes in the support structure, the model is forced to change, but supporting piles have been in construction, and therefore need to be checking the supporting piles.

(3) The theory of computation and the main results of internal bracing.
1) Model:
The internal force of the support structure is calculated per unit length along the surrounding fit according to

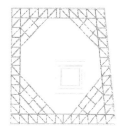

Figure 2. Gussets layout system.

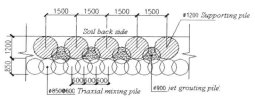

Figure 3. The optimized solution waterproof curtain.

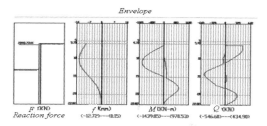

Figure 4. Supporting structure internal force and deformation envelope.

the theory of beams on elastic foundation. The internal force calculation according to the excavation and then construction of internal structure. During the excavation phase, calculations must be included in the initial displacement of the structure and deformation of the support, according to "supporting after deformation" the principles of structural analysis calculations.

2) The main calculation parameters are:
Soil natural gravity——according to the survey report values.
Reinforced concrete support ——the main support is 800×800 mm, the connecting rod is 700×600 mm.
Soil spring——according to the survey report values.
Ground overload——20 kN/m^2.
Lateral load——clay soil: water and soil together, sandy soil: water and soil are calculated separately.

3) The deformation and internal forces results of the supporting structure.
The supporting structure's internal force and deformation envelope are shown in Figure 4.

According to the results of internal forces, and after checking, piles of piles with anchor supporting structures meet the force requirements.

4) Stability checking of the supporting structure.
5) According to leading software calculations, the safety factor of stability $K = 2.617 > 1.3$, and the factor of safety against overturning $K = 1.481 > 1.2$, also meet the requirements.

4.3 The construction effect of supporting structures

(1) From August 2012 to the start of construction, the earthwork excavation began in March 2013, during Nanchang's rainy season, in order to reduce the impact of excavation on the surrounding environment. Monitoring technologies were used to grasp the deformation of the foundation pit structure and the supporting structure, so as to ensure the construction of information technology met the structural safety levels and environmental requirements.

(2) During the earthwork excavation process, the horizontal displacement of the top of slope, the horizontal displacement of the pile top, the deep horizontal displacement of the supporting piles, the horizontal displacement of the soil, the ground settlement of the foundation pit, and the deformation of surrounding buildings etc., were within the controllable range, to ensure the safety of the pit and the surrounding environment.

(3) During the earthwork excavation process, the foundation of the basic anhydrous, piping phenomenon does not appear in the pit bottom, and the slope was stable.

5 CONCLUSION

Nanchang is undergoing rapid economic development and urban construction. In order to meet the needs of the construction, deep excavation technology continues to develop, and a variety of deep excavation technology applications are constantly deepened. Through engineering examples, we can, from the following aspects, develop deep foundation pits in the future.

(1) Promotion of supporting structures in deep foundation pits.

The supporting structures currently used mostly in the Nanchang area are pile anchors and soil nail walls, including composite nailing walls. The soil nail wall is economic, of quick construction and has other advantages, but because of its displacement control is relatively weak, and although the pile with anchor displacement control is better, but with a higher cost, the prestressed anchor tensile force is difficult to guarantee in water-rich sands. At the same time, the two supporting structures have adverse effects on the surrounding land to be developed, so the piles with internal bracing for deep foundations in the application of water-rich sands, are also worth promoting.

(2) Improved construction technology.

Because of the presence of large numbers of water-rich sands in the Nanchang area, anchor use will be greatly restricted; the development of suitable construction technologies for water-rich sand, can expand the anchor applications.

(3) Application monitoring and information technology[9–10].

The deep foundation fit is located in the complex mechanical properties' stratum, due to the existence of the role of "space effect", as well as groundwater, the surrounding environment, such as preloading the ground, internal force calculation and estimation of soil deformation with the actual situation are quite different, so the design of foundation pits at this stage, to a great extent, depends on experience. Therefore, in the pit construction process, only by taking and monitoring, and using information technology to comprehensively monitor the supporting structure of the foundation pit, the soil around the pit, and the adjacent buildings, will a comprehensive and dynamic understanding of the degree of safety and the impact on the surrounding environment of foundation pit engineering be achieved. Through monitoring and discovering hidden safety problems, in the event of unusual circumstances timely feedback, and to take the necessary engineering emergency measures to prevent the pit accident, personal injury and avoid major economic losses.

(4) Foundation pit supporting technology and environmental harmony.

The issue of the foundation pit support design harmoniously fitting in with the environment, is one of the prominent problems in foundation pit support design. Because the foundation pit engineering is temporary, the use of anchor support technology will break through the land red line, and it will extend into the adjacent development project thereby affecting the project development; the prestressed anchor tensile force is difficult to guarantee in water-rich sands and easily causes damage to the surrounding environment during construction. Therefore, piles with internal bracing in the water-rich sand layers of the Nanchang region can be harmonized to support both technology and the environment.

ACKNOWLEDGEMENT

Project: school-level natural science research fund project (ZL-2010-02)

Author introduction: Li Yi-Hui (1981–), female, master's degree, structural engineering;

REFERENCES

LiuGuobin, Wang Weidong. Pit Engineering Handbook. Beijing: China Architecture & Building Press, 2009.

Yang Zhiyin, Zhang Jun. Development and application of supporting techniques for deep foundation pit excavation in

Shenzhen region. Chinese Journal of Rock Mechanics and Engineering. 2006, 25: 3377~3383.

Zheng Chenmin, Wang Zenghui, Zhang xin,Economic analysis of double-row piles in deep foundation pits in soft soils in coastal areas of Fujian Province.Chinese Journal of Geotechnical Engineering. Vol. 32, Supp. 1, July 2010: 317~320.

Li Fangzheng, Application of Technology of J et al. Grouting in Waterproof Engineering of Deep Foundation Pit of Subway Station. Journal of Highway and transpotation and Development. Vol. 20, No. 1, 2003:21~23.

Fu Wengang, Yang Zhiyin, Liu Junyan. Some theoretical questions of composite nailing wall and discussion on technical code for composite nailing wall in retaining and protection of excavation. Chinese Journal of Rock Mechanics and Engineering. Nov., 2012:2291~2304.

Mei Xiaoli,Chen Jie,Liu Wei.Construction Technology of Ground and Foundation for Super-large Deep Foundation Excavation of Some Engineering.Construction Technology. Vol. 43, 1, 2014:53~56.

Zhou Yong.Construction Technology of Deep Foundation Pit of Railway Underground Tunnel in Coastal High Groundwater Level Area.Journal of Railway Engineering Society. Mar 2014:82~86.

Shi Yiqing, Application of Supporting Technology Combined with Section SteelInclined Brace in Deep Foundation Excavation. Construction Technology. Vol. 43(7), 2014:18~21.

Zeng Xiuhua. Information on the deep excavation construction monitoring techniques. Guangxi Quality Supervision Guide Periodical, 2008, (7):192~194.

Zhou Yong, Guo Nan, Zhu Yanpeng. Construction Monitoring and Numerical Simulation of Deep Excavation of Century Avenue Metro Station in Lanzhou. Journal of Railway Engineering Society. Jan 2013:83~88.

Author index

Ao, L.P. 155

Bai, W.D. 177
Bao, G. 69
Bao, Y. 149
Bao, Y.H. 69
Bao, Z.B. 13

Chai, D. 263
Chen, D.W. 85
Chen, J.F. 223
Chen, Q. 223
Chen, Q.L. 17
Chen, Y.F. 47
Chen, Z.M. 63
Cheng, C. 21, 27
Chua, H. 59
Cong, X.N. 231

Deng, Y.J. 53
Ding, S.R. 195, 201
Dong, B. 239
Du, Q.J. 169, 271
Du, X. 101
Du, X.L. 149

E, X.Y. 159

Feng, X. 149
Feng, Y. 189

Gao, W.J. 163
Gao, W.L. 53
Ge, M.W. 39
Gu, J.J. 17
Guan, Q.Q. 17
Guo, Y.J. 159

Han, G.Y. 129
Hao, S.H. 159
He, Z.Y. 281
Hou, J. 169, 271
Hu, B.Q. 75, 79
Hu, G. 75, 79
Hu, W.J. 159
Huang, L.Y. 129
Huang, M.X. 69
Huo, Y.H. 195, 201

Jia, J. 21, 27
Jiang, B. 159

Jiang, S. 181
Jiang, W.Q. 177
Jin, D.C. 13

Kan, C.W. 59

Li, D. 217, 223
Li, G.H. 189
Li, J.L. 85
Li, L.Q. 93
Li, S.D. 47
Li, S.P. 169
Li, W.Z. 239
Li, X.H. 85
Li, X.N. 169
Li, Y. 159
Li, Y.F. 217
Li, Y.H. 281
Li, Y.Y. 247
Li, Z.J. 145
Liang, S.S. 27
Liao, X.L. 3
Lin, X. 27
Liu, D.C. 21, 27
Liu, G.G. 63, 145
Liu, J.J. 7
Liu, L.N. 109
Liu, P. 181
Liu, W.L. 109
Liu, X.G. 189
Lu, W.H. 255
Luo, J.S. 53
Luo, Y.M. 163

Ma, F.S. 101
Ma, H.D. 93
Meng, L. 181
Miao, R.R. 17

Ning, P. 17

Ou, X.M. 251

Pan, Y.L. 59
Pan, Z.H. 123
Pei, Z.Y. 119

Qi, J.L. 145
Qin, H.J. 239
Qin, J. 145

Qin, Y. 101
Quan, J.T. 123

Ruan, L. 123

Shen, Y. 93
Shi, L.N. 169
Shi, S.M. 149
Sun, S.H. 33

Tang, Q.N. 231
Tang, W.X. 47
Tian, S.L. 17
Tong, X. 123

Wang, C.L. 135
Wang, H. 129
Wang, J. 21, 277
Wang, J.C. 169, 271
Wang, L. 13, 21
Wang, M.L. 17
Wang, Q. 263
Wang, S. 39
Wang, W.F. 181
Wang, X.P. 281
Wang, Z. 231
Wang, Z.J. 119
Wang, Z.Y. 93
Wei, J.M. 79
Wen, X.S. 123
Wu, J.Y. 149
Wu, L. 163
Wu, R.H. 63
Wu, X.X. 149
Wu, Y. 255

Xi, D.M. 139
Xiang, X.Q. 255
Xiao, H. 119
Xiao, H.B. 33
Xiao, H.H. 63
Xiao, H.W. 39
Xiu, Z.L. 277
Xu, J. 177, 207, 211
Xu, J.X. 53
Xu, R.X. 271
Xu, S.H. 53
Xu, Y.B. 53
Xue, J.B. 231

Yang, H.S. 163
Yang, L. 13

Yang, Q. 123
Yang, X.G. 129
Yuan, Y.J. 101

Zhang, D. 139
Zhang, F.H. 69
Zhang, G.F. 169
Zhang, G.L. 113
Zhang, J. 33, 177

Zhang, L. 129
Zhang, L.S. 189
Zhang, Q. 251
Zhang, Q.L. 17
Zhang, W.Y. 13
Zhang, X. 251
Zhang, X.H. 53
Zhang, X.L. 251
Zhang, Z.H. 75

Zhang, Z.Q. 79
Zhao, J. 21, 27
Zhong, G.S. 155
Zhou, B.H. 101
Zhou, C.D. 33
Zhou, X.P. 263
Zhou, Y. 53
Zhu, J.G. 135
Zhu, W.Y. 239